Soil Vapor Extraction *Using* Radio Frequency Heating

Resource Manual *and* Technology Demonstration

Edited by

Donald F. Lowe
Rice University, Houston, TX

Carroll L. Oubre
Rice University, Houston, TX

C. Herb Ward
Rice University, Houston, TX

Authors
David E. Daniel
Raymond C. Loehr
Matthew T. Webster
Raymond S. Kasevich

CRC Press
Taylor & Francis Group
Boca Raton London New York

CRC Press is an imprint of the
Taylor & Francis Group, an **informa** business

Although the information described herein has been funded wholly or in part by the United States Department of Defense (DOD) under Grant No. DACA39-93-1-0001 to Rice University for the *Advanced Applied Technology Demonstration Facility for Environmental Technology Program* (AATDF), it may not necessarily reflect the views of the DOD or Rice University, and no official endorsement should be inferred.

CRC Press
Taylor & Francis Group
6000 Broken Sound Parkway NW, Suite 300
Boca Raton, FL 33487-2742

First issued in paperback 2019

© 2000 by Taylor & Francis Group, LLC
CRC Press is an imprint of Taylor & Francis Group, an Informa business

No claim to original U.S. Government works

ISBN-13: 978-1-56670-464-9 (hbk)
ISBN-13: 978-0-367-39918-4 (pbk)

Library of Congress Cataloging-in-Publication Data

Soil vapor extraction using radio frequency heating : resource manual and technology
    demonstration / edited by Donald F. Lowe, Carroll L. Oubre, and C. Herb Ward ;
    authors, David E. Daniel ... [et al.].
        p.   cm. — (AATDF mongraphs)
    Includes bibliographical references and index.
    ISBN 1-56670-464-2 (alk. paper)
    1. Soil vapor extraction. 2. Induction heating. 3. Radio frequency. 4. Soil
remediation—Technological innovations. I. Daniel, David E. (David Edwin), 1949–
II. Lowe, Donald F. III. Oubre, Carroll L. IV. Ward, C. H. (Calvin Herbert), 1933–
V. Series.
TD878.5 .S65 1999
628.5'5—dc21   99-030444
                                                                                    CIP

Library of Congress Card Number 99-030444

Visit the Taylor & Francis Web site at
http://www.taylorandfrancis.com

and the CRC Press Web site at
http://www.crcpress.com

Foreword

One of the most widely used techniques for treating soils contaminated with volatile organic compounds is soil vapor extraction (SVE). Soil vapor extraction can also be applied to semivolatile compounds if the soil is heated to increase the vapor pressure of the contaminants. The soil can be heated *in situ* by injecting hot air or steam, by direct resistive heating, or by applying electromagnetic energy in the radio frequency (RF) range. Although RF-SVE systems have been tried in several previous field demonstrations, success has been variable, and questions remain concerning the viability and cost-effectiveness of these treatment systems.

To answer these questions, the DOD-funded Advanced Applied Technology Demonstration Facility (AATDF) at Rice University provided funding to the University of Texas at Austin (UT) in 1995 for the project, Soil Vapor Extraction with Radio Frequency Heating (RF-SVE). Brown & Root Environmental was the prime contractor for the field demonstration. KAI Technologies, Inc. provided the RF heating equipment.

The project team carried out the RF-SVE demonstration at Kirtland Air Force Base (Kirtland AFB) in Albuquerque, NM. Mr. Christopher B. DeWitt was the Chief of the Environmental Restoration Branch at Kirtland and Mr. Jerroll G. Sillerud was Kirtland AFB's project contact for the AATDF project team. The field work at Kirtland AFB began on June 24, 1996, and ended on July 10, 1997. The work included 87 days of continuous RF-SVE operation starting on December 3, 1996.

The Principal Investigators for the RF-SVE project were Dr. David E. Daniel, Dr. Raymond C. Loehr, Dr. John A. Pearce, Dr. Gary A. Pope, and Dr. Richard L. Corsi with the University of Texas at Austin. The project team that carried out the demonstration at Kirtland AFB consisted of the following personnel:

- Dr. David E. Daniel, Dr. John J. Bowders, and Mr. Matthew T. Webster with the University of Texas at Austin;
- Mr. Clifton F. Blanchard, Mr. Ches Lyon, Mr. Brian Wolfe, Mr. Brad Sumrall, and Ms. Laura Whitt with Brown and Root Environmental;
- Mr. Raymond S. Kasevich, Mr. James C. Gardner, Mr. Daniel M. Jackson, and Mr. Stephen L. Price with KAI Technologies, Inc.

Other investigators included Eng-Chew Ang, Neil Deeds, Karl Huntzicker, Sarah Knight, Jean-Philippe Nicot, Thomas Groll, Jerry Garrard, Dustin Poppendieck, and J.P. Tan with the University of Texas at Austin.

The field data, calculated results, mass and energy balances, conclusions, and recommendations from the RF-SVE demonstration were initially reported in the two-volume Final Technical Report (FTR) by the team from the University of Texas at Austin. Fluor Daniel GTI prepared the hypothetical designs and project costs in their Technology Evaluation Report (TER).

This monograph also includes, as guides to future studies by others, the basic principles of RF heating of soils as well as the necessary databases, equations, and example calculations.

Other than the Monograph authors, the major contributors to the completion of the FTR and TER were as follows:

- Dr. John A. Pearce – University of Texas at Austin
- Dr. Gary A. Pope – University of Texas at Austin
- Dr. John J. Bowders – University of Missouri
- Mr. Todd E. Weese – Fluor Daniel GTI
- Mr. John F. "Jay" Dablow III – Fluor Daniel GTI

Besides serving as Editor, Dr. Donald F. Lowe of Rice University provided extensive technical input to the synthesis of the Monograph from the FTR and TER.

The authors worked closely with the AATDF staff and a team of Project Shepherds who participated in both design and review of the FTR, TER, and Monograph. The Project Shepherds were

- Dr. Paul C. Johnson — Arizona State University
- Mr. Steve Shoemaker — DuPont Company
- Mr. Michael C. Marley — KAI Technologies, Inc.
- Mr. Raymond S. Kasevich — KAI Technologies, Inc.
- Mr. Mark A. Johnson — Dahl & Associates, Inc.

The AATDF Technology Advisory Board (TAB) and other Advisory Committees provided valuable assistance at the start and during the project in such areas as technology selection, directional changes in field work, technology transfer, and reporting. Mr. Richard Conway and Ms. Karen Duston played pivotal roles in facilitating the publication process.

The AATDF Program was funded by the United States Department of Defense under Grant No. DACA39-93-1-0001 to Rice University. The Program was given oversight by the U.S. Army Engineer Waterways Experiment Station in Vicksburg, MS.

AATDF Monographs

This monograph is one of a 10-volume series that record the results of the AATDF Program:

- Surfactants and Cosolvents for NAPL Remediation: A Technology Practices Manual
- Sequenced Reactive Barriers for Groundwater Remediation
- Modular Remediation Testing System
- Phytoremediation of Hydrocarbon-Contaminated Soil
- Steam and Electroheating Remediation of Tight Soils
- Soil Vapor Extraction Using Radio Frequency Heating: Resource Manual and Technology Demonstration
- Subsurface Contamination Monitoring Using Laser Fluorescence
- Reuse of Surfactants and Cosolvents for NAPL Remediation
- Remediation of Firing-Range Impact Berms
- NAPL Removal: Surfactants, Foams, and Microemulsions

Advanced Applied Technology Demonstration Facility (AATDF)
Energy and Environmental Systems Institute MS-316
Rice University
6100 Main Street
Houston, TX 77005-1892

U.S. Army Engineer
Waterways Experiment Station
3909 Halls Ferry Road
Vicksburg, MS 39180-6199

Rice University
6100 Main Street
Houston, TX 77005-1892

Preface

Following a national competition, the Department of Defense (DOD) awarded a $19.3 million grant to a university consortium of environmental research centers led by Rice University and directed by Dr. C. Herb Ward, Foyt Family Chair of Engineering. The DOD Advanced Applied Technology Demonstration Facility (AATDF) Program for Environmental Remediation Technologies was established on May 1, 1993 to enhance the development of innovative remediation technologies for DOD by facilitating the process from academic research to full-scale utilization. The AATDF's focus was to select, test, and document performance of innovative environmental technologies for the remediation of DOD sites.

Participating universities included Stanford University, the University of Texas at Austin, Rice University, Lamar University, University of Waterloo, and Louisiana State University. The directors of the environmental research centers at these universities served as the Technology Advisory Board. The U.S. Army Engineer Waterways Experiment Station managed the AATDF grant for DOD. Dr. John Keeley was the Technical Grant Officer. The DOD/AATDF was supported by five leading consulting engineering firms: Remediation Technologies, Inc.; Battelle Memorial Institute; GeoTrans, Inc.; Arcadis Geraghty and Miller, Inc.; and Groundwater Services, Inc., along with advisory groups from the DOD, industry, and commercialization interests.

Starting with 170 preproposals that were submitted in response to a broadly disseminated announcement, 12 projects were chosen by a peer review process for field demonstrations. The technologies chosen were targeted at DOD's most serious problems of soil and groundwater contamination. The primary objective was to provide more cost-effective solutions, preferably using *in situ* treatment. Eight projects were led by university researchers, two projects were managed by government agencies, and two others were conducted by engineering companies. Engineering partners were paired with the academic teams to provide field demonstration experience. Technology experts helped guide each project.

DOD sites were evaluated for their potential to support quantitative technology demonstrations. More than 75 sites were evaluated to match test sites to technologies.

Following the development of detailed work plans, carefully monitored field tests were conducted and the performance and economics of each technology were evaluated.

One AATDF project designed and developed two portable Experimental Controlled Release Systems (ECRS) for testing and field simulations of emerging remediation concepts and technologies. The ECRS is modular and portable and allows researchers, at their sites, to safely simulate contaminant spills and study remediation techniques without contaminant loss to the environment. The completely contained system allows for accurate material and energy balances.

The results of the DOD/AATDF Program provided DOD and others with detailed performance and cost data for a number of emerging, field-tested technologies. The program also provided information on the niches and limitations of the technologies to allow for more informed selection of remedial solutions for environmental cleanup.

The AATDF Program can be contacted at: Energy and Environmental Systems Institute, MS-316, Rice University, 6100 Main, Houston, TX, 77005, phone 713-527-4700; fax 713-285-5948; e-mail <eesi@rice.edu>.

The DOD/AATDF Program staff include

Director:
    Dr. C. Herb Ward
Program Manager:
    Dr. Carroll L. Oubre
Assistant Program Manager:
    Dr. Kathy Balshaw-Biddle
Assistant Program Manager:
    Dr. Stephanie Fiorenza

Assistant Program Manager:
    Dr. Donald F. Lowe
Financial/Data Manager:
    Mr. Robert M. Dawson
Publications Coordinator/Graphic Designer:
    Ms. Mary Cormier
Meeting Coordinator:
    Ms. Susie Spicer

This volume, *Soil Vapor Extraction Using Radio Frequency Heating*, is one of a ten-monograph series that records the results of the DOD/AATDF environmental technology demonstrations. Many have contributed to the success of the AATDF program and to the knowledge gained. We trust that our efforts to fully disclose and record our findings will contribute valuable lessons learned and help further innovative technology development for environmental cleanup.

Donald F. Lowe
Carroll L. Oubre
C. Herb Ward

Authors and Editors

David E. Daniel is Professor and Head of the Department of Civil and Environmental Engineering at the University of Illinois at Urbana-Champaign, where he teaches and conducts research on waste containment and remediation of contaminated sites. He earned B.S., M.S., and Ph.D. degrees in Civil Engineering from the University of Texas at Austin (UT) and taught on the faculty at UT from 1980 to 1996. Dr. Daniel has worked on a variety of projects ranging from clay liners and vertical containment walls to remediation of contaminated sites using reactive materials or heat-enhanced soil vapor extraction. Dr. Daniel's research has been sponsored by the National Science Foundation, U.S. Environmental Protection Agency, Gulf Coast Hazardous Substance Research Center, Department of Energy, Department of Defense, Chemical Manufacturers' Association, and others. He has served as a consultant for over 100 projects involving waste containment or remediation of contaminated sites.

Raymond C. Loehr has decades of experience with the use of land as a waste management alternative, with bioremediation of sludges and contaminated soil, and with the practical application of research results. He has served as an advisor to industry, consulting engineering firms, and law firms on the use of waste management processes for environmental problems.

His research and innovative technological development studies have been widely recognized, including his being elected to the National Academy of Engineering in 1983. His many awards include the Rudolph Hering Award from the American Society of Civil Engineers in 1969, selection as the Water Conservationist of the Year by the Kansas Wildlife Federation in 1967, being a Senior Fulbright-Hays Scholar in 1979, and receiving the G. Brooks Earnest Award in 1991, the Claude Hocott Award for research excellence in 1991, the Rachel Carson Award in 1995, the Gordon M. Fair Award in 1996, and the Thomas R. Camp Medal in 1997, each from different national professional societies.

In addition, he has had major positions as the chair of committees of the National Research Council, Department of Defense, and the U.S. Environmental Protection Agency. Of special note is the fact that he served for 20 years as a member of the Science Advisory Board of USEPA, including serving as chair of the entire SAB from 1988 to 1993. A civil engineering graduate of Case Institute of Technology, he received a M.S. degree from that institution and Ph.D. in Environmental Engineering from the University of Wisconsin. He currently is the H.M. Alharthy Centennial Chair of Environmental and Water Resources Engineering and Head, Environmental Solutions Program at the University of Texas at Austin.

Matthew T. Webster is a Research Associate in the Environmental and Water Resources Engineering Program in the Department of Civil Engineering at the University of Texas at Austin. Mr. Webster obtained a B.S. in Environmental Engineering from Rensselaer Polytechnic Institute in 1992 and a M.S. in Environmental Health Engineering from the University of Texas at Austin (UT) in 1994. Mr. Webster has worked with Dr. Raymond C. Loehr on projects related to solidification, stabilization, bioremediation, and the determination of environmentally acceptable endpoints in soil. He helped manage this RF-SVE project.

Raymond S. Kasevich, founder of KAI Technologies, Inc., has 30 years of corporate research and development experience in electromagnetic science and engineering applications. These applications cover a wide range of frequencies from full-scale radio-frequency oil recovery and environmental systems to medical catheter systems for microwave hyperthermia. He also has 20 years of university teaching experience in electrical engineering. He holds 25 patents and has published numerous papers in professional journals. His education includes a M.E. from Yale University in 1963 with undergraduate studies in Electrical Engineering at Case Western Reserve University and the University of Hartford. He received a Ford Foundation Grant for Ph.D. studies at the University of Michigan while he was working part-time at the Radiation Laboratory in Ann Arbor on network synthesis problems. He continued his Ph.D. studies at MIT in the Department of Physics. Currently, Mr. Kasevich is President and CEO of KAI Technologies, Inc. and KASE Energy, LLC.

Donald F. Lowe is an Assistant Program Manager with AATDF at Rice University where he has been a project manager of four projects involving the field demonstration of innovative technologies. Dr. Lowe has a Ph.D. in Metallurgy-Chemical Engineering from the University of Arizona, a M.S. in Metallurgical Engineering from the University of Wisconsin at Madison, and a B.S. in Mining Engineering from the University of North Dakota. In his capacity as project manager for AATDF, Dr. Lowe provided the necessary managerial guidance and technical expertise to bring each project through a successful demonstration. He has also been an active participant in the preparation of the reports for each project. Since 1986, Dr. Lowe has been involved as a technical manager, proposals manager, and senior engineer with several environmental firms. His management responsibilities included cost estimation of remediation projects and economic feasibility studies for numerous processes. He also has provided technical guidance for many remediation projects. For approximately 25 years before 1986, Dr. Lowe was a research supervisor or research engineer with four primary mining or metals recycling companies. He has several patents and publications that are related to metals extraction and recycling processes.

Carroll L. Oubre is the Program Manager for the DOD/AATDF Program. As Program Manager he is responsible for the day-to-day management of the $19.3 million DOD/AATDF Program. This included guidance of the AATDF Team, overview of the 12 demonstration projects, assuring project milestones are met within budget, and that complete reports of the results are delivered.

Dr. Oubre has a B.S. in Chemical Engineering from the University of Southwestern Louisiana, a M.S. in Chemical Engineering from Ohio State University, and a Ph.D. in Chemical Engineering from Rice University. He worked for Shell Oil Company for 28 years with his last position as Manager of Environmental Research and Development for Royal Dutch Shell in England. Prior to that, he was Director of Environmental Research and Development at Shell Development Company in Houston, TX.

C. H. (Herb) Ward is the Foyt Family Chair of Engineering in the George R. Brown School of Engineering at Rice University. He is also Professor of Environmental Science and Engineering and Ecology and Evolutionary Biology. Dr. Ward has undergraduate (B.S.) and graduate (M.S. and Ph.D.) degrees from New Mexico State University and Cornell University, respectively. He also earned the M.P.H. in Environmental Health from the University of Texas.

Following 22 years as Chair of the Department of Environmental Science and Engineering at Rice University, Dr. Ward is now Director of the Energy and Environmental Systems Institute (EESI), a university-wide program designed to mobilize industry, government, and academia to focus on problems related to energy production and environmental protection. Dr. Ward is also Director of the Department of Defense Advanced Applied Technology Demonstration Facility (AATDF), a distinguished consortium of university-based environmental research centers supported by consulting environmental engineering firms to guide selection, development, demonstration, and commercialization of advanced applied environmental restoration technologies for the DOD. For the past 18 years he has directed the activities of the National Center for Ground Water Research (NCGWR), a consortium of universities charged with conducting long-range exploratory research to help anticipate and solve the nation's emerging groundwater problems. He is also Co-Director of the EPA-sponsored Hazardous Substances Research Center/South & Southwest (HSRC/S&SW), whose research focus is on contaminated sediments and dredged materials.

Dr. Ward has served as President of both the American Institute of Biological Sciences and the Society for Industrial Microbiology. He is the founding and current Editor-in-Chief of the international journal *Environmental Toxicology and Chemistry.*

AATDF Advisors

Mr. M. R. (Dick) Scalf (retired)
U.S. EPA
Robert S. Kerr Environmental Research
  Laboratory
Ada, OK

Mr. Terry A. Young
Executive Director
Technology Licensing Office
Texas A&M University
College Station, TX

Mr. Stephen J. Banks
President
BCM Technologies, Inc.
Houston, TX

Consulting Engineering Partners

Remediation Technologies, Inc.
Dr. Robert W. Dunlap, Chair
President and CEO
Concord, MA

Parsons Engineering
Dr. Robert E. Hinchee
(Originally with Battelle Memorial Institute)
Research Leader
South Jordan, UT

GeoTrans, Inc.
Dr. James W. Mercer
President and Principal Scientist
Sterling, VA

Arcadis Geraghty & Miller, Inc.
Mr. Nicholas Valkenburg and
Mr. David Miller
Vice Presidents
Plainview, NY

Groundwater Services, Inc.
Dr. Charles J. Newell
Vice President
Houston, TX

Industrial Advisory Committee

Mr. Richard A. Conway, Chair
Senior Corporate Fellow
Union Carbide
S. Charleston, WV

Dr. Ishwar Murarka, Associate Chair
Electric Power Research Institute
Currently with Ish, Inc.
Cupertino, CA

Dr. Philip H. Brodsky
Director, Research & Environmental
  Technology
Monsanto Company
St. Louis, MO

Dr. David E. Ellis
Bioremediation Technology Manager
DuPont Chemicals
Wilmington, DE

Dr. Paul C. Johnson
Arizona State University
Department of Civil Engineering
Tempe, AZ

Dr. Bruce Krewinghaus
Shell Development Company
Houston, TX

Dr. Frederick G. Pohland
Department of Civil Engineering
University of Pittsburgh
Pittsburgh, PA

Dr. Edward F. Neuhauser, Consultant
Niagara Mohawk Power Corporation
Syracuse, NY

Dr. Arthur Otermat, Consultant
Shell Development Company
Houston, TX

Mr. Michael S. Parr, Consultant
DuPont Chemicals
Wilmington, DE

Mr. Walter Simons, Consultant
Atlantic Richfield Company
Los Angeles, CA

Acronyms and Abbreviations

AATDF	. Advanced Applied Technology Demonstration Facility
AIR 3D	proprietary three-dimensional airflow mode (Envirogen, Inc.)
aq	aqueous phase compound
ASTM	American Society for Testing and Materials
Av, av	Average, average
BDL	below detection limit
bgs	below ground surface
BH	borehole
BRE	Brown & Root Environmental
cc	counter current
cm	centimeter
coc	cocurrent
°C	degree Celsius
D	Darcy
DOD	Department of Defense
DOE	Department of Energy
DRO	diesel range organic compounds
dwb	dry weight basis
EESI	Energy and Environmental Systems Institute (Rice University)
EOS	equation of state
EPA	Environmental Protection Agency (United States)
EWRE	Environmental and Water Resources Engineering Program (Civil Engineering Department, the University of Texas at Austin)
FID	flame ionization detector
FRTR Guide	*Guide to Documenting Cost and Performance for Remedial Technologies,* The Federal Remediation Technologies Roundtable (FRTR)
ft	feet
FT-14	Manzano Fire Training Pit #14, Kirtland Air Force Base
g	g, grams, or (g), gas phase compound
GAC	granular activated carbon
gal	gallon
g/g-mole	grams per gram-mole
GC	gas chromatograph
GHz	gigahertz
GRO	gasoline range organic compounds
GWT	groundwater table
h	hour
HPO	hydrous pyrolysis oxidation
ID	inside diameter
IITRI	Illinois Institute of Technology Research Institute
IPH	immiscible phase hydrocarbon
ISM	Industrial, Scientific, and Medical Frequencies
J	Joule
K	Kelvin
k	thermodynamic equilibrium constant
k	heat conductivity, kW/m-°C
KAFB	Kirtland Air Force Base (Albuquerque, NM)

KAI	KAI Technologies, Inc.
kg	kilogram
kHz	kilohertz (one thousand hertz)
kJ	kiloJoule
KVA	kilovoltampere
kW	kilowatt
kWh	kilowatthour
(l)	liquid phase compound
L	liters
LF	linear feet
lb	pound
LLNL	Lawrence Livermore National Laboratory
LLTD	low temperature thermal desorption
Lpd	liters per day
Lpm	liters per minute
m	meter
MDFIT	proprietary two-dimensional airflow model (Envirogen, Inc.)
mg/kg	milligrams per kilogram
MHz	megahertz (one million hertz)
min	minute
mL	milliliter
mm	millimeter
mm Hg	millimeters of mercury
mol	gram mole
MOU	Memorandum of Understanding
MPD	main disconnect panel
MRO	mineral range organic compounds
MW	microwave
N.A.	not applicable
NEC	Numerical Electromagnetic Code
nm	not measured
NO-e-SYS	computer program from Fortner Software used to calculate heat balances and temperature profiles
Ω	ohm
O&M	operation and maintenance
OD	outside diameter
OSHA	Occupational Safety and Health Administration
P	pressure
P/T	pressure and temperature monitoring well
PID	photoionization detector
PITT	partitioning interwell tracer test
PTFE	Teflon™
PVC	polyvinyl chloride
PVT	pressure, volume, and temperature (properties)
Qty	Quantity
RF	radio frequency
RF-SVE	enhanced soil vapor extraction using radio frequency heating
roi	radius of influence
scfm	standard cubic feet per minute
SD	standard deviation
sec	second

SITE	Superfund Innovative Technology Evaluation
slpm	standard liters per minute
std.	standard
SVE	soil vapor extraction
SVOC	semivolatile organic compound
t_c	temperature in Celsius: $t_c = t_K - 273.15$
t_K	temperature in Kelvin: $t_K = t_c + 273.15$
T	temperature
TBP	true boiling point
TCE	trichloroethylene
TDR	time domain reflectometer
TPH	total petroleum hydrocarbon
TT	tracer-test well
USGS	United States Geological Survey
UT	the University of Texas at Austin
UTCHEM	three-dimensional, multicomponent, multiphase flow and transport simulator – the University of Texas at Austin
UTCOMP	compositional simulator — the University of Texas at Austin
VOC	volatile organic compound
WBS	work breakdown structure
XRD	X-ray diffraction

Contents

Appendices

Figures

Tables

Executive Summary

OBJECTIVES

One of the most widely used techniques for treating soils contaminated with volatile organic compounds is soil vapor extraction (SVE). Soil vapor extraction can also be applied to semivolatile organic compounds (SVOCs) if the soil is heated to increase the vapor pressure of the contaminants. The soil can be heated *in situ* by injecting hot air or steam, by direct resistive heating, or by applying electromagnetic energy in the radio frequency (RF) range. Although RF-SVE systems have been tried in several previous field demonstrations, success has been variable, and questions remain concerning the viability and cost-effectiveness of RF-SVE treatment systems.

The objectives of the RF-SVE demonstration were to:

- Heat a 28-m^3 block of soil at a depth of 3.05 m (the *design* volume) to a temperature of 150°C at the center of the block and 130 to 150°C at the perimeter
- Determine and document the reduction in SVOC contaminants that occurred as a result of the RF-SVE treatment
- Achieve a significant reduction in SVOC concentration

The targeted SVOCs in the *design* volume consisted of diesel range organic (DRO) compounds in the range of C_{10} to C_{21} hydrocarbons.

The objectives of the technology evaluation were to:

- Develop one or more engineering designs, using the RF-SVE technology
- Examine magnitude of costs for RF-SVE
- Evaluate the effect on costs of changes in performance and design
- Compare the RF-SVE costs to similar and alternative technologies

In addition, the basic scientific and engineering considerations involved in RF heating of soil are discussed in detail in this monograph to facilitate future studies by others. Necessary databases, equations, and example calculations for RF heating are included as guides. Partitioning interwell tracer tests (PITT) are described in detail from both theoretical and execution aspects.

FIELD DEMONSTRATION

The RF-SVE demonstration took place at Kirtland AFB, Albuquerque, NM. The site was a former fire-training pit that contained a variety of fuels, oils, and lubricants. The total petroleum hydrocarbon (TPH) concentrations ranged from 500 to 29,900 mg/kg of dry soil, and the contaminants were located from the ground surface to a depth of about 12 m. The peak concentrations occurred at a depth of 4 to 5 m. About 20% dry weight basis (dwb) of the hydrocarbons were in the DRO range of C_{10} to C_{21} and were the targeted contaminants. Less than 1% dwb of the hydrocarbons were gasoline range organic (GRO) components (less than C_{10}). The remaining TPH components were above C_{21} hydrocarbons and were defined as motor-oil range organic (MRO) compounds.

The RF-SVE treatment system consisted of two RF antennae and three SVE well screens. The RF antennae were located on two sides of the central SVE well screen and were centered vertically within the *design* volume. The two dipole antennae were operated at 27 MHz. Power was alternated between the antennae so that only one antenna produced energy at any given time. The power was initially applied at 2 kilowatts (kW), but was gradually increased to about 14 kW. Before being released to the atmosphere, the SVE off-gas was treated in an air-cooled condenser, a secondary condenser (knock-out drum), and three carbon filters to remove water and organic vapors. Monitoring wells were drilled and used for soil sampling and analysis and for data gathering.

Before the RF-SVE treatment, borings were made, and soil samples were collected for testing of soil characteristics and contaminant type, concentration, and distribution. Pre-RF-SVE testing included a partitioning interwell tracer test (PITT) and several *in situ* respiration tests. The SVE system was initially operated without RF heating to enable definition of the enhancement that RF heating provided. At the end of the RF-SVE treatment, additional soil borings were made, and soil samples were obtained and analyzed to determine changes in soil characteristics and contaminant types, concentrations, and distributions. Finally, *in situ* respiration and partitioning interwell tracer tests were also performed at the end of the RF-SVE demonstration. The basics of RF heating are presented to allow translation of the results to other situations.

DEMONSTRATION RESULTS

The SVE system was initially operated without RF heating. The initial off-gas analysis revealed that the constituents removed were within the GRO range, not the DRO range. During RF-SVE operation, the RF heating occurred in three phases:

1. Heating to 94°C (the boiling point of water at the elevation of the site)
2. Vaporization of soil moisture and
3. Heating the soil to temperatures greater than 94°C

The goal established for heating of the 28-m³ *design* volume (150°C in the center, 130 to 150°C at the perimeter) was not quite met during this RF-SVE demonstration. Temperatures reached a maximum of about 140°C at the center of the *design* volume and 110 to 120°C at the perimeter of the *design* volume. Since the airflow was from the periphery of the *design* volume toward a central extraction well, the SVE air-water vapor leaving the central extraction well contained a considerable amount of available heat, about 37% of the total RF energy added. This helped account for the lower peripheral temperatures. If the airflow had been from the central well toward three or four peripheral extraction wells, a portion of the water vapor in the air could have condensed in the cooler soil. This condensation would have released heat that would have increased the wet soil's temperature.

The goals for determining and documenting the removal of chemicals from the *design* volume by achieving a significant reduction in the DRO concentration were met. Substantial DRO mass was removed from the site soil. SVE alone resulted in the removal of chemicals in the GRO range (lighter than C_{10} compounds). Analyses of DRO condensate samples and of the site soils indicated that chemicals as heavy as C_{21} were removed during RF-SVE treatment. Thus, RF heating clearly enhanced the removal of DRO compared to SVE alone.

Initially, the RF-SVE demonstration focused on a 28-m³ *design* volume. The *design* volume was a 3.05-m cube of soil at a depth of 3.05 m. However, the RF heating and SVE systems influenced a volume greater than the *design* volume. Hence, the 132-m³ *sampled* volume (4.6 m × 3.6 m × 8.0 m) and the 324-m³ *heated* volume (8.8 m × 8.8 m × 4.2 m) were also used to evaluate the removal of SVOCs and water and the temperature distributions. Since the *design* volume could not be readily isolated in the comparison, a portion of the *sampled* volume that included the *design* volume was also examined for chemical removal. This 50-m³ *extended design* volume measured 4.6 m × 3.6 m × 3.1 m at a depth of 3.1 m.

During RF-SVE operation, the RF system was powered for 87 days and delivered 17,000 kW-hours (61.1 million kJ) of energy to heat the soil. The *heated* volume of soil was used to estimate the heat balances for the RF-SVE phase of the demonstration. The *heated* volume was defined by the measured temperature profile and by the heat balance when RF heating ceased on February 28, 1997. For the 87 days, the distribution of heat was as follows:

- Heating dry soil mass: 17.1 million kJ (0.28 kJ/kJ$_{Input}$)
- Heating residual soil moisture: 5.3 million kJ (0.09 kJ/kJ$_{Input}$)
- Vaporizing soil moisture into SVE Air: 19.4 million kJ (0.32 kJ/kJ$_{Input}$)
- Heating SVE air: 2.9 million kJ (0.05 kJ/kJ$_{Input}$)
- Heat loss to the surroundings: 16.4 million kJ (0.27 kJ/kJ$_{Input}$)

The pre-demonstration sample analyses indicated that the *sampled* volume contained an initial DRO mass of 244 kg. The post-demonstration evaluation indicated that the residual DRO mass in the *sampled* volume was 128 kg. Thus, the RF-SVE treatment resulted in a 48% dwb or 116 kg decrease in the DRO mass within the *sampled* volume. In this volume, the average DRO content decreased from 1020 mg/kg of soil to 560 mg/kg. For the *extended design* volume, 70 kg, or 53% of the DRO mass, was removed by RF-SVE. In this volume, the average DRO content decreased from 1520 mg/kg to 690 mg/kg.

DRO condensate was also collected and used to determine the types and amounts of chemicals removed by the RF-SVE system. Measuring the volume of the DRO condensate indicated that about 100 kg was removed by RF-SVE. Analyses of condensate samples showed that the composition of chemicals shifted from GRO components during the SVE-only phase to DRO components during RF-SVE operation. The chemical composition of the condensate collected near the end of the RF-SVE demonstration consisted of the full range of DRO compounds (C_{10} to C_{21} hydrocarbons).

Analyses of the soil within the 132-m^3 *sampled* volume indicated that at least 300 kg of TPH mass was removed from the soil. The partitioning interwell tracer tests confirmed a similar loss of TPH mass at 300 kg but from a larger PITT volume of soil (about 380 m^3).

The 184-kg difference between the decrease in TPH and DRO masses in the *sampled* volume was probably caused by one or more of the following reasons:

- Errors introduced by soil sampling, handling, holding, subsampling, and analyses
- Movement of DRO and TPH compounds out of the *sampled* volume
- Hydrous pyrolysis oxidation of DRO and TPH compounds
- Biooxidation of DRO and TPH compounds

Obtaining mass closure was also complicated by incomplete off-gas data and by the fact that the *sampled* volume was an open system. The RF-SVE process could have affected soil beyond the limits of the *design* volume. Further, the system was not operated until a treatment endpoint for the site was reached (as it likely would have been in an actual remediation project). Rather, the demonstration was completed within the constraints of time and budget.

Recommendations resulting from this study include the following:

- The RF-SVE system should be optimized, with air and heat flow coordinated for maximum efficiency
- A design tool should be developed that links RF heating mechanisms with an SVE-transport model
- Further consideration should be given to residual chemicals and the risks that they pose, to provide assistance in defining appropriate endpoints for RF-SVE treatment
- Consideration should be given to using RF-SVE for the following applications:
  - Enhanced SVE of SVOC chemicals that are not readily removed by SVE alone
  - Spot heating of small areas
  - Thermally enhanced biodegradation and
  - Enhanced removal in soils of low air or water conductivity

HYPOTHETICAL RF-SVE DESIGN FOR FULL-SCALE REMEDIATION PROJECTS

Three hypothetical design scenarios were evaluated to study the economics and design sensitivities of the RF-SVE process. The scenarios addressed full-scale applications of RF-SVE that remove SVOCs from contaminated, unsaturated (vadose zone) subsurface soil. The field demonstration experience at Kirtland AFB formed the basis for the designs and economic evaluations.

The equipment and component designs for these case studies were based on the RF-SVE systems operating continuously over the entire area of each site. The designs for these cases addressed either a 4300 m² (~1 acre) site or an 850 m² site. Targeting a site with a smaller area with spot contamination could be more cost-effective. The design also used 500 mg/kg for the regulatory cleanup level of SVOCs.

The design of the RF heating system and its configuration for a particular application depend on the following factors:

- The operating frequency of the RF generator
- The design and number of antennae
- The soil bulk electrical properties
- The nature of the contaminants within the site
- The target temperature for RF heating
- The desired rate of contaminant removal

The RF antenna delivers energy to dielectric materials, such as soil, by electromagnetic radiation. Hence, RF heating is not directly dependent upon heat conduction and fluid convection. The antenna heats the soil quickly and efficiently because energy coupling occurs at the molecular level. During RF heating, the applied electromagnetic field interacts directly with the free charge and the permanent and induced electric dipoles present in the soil. RF heating can increase the soil temperature from a few degrees above ambient to more than 250°C, depending upon the application and the type of soil and its moisture content.

The soil's bulk electrical properties, the dielectric constant and conductivity, relate directly on the macroscopic level to the soil's ability to absorb RF energy and to heat. The dielectric constant and conductivity determine the wavelength of the RF energy in the soil, strongly influencing the design of the antenna for optimal heat transfer. The conductivity of the soil is directly proportional to the soil's ability to absorb RF energy. The mineral composition of the soil and the RF operating frequency affect the bulk electrical properties of soil. Many types of material, such as organic and inorganic matter and water, make up a soil. Each material has its own unique dielectric properties.

A generator conveys RF energy via coaxial cabling to the antenna. The matching network (tuner) allows the RF generator to provide maximum power to the antenna. Therefore, the combination of the optimally designed tuner and antenna helps insure the maximum transfer of RF energy to the soil.

Three methods of contaminant removal could occur in RF-SVE remediation of an SVOC-contaminated site. These methods are SVOC-vapor extraction with air, thermal oxidation at temperatures near 100°C, and biooxidation at low temperatures (25 to 35°C). In most RF-SVE applications, all three will occur with vapor extraction and thermal oxidation predominating. However, thermal oxidation and biooxidation require sufficient soil moisture and dissolved oxygen in the moisture before SVOC oxidation can occur.

If the SVOCs contain high boiling point compounds with vapor pressures less than 1 mm Hg at and above 100°C, using SVE can require soil temperatures greater than 150°C. Therefore, most of the soil moisture must be removed before the soil temperature can increase above 100°C and the heavier SVOCs can be extracted. If the air and vapor mixture is extracted from a central well, the soil moisture will decrease in a continuously moving pattern toward the periphery of the treatment area. Removing water and heating to a design temperature of 150°C may require a relatively large number of RF generators and power. Operating at 100°C and relying on thermal oxidation require fewer RF generators and less power consumption. Biooxidation could further decrease the number of generators required for soil treatment. However, longer treatment times are usually needed, and the bacteria might not easily oxidize the larger, more complex SVOC molecules.

When designing the RF-SVE system, the antenna should be at the center of the specific treatment area. Based on preliminary assessment, it is believed that the SVE air-vapor extraction wells should be at the periphery of the treatment area in a 3- or 4-spot pattern. If required by soil characteristics like air permeability and high airflow channeling, a SVE well near the antenna well can also be used to inject air at the same rate as the total airflow from all extraction wells. This mode of operation will minimize the amount of air being drawn from outside the treatment volume. A central injection well will create a positive flow of air toward the extraction wells.

The heating pattern for a particular soil application depends on such factors as:

- RF operating frequency and power level
- Length and number of antennae
- Antenna position and orientation and
- Mutual coupling between the antennae, electrical phasing, and soil properties

The volume that each antenna will heat depends upon the design of the antenna (length, RF frequency) and soil properties (water and mineral contents, conductivity). The heat radiates outward from the antenna toward the periphery of the volume. Concentric zones of gradually increasing temperatures are established, and heating continues until the desired temperature is achieved. The rate of heating is dependent upon the kilowatt output of the antenna, SVE airflow rate, soil characteristics (thermodynamic properties and water content), and heat loss from the treatment volume. Smaller volumes will have higher heat losses. Low RF antenna outputs can increase treatment time and the heat loss from the treatment volume. Conversely, if the RF output is too high, overheating can damage the antenna. Furthermore, a high RF output can produce a condensing steam-vapor front around the antennae. This front could saturate the soil with water and contaminant and block the airflow through the central portion of the soil. The excess fluid in the soil could also flow below the contaminated zone.

High airflow rates can remove excessive amounts of heat as water vapor (humidity) that will slow the soil heating process. When the air is injected into a central well and extracted from peripheral wells, it is believed that this heat loss can be minimized. When treating the soil by the thermal oxidation or biooxidation modes, the airflow should be minimized. However, the airflow should be sufficient to maintain aqueous-phase oxygen contents that will sustain the oxidation reactions. If removal of water is the goal to achieve soil temperatures approaching 150°C, the airflow and its humidity leaving the extraction wells must be optimized to maximize the removal rate of water after heating the wet soil to 100°C.

ECONOMIC CONSIDERATIONS

The following three hypothetical design cases were considered in this technology evaluation:

Case 1 and Case 2
- Heated wet soil from 15 to 100°C (the boiling point of water)
- Maintained temperature at 100°C and removed water and light-end SVOCs by SVE
- Heated dry soil from 100 to 132°C and removed heavy-end SVOCs by SVE

Case 3
- Heated wet soil from 15 to about 100°C
- Maintained temperature near 100°C and removed SVOCs by a combination of SVE, thermal oxidation in the heated zone, and biooxidation near the perimeter of the heated zone

Cases 1 and 2 were for different treatment volumes, 24,800 m³ and 4900 m³, respectively. Since water was essential for the thermal oxidation and biooxidation processes in Case 3, the amount of water that was removed from the soil was minimized. In all cases, the SVOC contamination extended from ground surface to 5.8 m below ground surface (bgs).

The estimated unit remediation costs for the three hypothetical cases for the RF-SVE technology areas follow:

Case	Volume, m³	$ per m³	$ per yd³	$ per Ton[a]
1	24,800	182	139	93
2	4900	288	222	147
3	4900	209	161	107

[a]Using a density of 1.50 ton (3000 lbs) per yd³.

In comparison, the estimated treatment costs associated with above-ground, low-temperature, thermal desorption (LTTD) of excavated soil containing SVOCs are as follows:

Process	$ per m³	$ per yd³	$ per Ton[a]
Thermal treatment	80–195	60–150	40–100
Excavation-soil replacement	20–40	15–30	10–20
Total cost	100–235	75–180	50–120

[a]Using a density of 1.50 ton (3000 lbs) per yd³.

However, LTTD treatment of soil is limited by the depth of the soil excavation that is required to remove the contaminated soil. As the depth of the contaminated soil increases, the unit cost for excavating the soil will increase as more over-burden must be removed. Generally, the application of RF-SVE is not limited by the depth of contamination.

The cost estimates for the hypothetical designs assume the mobile RF units will be leased. The capitalization of the RF units will be a more cost-effective option. For Cases 2 and 3, capitalizing the RF units could decrease the estimated unit costs to the following values:

Case	Volume, m³	$ per m³	$ per yd³	$ per Ton[a]
1	24,800	109	84	56
2	4900	216	166	110
3	4900	185	142	94

[a]Using a density of 1.50 ton (3000 lbs) per yd³.

Because of the high cost for leasing the RF heating units, shortening the amount of time for remediation or decreasing the number of RF units can significantly decrease costs. Conversely, RF units with higher energy output could be more cost-effective by decreasing the lease or capital cost.

The RF-SVE technology appears to be less cost-effective for small volume projects. According to previous pilot-scale studies, the volume of contaminated soil should not be less than 350 m³ (455 yd³). The primary reasons for this limitation are the high mobilization and capital cost associated with installation of the SVE and vapor abatement systems and the high O&M costs associated with the RF trailer units. The economics of scale come into effect for larger systems that can justify the remediation costs per unit volume of soil (dollars per m³).

In comparison, the estimated costs for treating contaminated vadose-zone soil using similar heating technologies are estimated as follows:

Process	$ per m³	$ per yd³	$ per Ton
Steam injection with SVE	105–250	80–200	53–133
Hydraulic fracturing with steam injection	144	110	74
Six-phase soil heating with SVE	112	86	57

Several low-cost, low-temperature technologies can also be used to treat vadose-zone soil to remove organic contaminants. These are bioventing, biooxidation, and conventional SVE. However, technologies relying on biodegradation will take years to complete and usually cannot treat sites with high contaminant concentrations and complex, high molecular-weight compounds. Furthermore, SVE can only remove volatile organic compounds (VOCs) not SVOCs from the soil.

POTENTIAL APPLICATIONS

The RF-SVE technology appears to be less cost-effective when applied for remediation of entire contaminant plumes. This is due primarily to the high operation and material (O&M) costs associated with the lease of the RF trailer units over long periods. The cost is also impacted by the large quantity of heat energy needed to remove water and hydrocarbon contaminants from large volumes of soil. RF-SVE appears to be more suited for small "hot spot" contaminant removal.

Although the RF-SVE technology can be used in virtually any soil, the methodology could pose problems in soils containing cobbles because thermal "shadows" could develop behind the cobbles. RF heating could become ineffective in extremely dry soils since some water is needed for RF heating to occur. Moist permeable soils have an increased ability to absorb RF energy. Conversely, very high moisture levels can limit the effectiveness of RF heating by shrinking the radiative heating pattern and increasing the amount of energy required to heat and remove moisture where high heating applications (greater than 100°C) are needed.

As with any technology, limitations exist for RF heating. However, it has an important niche when soil volumes to be treated are relatively small, contaminants are of higher molecular weights, and soils are moist and permeable.

CHAPTER 1

Introduction

This monograph was prepared to provide detailed technical information obtained from a recent technology demonstration of enhanced soil vapor extraction (SVE) using radio frequency (RF) heating. A pilot-scale demonstration was performed under the Advanced Applied Technology Demonstration Facility (AATDF) program, which was established at Rice University by the Department of Defense (DOD). The demonstration site was within a former fire-training facility at Kirtland Air Force Base (AFB) in Albuquerque, NM. This monograph also provides cost and performance analysis that may foster potential future use of this technology on other sites.

The overall objective of the RF-SVE demonstration was to provide information for a full-scale engineering design of several potential RF-SVE applications. These designs use the results from the demonstration to examine predicted performance, magnitude of costs, and modifications to the design that may decrease cost. The monograph also provides information on the economic analysis of the costs, site characteristics favorable for the technology, and a comparison of the applicability of this technology to other applications.

This monograph also had an objective to provide detailed information of both a theoretical and practical nature to facilitate future testing by others of RF-SVE treatment of soils.

1.1 PROCESS DESCRIPTION

Radio frequency (RF) refers to those frequencies between 500 kilohertz (kHz) and 500 megahertz (MHz). The ISM (Industrial, Scientific, and Medical) frequencies used for RF heating are 6.78, 13.56, 27.12, and 40.68 MHz. Radio frequency heating is similar to microwave (MW) heating that uses 500 MHz to 500 gigahertz (GHz). The lower frequency for RF operating systems means that electromagnetic energy penetrates several meters into a material, as opposed to only several centimeters for MW. The four subsystems (Figure 1.1) required for an operational RF heating system for soil decontamination are

- Antenna, or applicator, array in the soil
- RF power generator with an impedance-matching network and a control system
- Vapor removal system consisting of extraction wells and contaminant containment (impervious cover) and
- Off-gas vapor treatment system

Of these components, the RF antenna is the critical item that drives the design and constrains the other components.

Electromagnetic energy heats materials quickly and efficiently because energy deposition occurs through interactions between the applied electromagnetic field and molecules and atoms present

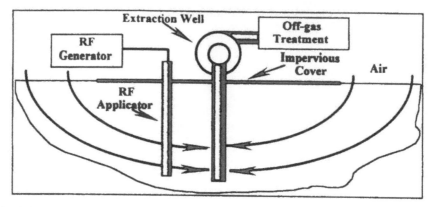

Figure 1.1   Schematic of RF enhanced soil vapor extraction system.

in the material. Electromagnetic heating does not rely on surface heat transfer to heat the material. The dominant heating mechanisms depend on material composition and usually differ for RF and MW. These mechanisms are discussed in Section 3.2. The electrical properties of the soil determine the amount of RF power needed to heat the soil and the propagation characteristics of the electromagnetic energy in the soil. The volume distribution of the electromagnetic energy is determined by the geometry of the RF antenna and soil electrical properties. The actual heating pattern is significantly affected by the heat and mass transfer through thermal conduction and fluid convection that occur in the soil as it is heated. The temperatures that can be achieved using a RF system depend on the antenna, the RF generator, the type of soil, and the soil moisture content.

1.2 SITE IDENTIFICATION

1.2.1   Introduction

It is extremely important that any RF-SVE demonstration is conducted at a site that is compatible to the technology. Furthermore, the site must lend itself to critical monitoring of system performance. For the RF-SVE demonstration, an extensive search and evaluation of field sites were undertaken. Several candidate sites were selected, and preliminary soil and contaminant data were obtained. Based on the results of a careful analysis of data, a fire-training site (FT-14) at Kirtland AFB, Albuquerque, NM was selected for the RF-SVE demonstration.

1.2.2   Selection Decision

The prominent site characteristics at FT-14, Kirtland AFB, were as follows:

- Sandy silt with some fine gravel
- Relatively uniform deposit
- Soil moisture at 3 to 10% dwb
- Groundwater table greater than 150 m bgs
- Contaminants from ground surface to about 12 m depth
- Total petroleum hydrocarbon (TPH) concentrations at 2000 to 16,000 mg/kg of dry soil
- Maximum TPH concentrations at ~4.5 m below ground surface
- Diesel range organic (DRO) compounds found in the range 11- to 18-carbon chains

These characteristics made the FT-14 site a likely choice for the RF-SVE demonstration. The following additional parameters made FT-14 a first-priority site:

- Contaminants reasonably contained laterally, simplifying mass balance efforts
- Site free of any potential RF interference, such as buried steel drums
- Base environmental management team cooperative and ready to act
- Established personnel at the base from the prime contractor for the demonstration work
- Routine site permitting
- Sufficiently dry and permeable soil to permit vapor extraction
- Majority of contaminants not extractable with SVE at ambient temperature, but significant quantities extractable at the elevated temperatures associated with RF heating

1.2.3 Alternative Sites

Preliminary borehole sampling and analyses were performed at Aberdeen Proving Grounds and Eaker AFB. The Aberdeen Proving Ground site was held in reserve until further borehole sampling and analysis could be completed at the Kirtland AFB, FT-14 site. The Eaker site was not appropriate for the RF-SVE demonstration because of tight clayey soils.

Several other sites were also evaluated; however, no field sampling or chemical analyses were performed for these sites. The sites included McClellan AFB, Fort Lee, Fort Jackson, and Beale AFB. The site details are provided in Table 1.1 to show factors that need to be considered.

1.3 PROJECT OVERVIEW

In situ RF heating was originally developed to extract petroleum from tar sand and oil shale in the mid 1970's. Interest in the technology shifted to remediation of contaminated soil in the mid 1980s. As referenced in Chapter 2, several pilot-scale demonstrations of soil vapor extraction enhanced with RF heating (RF-SVE) have verified the feasibility of the concept. Since the results of these past demonstrations have been variable, further pilot-scale demonstration and testing were deemed necessary.

The demonstration at Kirtland AFB used the RF technology in conjunction with conventional SVE. The RF technology was used to heat the soil. Soil heating improved organic contaminant volatilization that then allowed for enhanced contaminant removal via conventional SVE using a vacuum drawn from one or more extraction wells.

The RF heating system used in this demonstration was a remotely operated, computer-controlled module consisting of:

- RF generator
- Matching network, or tuner
- One or more antennae

The RF heating technology essentially delivers energy to materials such as soil by electromagnetic radiation and diffusion. The RF generator produces the RF energy that is applied to the antennae via coaxial cabling, the transmission lines. The antennae, usually placed in fiberglass-cased boreholes, couple the RF energy into the soil.

The design of the RF-SVE system focused on a *design* volume that was a 28-m^3 cube of soil at a depth of 3.05 m. The RF-SVE system consisted of two antennae and three SVE well screens in or near the center of the *design* volume. Before being released to the atmosphere, the SVE off-gas was treated in an air-cooled condenser, a knock-out drum to collect the condensate, and three carbon filters to remove water and organic vapors. Monitoring wells were drilled and used for soil sampling and for data gathering.

The demonstration project goals were

- Achieve a temperature of 150°C at the center of the *design* volume and 130°C at the periphery of the *design* volume.

Table 1.1 Summary of Potential RF-SVE Demonstration Sites

Site	Facility	Soil Conditions	GWT	Contaminants	Comments
Kirtland Air Force Base, Albuquerque, NM	Fire-training pit, FT-14	Weathered granite, sandy silt, gravel	>80 m	GRO & DRO compounds with TPH up to 16,000 mg/kg	Site selected for RF-SVE demonstration
Aberdeen Proving Grounds, Baltimore, MD	Storage warehouses & #2 fuel oil tank area	Sand, silty sands, and silty clays	~5.5 m	#2 fuel oil up to 6,000 mg/kg	Reserve site
Eaker Air Force Base, Blytheville, AR	Spill sites #1 and #2, fuel oil spills	Stiff silty clay, silty clay, fine silty sand, fine sand	~3 m	Samples from 2 boreholes, TPH varied from 300–9,000 mg/kg, moisture from 17–33% dwb	Soil too tight for effective SVE, high water content
McClellan Air Force Base, Sacramento, CA	T-16, T-18, TRL-19 oil spill sites	Silty sand with some clay	~30 m	TPH-D (Diesel) @4,900 mg/kg	MOU finalized in May, 1996
Beal Air Force Base, Marysville, CA	Underground pipeline leak near runway	Sand, clay, silty sand, silt, sand	>5.5 m	JP-7 jet fuel	Wrong contamination
Fort Lee, VA	Fire-training pit	Silty sand, silty clay, sandy silt, clays	4–5.5 m	Diesel fuel, TPH @ 1,350 mg/kg (3 holes near pit)	
Fort Jackson, SC	Above-ground oil leak	Sandy, silty clay	~5.5 m	Diesel fuel, TPH-heavy @ 13,000 mg/kg, highest concentration under bldg.	Site may be too crowded with utilities
Naval Air Station Moffett, San Francisco, CA				Heavy hydrocarbons speculated	
EPA SITE Program, Kansas, Nebraska, IA	Coal tar gasification sites, heavy hydrocarbon residuals			Heavy hydrocarbons	

Note: DRO, diesel range organic compounds; GRO, gasoline range organic compounds; TPH, total petroleum hydrocarbons; GWT, groundwater table.

- Determine and document the effect of RF-SVE on the removal of SVOCs that have a pure component vapor pressure of greater than 1 mm of Hg in the temperature range of 100 to 150°C.
- Achieve a significant reduction in SVOC concentration.

The target SVOCs were in the carbon range of C_{10} to C_{21}. The RF-SVE demonstration at Kirtland AFB specifically addressed DRO compounds. The target soil temperatures were selected to ensure removal of these SVOCs. To reach this temperature, however, most of the soil moisture had to be removed. Hence, the following three-step approach was used to achieve the goals:

- Heated the soil and the contained moisture and contaminants in the *design* volume to the boiling point of water. At the Kirtland AFB demonstration site, the boiling point was 94°C.
- While at 94°C, removed the moisture and low-boiling point SVOCs from the soil.
- After the moisture was removed from the soil, heated the soil toward 150°C and removed higher boiling point SVOCs.

The field work at Kirtland began on June 24, 1996 and ended on July 10, 1997. The fieldwork included 87 days of nearly continuous RF-SVE operation that started on December 3, 1996, and ended on February 28, 1997.

Before RF heating began, the SVE system was operated to compare the chemical removal of SVE to that of enhanced SVE using RF heating. During the 3 months of RF-SVE operation, 17,000 kWh of RF energy were delivered to the soil. Initially, the demonstration focused on the 3 m × 3 m × 3 m *design* volume. However, the RF heating and SVE systems influenced a volume greater than the *design* volume. Hence, the 9 m × 9 m × 4 m *heated* volume (324 m³) and 5 m × 4 m × 8 m *sampled* volume (132 m³) were also used to evaluate the temperature distributions and the removal of SVOCs and water. Since the *design* volume could not be readily isolated in the comparison, a portion of the *sampled* volume that included the *design* volume was also examined for chemical removal. This *extended design* volume measured 5 m × 4 m × 3 m (50 m³) at a depth of 3 m.

Data collected as part of the investigation included pre- and posttesting information, as well as data collected during the demonstration. Before the field demonstration, soil borings were made, soil samples were obtained, and chemical analyses were performed to determine the initial concentration of various constituents in the soil. A partitioning interwell tracer test (PITT) was also performed to help identify the initial mass of contaminants and to identify preferential pathways for airflow. *In situ* air permeability tests were performed, and *in situ* respiration tests were also completed. Other characterization tests (e.g., grain size and mineralogy of the soil) were also performed. During the demonstration, airflow rates, temperatures, and pressures at various points were measured, and air samples were collected and analyzed from the off-gas stream. After the demonstration, additional soil borings were made and chemical analyses performed on samples collected from the borings. A post-PITT test and a series of *in situ* respiration tests were performed.

Based on the demonstration results, designs were developed for hypothetical full-scale RF-SVE systems, configured based on knowledge gained. Several "typical" situations were considered and conceptual designs of RF-SVE systems developed. Cost estimates for the RF-SVE systems were then developed and compared with cost estimates for alternative technologies.

CHAPTER 2

Measurement Procedures

The following sections discuss the measurements necessary for characterizing the site conditions, evaluating the contaminants, and monitoring system operation and performance during pilot-scale demonstrations or full-scale remediation of RF-SVE projects.

2.1 MEASUREMENT PROCEDURES FOR CHARACTERIZING THE MATRIX

Extensive characterization of the geology and nature of the contaminants at the site is necessary in determining the applicability of RF heating as a remediation alternative. Table 2.1 summarizes the analyses and testing required for determining site parameters, characterizing the contaminants, and evaluating other remediation technologies to enhance the use of RF heating.

2.1.1 Contaminant

Before the application of any remedial technology, an understanding of the nature of the contaminants and any by-products is necessary. The installation of borings and collection of soil samples for laboratory analysis is necessary for evaluating the extent of contamination. Soils are analyzed for the presence of all contaminating semivolatile organic compounds (SVOCs). For example, the modified EPA Method 8015 can be used to measure DRO content. Trichloroethylene (TCE) and dichloroethane (DCA) can be measured using EPA Method 8010. EPA Method 418.1 can be used to quantify petroleum hydrocarbons if more than one fuel contaminant is present. Partitioning interwell tracer tests (PITT) can also be used to characterize organic contaminants in the vadose zone (see Section 3.7.8).

2.1.2 Geology

Barriers such as building foundations, underground storage tanks, previous borings or wells, and underground utility tunnels or trenches must be identified and avoided if possible. These features may provide preferential pathways and cause a short-circuiting during venting.

Soil borings should be continuously logged to identify permeable zones, fractures, and structural features. Cores should be examined for bedding or changes in permeability with increases in the silt content. More than one continuously logged soil boring may be necessary due to the potential for clay or impermeable zones that may be discontinuous or limited in lateral extent. The distribution of these impermeable zones will help provide an understanding of contaminant migration.

Table 2.1 Summary of Pre-Testing Analyses

	Analysis	Method
Contaminant characteristics	Volatile organic compounds[a]	EPA 8010[1]
	Semivolatile organic compounds[a]	EPA 8270B[1]
	Diesel range organic compounds[a]	Modified EPA 8015[1]
	TPH[a]	EPA 418.1[1]
	PITT	Section 3.7.8
Geology	Soil samples[a]	Visual
Remediation feasibility	Soil vapor extraction test[a]	Section 2.1.3[2,3]
(field tests)	Air permeability tests[a]	Section 2.1.3[2,3]
	Respiration tests[a]	Section 2.1.3[2,3]
Geotechnical	Particle size analysis[a]	ASTM D1140, D422[4]
	Moisture content[a]	ASTM D2216[4]
	Soil mineral analyses[a]	XRD
	Dry bulk density[a]	API RP40[5]
	Air permeability	API RP40[5]
	Effective porosity	API RP40[5]
	Pore fluid saturation	API RP40[5]
Oxidation potential	PH	SW-846[1]
	Alkalinity	EPA 350.1[1]
	Total kjeldahl nitrogen	EPA 350.2[1]
	Total phosphorous	EPA 365.4[1]
	Iron	EPA 6010a[1]
	Total organic carbon	EPA 9031[1]
	Total heterotrophic bacteria	EPA 9215[1]
	Contaminant utilizing bacteria	EPA 9215[1]

[a] Critical Items for RF-SVE Technology
[1] (EPA, 1995); [2](BRE, 1996b); [3](ENVIROGEN, 1996); [4](ASTM, 1996); and [5](API, 1996).

2.1.3 Remediation Feasibility Testing

Before the treatment plant is designed, feasibility testing is conducted to determine the applicability of SVE to increase heat distribution in the subsurface. An *in situ* air permeability test is used to measure achievable flows vs. applied vacuum and to determine the radius of pressure influence.

Field testing of a SVE system consists of applying a vacuum to an extraction well via a vacuum pump. The induced vacuum creates a vacuum gradient surrounding the extraction point. Monitoring points [typically located 1.5 m (5 ft), 3 m (10 ft), and 4.6 m (15 ft) from the extraction point] are sealed with specially designed seals that allow the induced vacuum to be measured at each monitoring point. The extraction point is sealed, and an averaging pitot tube or totaling meter is mounted on the extraction well to measure the air volume removed from the subsurface. Vacuum readings are taken at 15-min intervals from each monitoring point. Airflow, induced vacuum, and VOC content of the effluent air will be taken at the extraction point during each interval. The test is conducted until the pressure and flow readings reach static conditions. The measured test parameters at the soil-gas monitoring points are as follows:

- Soil-gas pressure or vacuum
- Percent oxygen
- Percent carbon dioxide
- Volatile organic compound content at the exhaust-gas wellhead

In situ respiration tests involve the injection of air into the contaminated, vadose-zone soil to stimulate oxidation processes that occur at ambient temperatures or at elevated temperatures. Similar tests are also conducted in noncontaminated soils to measure background oxygen uptake rates. In these tests, air that contains ambient levels of oxygen is introduced to the subsurface using either vacuum extraction or air injection. Air injection minimizes volatilization of compounds to the surface and provides a low cost, effective means of removing contaminants. The testing consists of placing narrowly screened soil gas monitoring probes into contaminated and uncontaminated soils in the vadose zone and aerating the soils with air that contains an inert tracer gas, such as helium, for various periods of time. Before air injection begins, subsurface oxygen and carbon dioxide concentrations are measured using field instruments. The air-helium mixture is injected into the soil for a period of 24 h. Following the air injection, carbon dioxide, oxygen, and helium in the soil gas are monitored over a 40- to 80-h period. Once the *in situ* test has been completed, subsurface gas concentrations will be plotted vs. time to calculate oxygen uptake rate in the soil. Using the oxygen uptake rates, it is possible to calculate a hydrocarbon degradation rate.

2.1.4   Geotechnical

Physical and chemical properties of soil, such as mineral analyses, particle size analysis, dry bulk density, and moisture content, should also be obtained. Air permeability and pore fluid saturation play an important role in SVE design and the potential effectiveness of chemical removal potential by SVE.

2.1.5   Oxidation Potential

Soil samples can be submitted to a laboratory to characterize the site conditions for potential hydrous–pyrolysis oxidation and biooxidation. Pyrolysis oxidation can occur only at temperatures near 100°C and with sufficient soil moisture. Biooxidation rates are sustainable only if favorable temperatures, soil moisture content, and sufficient quantities of inorganic nutrients, nitrogen and phosphorus, are present in the soil. Stable physical and chemical conditions must occur for the development of an active biomass. Tests should also be conducted to measure the effect of RF waves on the viability of the biomass. KAI Technologies, Inc. is currently completing a demonstration in Alaska of enhanced biodegradation of fuel oil using RF heating.

2.2 MEASUREMENT PROCEDURES FOR
MONITORING OPERATING PARAMETERS

A summary of measurement procedures that would be utilized during the operation of the RF-SVE system is presented in Table 2.2.

2.2.1   Vapor

Vapor samples are collected from the influent stream of the vapor treatment equipment for determining the mass of contaminant removed from the soil. From the effluent stream of the vapor treatment equipment, vapor samples must be collected to show regulatory compliance.

Samples can be collected using Tedlar® bags and adsorbent media like Tenax® or charcoal tubes. An adsorbent medium is necessary for sampling the heavier, less volatile hydrocarbons that can adhere to the sides of the Tedlar bag and cannot be analyzed. Due to the probable presence of moisture in the vapor stream, a condensate trap should be used to remove the moisture, and some of the heavier hydrocarbons, before entering the adsorbent.

During sample collection, the total volatile organic concentration in the vapor stream should be measured. Measurements should be performed using a flame ionization detector (FID) to detect both the aromatic and aliphatic fractions of hydrocarbons.

Table 2.2 Summary of Monitoring Parameters

	Analysis	Method
Vapor	C_2–C_6 hydrocarbons	Modified EPA 8015
	C_7–C_{16} hydrocarbons	Modified EPA 8015
	$C>_{16}$ hydrocarbons	Modified EPA 8015
	Volatile organic compounds	EPA 8010
	Total volatile VOC (field analysis)	Section 2.2.1
Condensate	TPH	EPA 418.1
	Volatile organic compounds	Modified EPA 8015
	Semivolatile organic compounds	EPA 8270B
	Diesel range organic compounds	EPA 8010
RF system operation	Power	kW meter
	Soil temperature in °C	Fiber optic thermometer
	Airflow Rates	Annubar flow meter
	Pressure (centimeters of water)	Magnahelix pressure gauge
Soil	TPH	EPA 418.1
	Volatile organic compounds	EPA 8010
	Semivolatile organic compounds	EPA 8270B
	Diesel range organic compounds	Modified EPA 8015
Oxidation potential	In situ respiration tests	O_2 and CO_2 meters
	Contaminant utilizing bacteria	EPA 9215
	Total heterotrophic bacteria	EPA 9215

2.2.2 Condensate

The condensate collected in the air–water separator consists of immiscible phase hydrocarbons (IPH) and water. Following the removal of most of the soil moisture, the majority of the condensate will be IPH with some water. The volume of the recovered IPH from the air–water separator must be tracked to measure the mass removal. Samples of the recovered IPH are collected and submitted to the laboratory to determine the hydrocarbon composition. Changes in the petroleum hydrocarbon composition and the increase in subsurface temperature can be monitored through time. At lower temperatures, the more volatile hydrocarbons represent a greater percentage of the contaminant volatilized and removed. With an increase in temperature, the heavier, less volatile hydrocarbons are removed.

2.2.3 RF Heating System Operation

The temperature of the soil is monitored using the fiber optic thermometer system to ensure the soil is properly heated to the target temperature. To monitor system performance, vacuum and airflow rates are measured at the SVE wellheads using an Annubar flow meter coupled with a magnahelix pressure gauge. The vacuum can be adjusted to balance the flow of air in the subsurface.

The operating characteristics of the RF unit are also measured to determine the overall efficiency of the conversion of AC power to heat energy. The measured values include

- RF "On-time" AC power in kilowatts (kW)
- AC energy (kW-hours)
- RF power to the antennae (kW)
- RF energy to the antennae (kWh)
- RF power delivered to the soil (kW)
- RF energy to the soil (kWh)
- Power delivered to the RF generator (monitored using a kWh meter)

2.2.4 Soil

The mass of contaminant removed from the soil by volatilization is calculated from the vapor and condensate samples analyzed during the operation of the RF-SVE system. This information is then compared to the analyses of pre- and postremediation soil samples to demonstrate to the regulatory agency that there was a reduction in the mass of contaminant. The soil sample results can also be compared to pre- and post-PITT results to verify the amount of contaminant mass that was removed from the site.

2.2.5 Oxidation Potential

During the collection of vapor samples, the oxygen and carbon dioxide concentrations in the vapor stream should be measured using oxygen and carbon dioxide meters. These measurements will monitor the oxygen consumed and carbon dioxide produced by the oxidation of SVOCs in the subsurface. At ambient and moderately elevated temperatures, biooxidation mechanisms can occur. At temperatures near 100°C, oxygen can be consumed by hydrous pyrolysis oxidation of organic compounds.

CHAPTER **3**

Summary of Technology Demonstration

3.1 INTRODUCTION

A field demonstration of RF-SVE technology to remove semivolatile organic compounds from subsurface soil was performed at Kirtland AFB in Albuquerque, NM. The RF-SVE project was conceived, developed, managed, and led by a research team at the University of Texas at Austin (UT). To complete the field demonstration, UT collaborated with Brown & Root Environmental (BRE), the prime contractor, and KAI Technologies, Inc. (KAI), the RF heating contractor.

The principal technical goal of the RF-SVE demonstration was to determine if RF heating could increase the removal of contaminants from soil relative to SVE. Generally, remediation by SVE is limited to lighter organic compounds whose vapor pressures are greater than 1.0 mm Hg at 20°C; for example, benzene at 76 mm of Hg and toluene at 22 mm of Hg. Figures 3.1, 3.2, and 3.3 indicate that heavier organic compounds (C_{12} to C_{21}) are not rapidly removed by SVE at ambient temperatures at ~20°C (~300 K). Figure 3.1 shows that the vapor pressures are considerably below 1 mm of Hg at 300 K. Figure 3.2 gives the concentration of the saturated organic compounds in air at increasing air temperatures. Figure 3.3 plots the air temperature and mass removal rates for each compound when the total airflow rate is 1.0 m³/min (35 scfm). Figure 3.3 suggests that at 300 K the removal rates for all but one alkane are considerably less than 10 kg/day. However, increasing the temperature toward 150°C (~425 K) can significantly increase their vapor pressures and removal rates. For example, the removal rate ranges from about 25 kg/day for heneicosane to about 2000 kg/day for dodecane. Therefore, heating a contaminated soil with RF energy can extend the range of chemicals available for removal by SVE. The data and calculations for these figures are listed in Appendix F.

3.2 RF HEATING

The RF heating technology delivers energy to soil by electromagnetic radiation and diffusion. The RF generator produces RF energy that is applied to the soil through coaxial cabling (transmission lines) to each antenna. The antenna, usually placed in a fiberglass-cased borehole in the KAI system, then radiates the RF energy into the soil. During RF heating, the applied electromagnetic field interacts directly with mobile free charges, permanent dipoles, or induced dipoles (or all three) that are present in the soil. The soil's bulk electrical properties, permittivity and conductivity, are directly related to the soil's ability to absorb RF energy and produce heat.

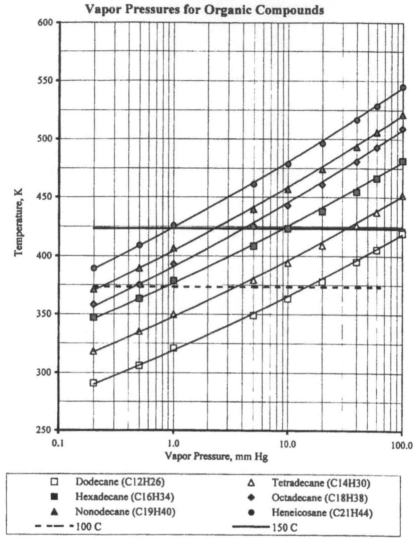

Figure 3.1 Vapor pressure–temperature curves for organic compounds.

3.2.1 RF Electromagnetic Antennae

Two RF antennae designs have been used to heat contaminated soils in field tests: bound-wave exciters and buried radiating antennae. The bound-wave exciters, proprietary to the Illinois Institute of Technology Research Institute (IITRI), were designed to essentially contain RF energy within a defined volume of soil (Dev, 1986, 1988, 1989; EPA, 1994a, 1995a; and Weston, 1992). One IITRI technique uses three lines of electrodes formed by cylindrical rods driven vertically into the ground, or placed inside boreholes (Figure 3.4). The cylindrical electrodes must be in electrical contact with the soil so that the central line acts as the "source" electrode, and the two outside lines act as "return" electrodes. Generator coupling to the soil is controlled by impedance-matching circuits to optimize heating. However, positioning or steering of the heated volume is limited.

The RF antenna used in the Kirtland AFB demonstration was patented by KAI Technologies, Inc. (DOE, 1994; EPA, 1994b, 1995b; and Kasevich, 1993, 1996). The antenna consisted of a radiating dipole formed at the end of a coaxial transmission line (Figure 3.5) that was placed in a

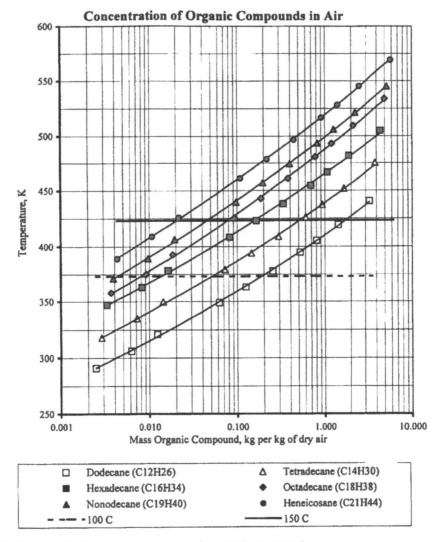

Figure 3.2 Air concentration–temperature curves for organic compounds.

buried fiberglass well casing. A properly configured RF system that uses multiple radiating antennae should provide controlled soil temperatures and a focused subsurface heating pattern.

The dipole antenna has several advantages over a triplate bound-wave exciter for a particular site. The antennae can be moved during heating to apply the RF field at varying depths and modified to match changing soil conditions. An alternate antenna can be activated for differing duty cycles to apply the RF power, facilitating dynamic control.

KAI's RF modules are capable of providing up to 25 kW of RF power at any one of the assigned ISM frequencies of 6.78, 13.56, 27.12, and 40.68 MHz. A heating pattern to fit almost any soil remediation application can be designed by considering the operating frequency, power level, antenna length, number of antennae, electrical phasing, and soil properties.

3.2.2   Key Components of Radio Frequency Heating System

KAI's mobile 25 kW RF system was used during the Kirtland AFB demonstration. This RF unit was designed specifically for demonstrations and pilot studies where an in-depth understanding of the RF process is required. This unit contained more diagnostic equipment plus storage and working space than necessary for commercial operations.

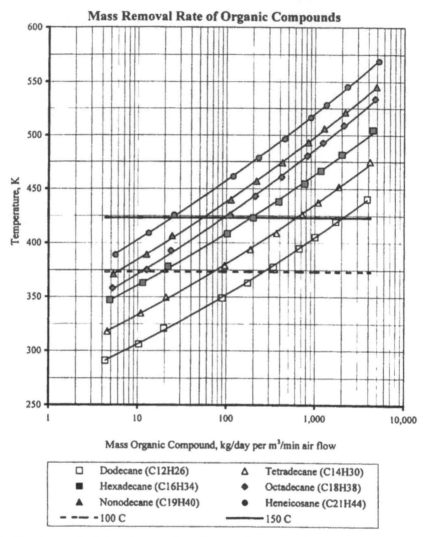

Figure 3.3   Mass removal rate-temperature curves for organic compounds.

The key component of the KAI RF heating system was an on-site and remotely operated computer-controlled RF module consisting of:

- Control and diagnostic center
- RF generator with matching network
- One or more antennae
- Vector voltmeter
- Several fiber optic thermometers

3.2.2.1   Control and Diagnostic Center

Computer control and remote operation of the RF heating system was provided by a rack-mounted computer system running KAI's proprietary RF heating process control software. The process control software monitored and controlled the 25-kW RF generator as well as the RF power supplied to the 2 RF antennae. The software was designed to assure safe operation of the RF equipment and to shut down the RF generator if an alarm condition, such as excessive power being reflected

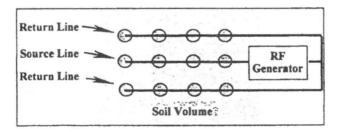

Figure 3.4 Top view of representative triplate bound wave-exciter electrodes, the IITRI system.

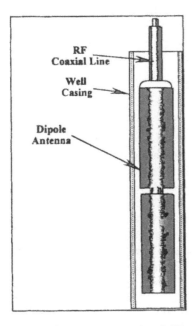

Figure 3.5 Dipole RF antenna, KAI Technologies, Inc.

back toward the generator, existed. The computer also stored data such as temperature, RF power, AC power, and antenna impedance for later analysis. During RF heating, the control and diagnostic computer communicated status information to or notified KAI personnel of an alarm condition via an alphanumeric pager. Carbon Copy® remote-control software by a laptop computer was used for monitoring and controlling the RF heating system off-site. The combination of computer control and remote operation minimized the amount of labor required to operate the RF heating system and permitted unattended round-the-clock operation. The pager, by providing KAI personnel with prompt notification of an alarm condition, allowed for rapid response and a quick resumption of RF operations.

3.2.2.2 Radio Frequency Generator and Tuner

The mobile RF generator could produce up to 25 kW at the ISM frequency of 27.12 MHz. The RF generator was divided into three stages:

- Exciter,
- Intermediate power amplifier, and
- Final power amplifier.

The exciter, a crystal oscillator, generated the initial 27.12 MHz signal that was used to drive the 1-kW intermediate power amplifier second stage. The intermediate power amplifier then drove the high-power final amplifier stage. Power from the RF generator was supplied to two 41.3-mm (1.625 in.) diameter 50-Ω coaxial cables. Each transmission line delivered the RF power to one of two switchable antennae.

A matching network, or tuner, was located between the RF generator and the transmission lines. This tuner insured that the RF generator saw matched load impedance, minimal reflected power, even if the impedance of the antenna and soil was mismatched to the impedance of the generator. The antenna radiation impedance was determined by the diameter and length of the dipole halves and the soil and well casing properties. As the soil was heated, its effect on the radiation impedance changed. The tuner adjusted to assure maximum transfer of the RF energy to the soil during these dynamic heating variations. The tuner transformed the antenna–soil impedance to a value nearly equal to the source impedance of the RF generator. This provided maximum power transmission to the soil and prevented damage to the generator.

Periodically, the RF generator was briefly cycled on and off to allow the control and diagnostic computer to switch between the RF antennae. A 25-kW, water-cooled, 50-Ω dummy load was used to test the RF system before connection to an antenna.

3.2.2.3 RF Antenna

There were two identical 88.9 mm (3.50 in.) diameter, 3.5 m (11.5 ft) long antennae employed to deliver RF energy to the soil. The Numerical Electromagnetic Code 4 (NEC 4) was used to calculate the input impedance of the antenna. NEC 4 uses the electrical properties (dielectric constant and conductivity) of the soil at 27.12 MHz, the antenna parameters (length and diameter), and the geometry of the borehole in the calculation. This code was developed by Lawrence Livermore National Laboratory. Properties of the soil samples from the Kirtland site were measured at field moisture content and after they had been dried in a microwave oven for a total of 11 min. The soil sample collected in June 1996 from a pit near the demonstration site contained 1% moisture dwb as received. NEC 4 calculated that the antenna would be 98.2% efficient in transferring RF energy to the soil in 1% moisture soil and 99.2% efficient in the dry soil. However, pre-demonstration borehole samples collected for analysis revealed an average of 7% moisture content. The dielectric constant and conductivity of the soil both increase with moisture level. This caused the antennae to be electrically longer than their half wavelength resonant design due to the shorter wavelength in the soil. This also lowered the predicted power transfer efficiency of the antennae from a projected level of 98 to 80%. This was confirmed during the demonstration when the average RF power transfer efficiency of the antennae was measured at 81%.

During the demonstration, the RF antennae were placed in the fiberglass-cased 17.78-cm (7-in.) OD boreholes with a tripod assembly. The antenna centers, or feed-points, were positioned at 4.57 m (15 ft) below grade. To allow power to be transmitted at high levels, the antennae and 41.3-mm (1.625 in.) diameter transmission line were pressurized with nitrogen gas. To prevent heat from building up in the two antennae boreholes, ambient temperature air was blown into each of the two RF boreholes. The air entered at 142 slpm or 5 scfm at 1.52 m (5 ft) below grade through a 2.54-cm (1-in.) diameter plastic pipe. The pipe from each borehole was connected to a 284-slpm (10 scfm) blower.

3.2.2.4 Vector Voltmeter

This instrument was used to measure the efficiency with which the antennae radiated RF energy during the demonstration. During RF heating, it measured the power incident to the antenna (forward power) and the power reflected from the antenna (reverse power) through a power dual-directional coupler located on the output of the tuner. The difference between the forward and reverse powers is called the return loss and is measured in decibels (dB). The return loss is directly related to power transfer efficiency. For example, a return loss of 10 dB means that the reflected power is 10 dB below the incident power (or 10% of the incident power is reflected).

3.2.2.5 Fiber Optic Thermometer

Fiber optic thermometry was used to measure temperatures of the antennae during RF heating. Fiber optic temperature probes were placed at the point of highest electric field intensity (i.e., near the feed point of each antenna).

3.2.3 Electrical and Magnetic Properties

The ability of RF energy to heat soils can be described on a macroscopic level by three bulk electrical properties of the soil: the electrical conductivity, the electric permittivity, and the magnetic permeability. Each material property relates a flux density to a field strength according to the three primary formulas:

$$\mathbf{J} = \sigma\mathbf{E} \tag{1}$$

$$\mathbf{D} = \varepsilon\mathbf{E} \tag{2}$$

$$\mathbf{B} = \mu\mathbf{H} \tag{3}$$

where $\mathbf{J}$ = current density, amps per square meter (A/m^2); σ = electrical conductivity, siemans (historically, mhos) per meter (S/m); $\mathbf{E}$ = electric field strength, volts per m (V/m); $\mathbf{D}$ = electric flux density, coulombs per m^2 (C/m^2); ε = electric permittivity, farads per m (F/m); $\mathbf{B}$ = magnetic flux density, tesla (T); μ = magnetic permeability, Henries per m (H/m); and $\mathbf{H}$ = magnetic field strength, ampere (turns) per m (A/m). The bold quantities and signifies vector fields.

The bulk electrical properties are determined by soil heterogeneity, the distribution and geometry of constituent soil types, and, ultimately, by microscopic interactions at the molecular and atomic levels. At the molecular and atomic level, applied external electric and magnetic fields induce changes in the "particle" charge distribution and its velocity. In this case, the term "particle" refers to the charge element under consideration, i.e., electrons, ions, atoms, molecules, or larger relatively homogeneous clusters of atoms and molecules. The three fundamental degrees of freedom of charge motion are translational, rotational, and vibrational motion. The electrical conductivity, σ, of a material describes the net translational motion of free charge (or charge carriers) in the material. The electric permittivity, ε, is determined by rotational and vibrational motion of bound charge in the material. There is no net translational motion of bound charge. The distinction between free and bound charge is significant in this respect. Magnetic permeability, μ, describes interactions between charge velocities, or "spins," i.e., the orbital motions of charge in molecules and atoms create magnetic dipoles which interact with each other and with external fields through magnetic forces.

The mechanism that determines RF heating in a particular soil depends on soil electrical properties. The electrical properties are frequency dependent in most materials, and the relative importance of each property is frequency dependent. Heating may be dominated by conduction losses, dielectric losses, or magnetic losses. As a practical matter, few materials exhibit significant heating due to magnetic permeability losses. Notable exceptions may be ferromagnetic soil constituents like hematite at least at MW frequencies. For the most part, soil heating in an RF field will be shared between dielectric and conductive losses. There is some confusion in the literature regarding these two coexisting phenomena; so a short discussion is warranted.

Electrical conductivity, σ, depends on the free charge volume density, n and p, and the charge mobility, v_n and v_p, m^2 per volt-second (m^2/V-s):

$$\sigma = n \, q \, v_n + p \, |q| \, v_p \tag{4}$$

where n = the number density of negative free charge (number per m³); p = the number density of positive free charge (number per m³); and q = the charge of an electron = -1.602×10^{-19} coulombs (C). The mobility is essentially a measure of the mean free path between collisions and collision mechanics in a material. That is, when an electric field is applied to a material, free charge is accelerated by electrostatic forces until it collides with other "particles" and exchanges momentum with them. The collision processes are **lossy**; that is, they dissipate energy. The net result of a statistically significant number of charge accelerations and collisions is that for a fixed electric field, the net average velocity, the "drift" velocity, u_d meters per second (m/s), determines the charge carrier mobility, v:

$$u_d = vE \tag{5}$$

where v = m² per volt-second (m²/V-s) and E = volts per m (V/m). The mobility of a charge carrier, and thus the electrical conductivity, is a function of temperature, among many other factors.

The electric permittivity indicates how much energy can be absorbed from an electric field by rotating or vibrating charge dipoles. In an externally applied electric field, a dipole does not have a net charge and is not accelerated. However, it does experience a torque, since the minus end has a force antiparallel to the electric field, and the positive end has a force parallel to the electric field. These forces result in the rotation or vibration of the dipole. Significant heating occurs when the rotational and vibrational dipolar motions are **lossy**. The standard example is water dipoles rotating in a microwave field and dissipating energy by viscous interaction. At the RF ISM frequencies of interest, the majority of dipole interactions are rotational rather than vibrational.

3.2.4 Heating Mechanisms

While collisions in translational motion of charge are inherently **lossy**, rotational motion may be lossless. That is, the rotating dipoles may not dissipate energy, only store it. These kinds of materials make good insulators. Examples are Teflon, polyethylene, glass, most ceramics, some alcohols, and most kinds of oils, which are essentially **lossless** in a RF field. These materials have an electric permittivity that has only a real value. On the other hand, **lossy** materials (materials that absorb energy from an electromagnetic field and dissipate it as heat) have a complex permittivity:

$$\varepsilon^* = \varepsilon_0 \, (\varepsilon' - j\varepsilon'') \tag{6}$$

where ε^* (F/m) represents the complex permittivity; ε_0 = the permittivity of free space (8.85 × 10^{-12} F/m); ε' is the relative real part; ε'', the relative imaginary part, is the loss factor; and j is the square root of minus one (−1). The relative real part, ε', is often called the "dielectric constant," but it is not constant since it depends, for example, on temperature, pressure, and RF frequency. The real part, ε', has a very strong influence on the wavelength of the electromagnetic field in the soil ($1/\lambda \approx 2 \, \pi \times f[\mu \, \varepsilon_0 \, \varepsilon']^{0.5}$), but if the soil is very **lossy**, the electrical conductivity and the imaginary part, ε'', also significantly affect the wavelength.

In a RF field, conductive and dielectric losses coexist. The difference between losses due to conductivity (translational motion) and those due to permittivity (rotational motion) cannot be discerned. The usual approach, therefore, is to define an "effective" conductivity, σ_{eff}:

$$\sigma_{eff} = \sigma + \omega\varepsilon''\varepsilon_0 \tag{7a}$$

or an effective loss factor, ε'':

$$\varepsilon'' = \varepsilon''_{pol} + \sigma/\omega\varepsilon_0 \qquad (7b)$$

where ω (radians per second, r/s) is the angular frequency ($\omega = 2\pi f$) and the two contributions (from rotation of polar particles, ε''_{pol}, and translational motion, σ) are combined into one term. Care must be taken regarding units, since ε'' has none and σ does.

3.2.5 Volume Power Generation

The volume power deposition term in the energy balance, Q_{gen}, watts per cubic meter (W/m³), is determined by the total losses and the square of the magnitude of the local electric field in the soil:

$$Q_{gen} = (\sigma + \omega\varepsilon''\varepsilon_0) |E|^2 \qquad (8a)$$

$$Q_{gen} = \omega\varepsilon_0 (\varepsilon''_{pol} + \sigma/\omega\varepsilon_0) |E|^2 \qquad (8b)$$

The two loss mechanisms in terms of overall heating cannot be separated since they coexist. Sometimes a σ/ω contribution in a plot of ε'' vs. frequency occurs such as illustrated in the measurements of fine builder's sand shown in Figure 3.6. In this figure, the relative imaginary permittivity (the loss factor, unit-less) is plotted as a function of frequency at a soil moisture content of 10% dwb and 20°C. On log–log axes, the σ/ω behavior of the curve is readily identifiable as a monotonically decreasing straight line. If dipolar rotational losses were significant over this frequency range, this curve would transition to a horizontal asymptote above some critical frequency. These measurements are discussed in more detail in the next section.

3.2.6 Important Implications for the Kirtland Air Force Base Site

If the goal of the soil treatment protocol is to heat the soil to or above the boiling point of water (94°C at Kirtland AFB), the soil must be completely dry before the boiling point can be exceeded. Very moist soils, for example, those with water content of 10% dwb, have significant concentrations of mobile ions. Therefore, the heating term in Equations 7 and 8 is dominated by the electrical conductivity due to the mobile ions, i.e., ionic conduction. During soil heating, water is vaporized. The ion mobility decreases to zero as the water is eliminated. The heating mechanism then shifts from primarily ionic conductivity to dominantly dielectric loss, possibly in combination with solid state conductivity. To achieve target temperatures greater than 94°C, a goal established for the Kirtland AFB site, there must be soil constituents present with significant dipole loss factors or measurable electrical conductivity not due to ionic currents. This was the case for the Kirtland

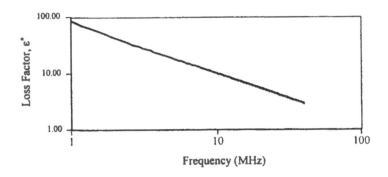

Figure 3.6 Loss factor and frequency plot for fine builder's sand at 10% dwb moisture and 20°C.

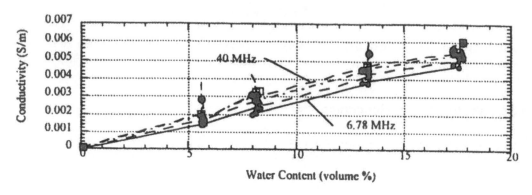

Figure 3.7   Electrical conductivity of fine sand (S/m) vs. water content (volume %) at RF frequencies of 6.78, 13.56, 27.12, and 40 MHz.

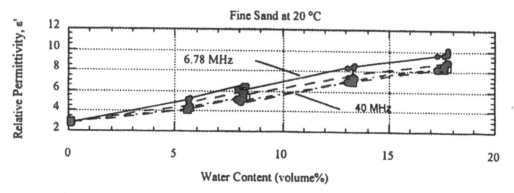

Figure 3.8   Real part of the relative permittivity, ε', for fine grain builder's sand at 20°C vs. water content (volume %) at RF frequencies of 6.78, 13.56, 27.12, and 40 MHz.

AFB soils (soil data in Section 3.4.5), and it was possible to heat them above 94°C. This is *not* at all the case for clean builder's sand (Figure 3.7). Hence, it would not be possible to exceed 94°C in that material with practically sized RF generators.

A clear understanding of soil bulk electrical properties is essential to the design of a RF heat enhancement protocol. If no significant **loss** occurs in dry soils, the heating will not exceed 100°C by more than the binding energy of bound water. That is, in **low loss** clays, the temperatures may exceed 100°C by a few degrees, depending on the water binding energy barriers which must be overcome by the RF heating, but no more.

3.2.7   Example of Low Loss Soil: Clean Fine Builder's Sand

The relative permittivity as a function of moisture content, w (% dwb), is shown in Figure 3.8. Sand is a nonhygroscopic material, and the highly linear nature of the relationship between ε' and water content suggests that a simple mixture formula would describe behavior of the permittivity. In contrast, the relationship in clay-containing soils is nonlinear. The relative permittivity of dry builder's sand is comparable to the Kirtland AFB surface soil sample at 20°C (compare to data in Section 3.4.5). The builder's sand data of Figure 3.8 are reasonably well fit by a line with an intercept of 2.75 and slope of 0.352:

$$\varepsilon' = 2.75 + 0.352 \, w \qquad\qquad (9)$$

At a reported soil moisture content of around 7% dwb in the deeper soils at the Kirtland AFB site, the real part of the permittivity should be around 5 (relative). This estimate was based on the assumption that the deeper soils are similar to the surface soil.

In wet sand, the losses are primarily due to ionic conduction rather than dipolar phenomena. Sand, like glass, leaches sodium ions into a solution. Consequently, the plot in Figure 3.8 is of measured conductivity (σ, S/m) rather than loss factor, ε''. As in the relative permittivity data, the conductivity data, as a group nearly independent of frequency, are reasonably fit by a single straight line:

$$\sigma = 9.8 \times 10^{-5} + 3.18 \times 10^{-4} \, w \qquad (10)$$

The plots for different frequencies nearly superimpose, further evidence that the loss is mostly conductive rather than mostly dielectric. There does appear to be a slight monotonic increase in conductivity with frequency, suggesting that some dielectric loss may be at work as well. However, because the dry conductivity is so small (9.8×10^{-5} S/m), for all practical purposes, the sand is essentially **lossless** when completely dry. The monotonic increase with frequency is not likely due to significant dielectric loss. At 27 MHz, the equivalent loss factor of the dry sand is about 0.065, which is much too small to be of any use.

3.3 DEMONSTRATION PROJECT APPROACH

This section provides a summary of the technical approach used by the project team to conduct the project. The entire RF-SVE demonstration project can be broken down into the activities that are described in Table 3.1.

The first phase of the demonstration project consisted of selecting the demonstration site. The site was to contain soil contaminated with chemicals that were not readily removed by SVE alone but could potentially be removed with thermally enhanced SVE using RF heating. The next step was to complete the characterization of soil within the selected site. To obtain soil samples, several boreholes were drilled, and test pits were excavated. The pre-demonstration characterization work, which was completed by UT, BRE, and KAI, is described below.

Borehole and Bulk Sample Analysis and Tests
- UT completed chemical analyses of borehole samples for water, gasoline range organic compounds (GRO compounds), diesel range organic compounds (DRO compounds or SVOCs), and total petroleum hydrocarbons (TPH).
- UT collected data on soil classification; mineralogy; geotechnical properties, for example, particle size distribution, dry density, specific gravity of solid, porosity, hydraulic conductivity, and air permeability; and geology of the site area.
- UT analyzed samples for thermal properties, such as thermal conductivity, density, and specific heat, and air permeability of the soil as a function of moisture content and temperature.
- KAI obtained information from a bulk sample on the soil's electrical properties, like conductivity and dielectric constant.

On-Site Pre-demonstration Tests
- UT and BRE determined the soil's *in situ* air permeability and penetration resistance.
- UT also completed a pre-demonstration partitioning interwell tracer test (PITT) program to estimate the contaminant distribution and initial concentration of DRO compounds (SVOCs).
- UT measured oxygen respiration rates in several SVE and pressure-temperature (P/T) wells during SVE-only operation. From these tests, the oxygen utilization rates were quantified, and the potential for biooxidation of the organic compounds within the RF-SVE site was inferred.
- The next step was to select and design the RF heating system, the SVE treatment system, and instrumentation for the field demonstration. The four major components of a typical RF-SVE treatment system are (1) a RF generator, a matching network, and one or more RF antennae;

Table 3.1   Components of the RF-SVE Project

Component	Purpose
Site selection	Select applicable site with appropriate contaminants and concentrations
Pre-demonstration soil sampling and analysis	Obtain representative samples for detailed site characterization and initial mass estimates for water and organic contaminants (mainly DRO compounds and TPH) inside and outside the *design* volume
Pre-demonstration partitioning interwell tracer test (PITT)	Demonstrate PITT in vadose zone as an independent check on the initial DRO–mass estimate
Pre-demonstration *in situ* respirometry test	Measure oxygen utilization rates and infer organic degradation by aerobic biooxidation
SVE only (no heat)	Assess removal rates of water and DRO constituents by SVE alone
RF heating with SVE	Decrease the water and DRO mass in *design* volume
Cool down period	Allow cooling of soil to facilitate assessment of contaminants after soil heating
Post-demonstration *in situ* respiration tests	Measure oxygen utilization rates and evaluate the effects of soil heating and water removal on inferred thermal oxidation and biooxidation processes
Post-demonstration soil sampling and analysis	Obtain representative samples for determination of residual masses of water and organic contaminants (mainly DRO compounds and TPH)
Post-demonstration PITT	Independent check on residual DRO mass

(2) the SVE system; (3) instrumentation system; and (4) off-gas treatment system. KAI provided a mobile heating system with a 25- kW RF generator. This mobile system was designed specifically for demonstrations and pilot studies where an in-depth understanding of the RF heating process was required. The design of the RF antennae was based on the pretests that measured the electrical properties of the bulk soil sample. Each component of the RF-SVE pilot study system was designed using information from past treatability and pilot-scale studies to optimize field performance of this demonstration.

The next phase of the project was completing the field demonstration activities using the RF-SVE system. Initially, the field demonstration focused on a selected *design* volume of soil within the fire-training pit. The *design* volume was a 3.05-m (10-ft) cube of soil (28.3 m³) at a depth of 3.05 m (10 ft) bgs. Once pre-treatment soil sampling and characterization were completed, the field operations for the demonstration were conducted in these major phases:

1. **SVE Without Heating:** Vapors were drawn from soil to determine the amount of contaminant removed in off-gas without RF heating. At the end of this venting period, air samples were collected to estimate SVE-only mass removal rate of organic compounds. Before the SVE-only operation stage, three respiration tests were performed to determine the possible contribution of aerobic biooxidation in the removal of the SVOCs.
2. **RF and SVE Operation:** The soil was heated with RF energy and maintained at an elevated temperature. Operating data for the RF system were continuously recorded. Water and contaminants were removed, the off-gas flow was measured, and off-gas concentrations were monitored and recorded. The temperature and vacuum pressures in the SVE, P/T, and PITT wells and the temperature of the air vapors at the SVE wellheads and before the carbon drums and air blower were also monitored. Condensed water and DRO liquid were collected, and the volumes measured.
3. **Cool Down:** After the RF system was turned off, SVE-only operation cooled the soil. After the cool-down phase, 13 respiration tests were performed. The measured oxygen utilization rates were used to infer possible degradation of SVOCs by thermal oxidation reactions and aerobic biodegradation.
4. **Post RF Testing/Data Collection:** Soil samples were collected to document actual subsurface conditions after the RF-SVE system was turned off. Variations in grain size and distribution of residual contamination in the soil, both within and outside the *design* volume, were noted.

Data from the demonstration were analyzed as the next step in the project. Several methods were used to monitor the RF-SVE system and evaluate the process performance. Temperatures and pressures were measured to track performance of the RF and SVE systems. Soil samples were obtained from the site before and after RF-SVE to determine the decrease in water, DRO, and TPH concentrations and mass in the soil. Samples of the DRO liquid were also analyzed for total DRO and normal paraffin (C_{10} to C_{21}) contents. PITTs were conducted before and after RF-SVE as another means to estimate the decrease in soil contaminants. *In situ* respirometry tests were conducted before and after RF-SVE to assess the activity of thermal- and biooxidation processes. Predicted and measured performance and energy consumption were then compared. Improvements in design and demonstration procedures were also recommended in this step.

Data were examined collectively and the overall performance of the demonstration and accuracy of the results were estimated based on field data that included the following:

- The initial mass of DRO compounds and water
- The mass removed by SVE "alone" operation
- The mass removed during RF-SVE operation
- The mass removed during cool-down
- The mass of DRO contaminants and water that remain after treatment

3.4 TEST SITE DESCRIPTION

3.4.1 Demonstration Site Location

The site selected to demonstrate the RF-SVE technology was a former fire-training pit (FT-14) located at Kirtland Air Force Base (AFB) in Albuquerque, NM (Figure 3.9). This abandoned training site consisted of two burn pits located about 18 m apart. The total area of the western pit was about 46 m^2, and the area of the eastern pit was about 93 m^2. Both pits were about 0.1 to 0.6 m deep, with earthen berms rising 0.3 to 0.6 m above the surrounding land surface. Each pit contained approximately 37 m^2 of soil darkened with carbonaceous material. The western pit was used for the RF-SVE demonstration.

3.4.2 Site Description Standard Terminology

The Federal Remediation Technologies Roundtable has developed a guidance document entitled *Guide to Documenting Cost and Performance for Remedial Technologies* (FRTR, 1995). Hereafter, this document will be referred to as the *FRTR Guide*. The purpose of the *FRTR Guide* was to foster the use of a consistent method to document and provide information regarding contaminated sites and media as well as the method in which cost and performance information is presented. The *FRTR Guide* will be referenced, and its terminology will be used to describe the demonstration site (see Table 3.2) and to describe and document typical design cost and performance evaluations.

3.4.3 Past Usage

Information is limited on the procedures used for fire training and the duration and frequency of the fire-training exercises. Fire-training procedures probably consisted of first applying water, then adding fuel to the pit, igniting the fuel, and extinguishing the fire with chemical foam.

3.4.4 Site Characteristics

The FT-14 site had many characteristics that made it desirable for the demonstration. These characteristics included soil particle size, air permeability, moisture content, and porosity. Furthermore, the site soil was weathered granite that contained desirable minerals for RF heat generation.

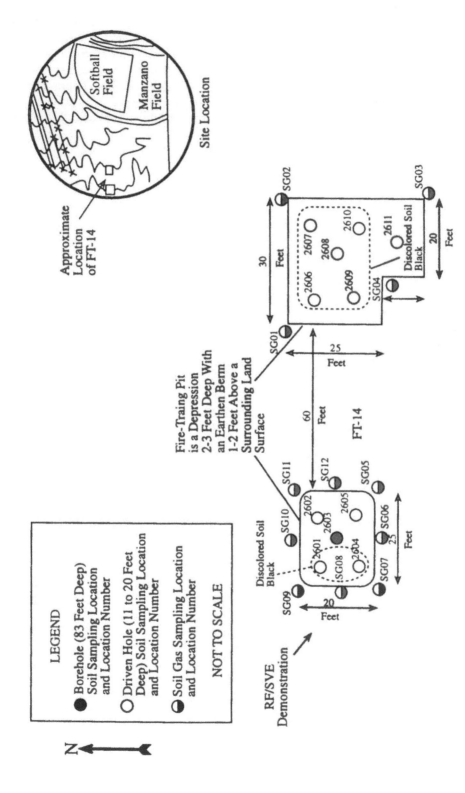

Figure 3.9 RF-SVE demonstration site at Kirtland AFB, Albuquerque, NM.

Table 3.2 Test Site Description Using FRTR Standard Terminology

Site Background
   Historical activity that generated contamination
      Former fire-training pit;
      SIC Code: 4581, Airport terminal services
   Management practice that contributed to contamination
      Fire/crash training area
Site Characteristics:
   Media treated
      Soil (*in situ*)
   Contaminants treated
      Nonvolatiles, nonhalogenated (such as TPH); semivolatiles (such as
      diesel fuels), polynuclear aromatic hydrocarbons (PAHs)
Treatment Systems:
   Primary treatment technology
      Thermally enhanced recovery with soil vapor extraction
   Supplementary treatment technology
      Post-treatment (air) – air condenser, activated carbon drums

The major soil characteristics are summarized in Table 3.3. The average mineral contents of the soil are listed in Table 3.4. The data in Table 3.4 were prepared using X-ray diffraction (XRD) methods.

Table 3.3 Geotechnical Properties of Site Soil

Property	Value	Method Used
Specific gravity of solids	2.70	ASTM D854[1]
Grain size analysis (dry weight basis)		
% Fines (< 0.074 mm)	10–15	ASTM D1140[1]
% Sand (size)	60–75	ASTM D422[1]
% Silt (size)	5–15	ASTM D422[1]
% Clay (size)	0–15	ASTM D422[1]
Moisture content (*Sampled* Volume)		
Mean	6.8% dwb	ASTM D2216[1]
Low	1.6% dwb	
High	13.1% dwb	
Dry density		
Mean	1780 kg/m³	UT Method (sleeve)
Low	1600 kg/m³	
High	2000 kg/m³	
Porosity		
Mean	0.34	UT Method
Low	0.31	
High	0.39	
Air permeabilities (field data)		
Mean	5 D	Section 2.1.3
Low	0.1 D	
High	100 D	
Air permeability (laboratory data)		
Low	0.1 D	UT Method (air perm)
High	10 D	
Average saturated hydraulic conductivity (laboratory data)	1×10^{-4} cm/s (0.1 D)	ASTM D5856[1]

[1](ASTM, 1996).

Additional characteristics made the soil desirable for the RF-SVE demonstration.

- The site was free of potential RF interference, such as buried iron pipes and steel drums.
- Site permitting for the demonstration was easily obtained.
- The soil deposit was relatively uniform.
- The groundwater table was greater than 150 m (>500 ft) below ground surface (bgs).

Table 3.4   Mineral Content of Soil at the Demonstration Site

Mineral	Chemical Composition	% dwb
Quartz	SiO_2	53
Plagioclase Feldspar	$1/2(Na_2O)1/2(Al_2O_3)3(SiO_2)$;	13
K-Feldspar	$1/2(K_2O)1/2(Al_2O_3)3(SiO_2)$;	7
Calcite	$CaCO_3$	5
Magnetite	Fe_3O_4	3
Chlorite	$3(MgO)2(SiO_2)2(H_2O)$	2
Kaolinite	$(Al_2O_3)2(SiO_2)2(H_2O)$	1
Hematite	Fe_2O_3	>1
Mixed layer illite-smectite		16
Illite	$1/2(K_2O)2(Al_2O_3)7(SiO_2)1/2(Al_2O_3)2(H_2O)$	25
Smectite	$(Al_2O_3)4(SiO_2)n(H_2O)$, where n = 2 or >2	75
Illite-mica mixture		2
Illite	$1/2(K_2O)2(Al_2O_3)7(SiO_2)1/2(Al_2O_3)2(H_2O)$	nm
Mica	$1/2(K_2O)(Al_2O_3)3(SiO_2)1/2(Al_2O_3)(H_2O)$	nm

- The SVOC contaminants were heavy oil-based hydrocarbons.
- The maximum depth of contamination was about 12 m (40 ft) bgs.
- The contaminants were reasonably contained (laterally), simplifying mass balance calculations.
- The maximum concentrations were located at 3 to 7 m (10 to 23 ft) bgs.
- Within the *sampled* volume, the maximum contaminant concentrations were 550 mg/kg GRO compounds, 5200 mg/kg DRO compounds, and 30,000 mg/kg TPH.
- The average concentrations within the *sampled* volume were 920 mg/kg DRO compounds and 6739 mg/kg TPH.
- The DRO hydrocarbons ranged from 10- to 21-carbon compounds. Peak gas chromatograms ranged from 15- to 18-carbon compounds.

3.4.5   Electrical Properties of the Soil

The complex electrical permittivity of soils is sensitive to moisture content and temperature, as well as the binding chemistry. A series of measurements was conducted on the surface soil sample from the Kirtland AFB site at the "as provided" moisture content about 0.8% dwb. Soils in the treatment zone were considerably wetter, with moisture contents from 7 to 8% dwb (Table A4, Appendix A). The measurements were conducted in a 4.13-cm (1.625-in.) coaxial chamber with flush center conductor. Impedance data were collected over the range 1 to 40 MHz with a Hewlett-Packard 4194A Impedance Analyzer, and the relative electric permittivity was determined using a bilinear transformation. The measurements showed the soil properties to be quite variable due to inclusions such as rocks and small pebbles. Nevertheless, the mean values were relatively well behaved functions, even though the standard deviations were large.

3.4.5.1   Methods

The RF antenna used for the measurements was a shielded open chamber built in a standard RS-225, 4.13-cm (1.625-in.) coaxial air line. The specific device used was described by Dong (1991) from a design originally presented by Harrington (1961). The center electrode was 1.7 cm (0.67 in.) in diameter. The sample holder was small enough that wave-guide propagation did not occur in the chamber and was suitable for fine pellet and liquidus materials, as well as solid material carefully fit to the chamber dimensions. Calibration was accomplished by measuring the impedance of slugs of PTFE (Teflon), plexiglass (Perspex™), and polystyrene. Impedance measurements were made at 201 sample points between 1 and 40 MHz.

The electric permittivity (ε^*, F/m) is a complex number, which is the product of the relative permittivity, $\varepsilon' - j\varepsilon''$, and the permittivity of free space, ε_0, or $\varepsilon^* = \varepsilon_0(\varepsilon' - j\varepsilon'')$, where j is the

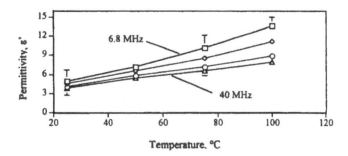

Figure 3.10 The temperature effect on the real part of the relative permittivity, ε', for Kirtland AFB soil at RF frequencies of 6.8, 13.6, 27.1, and 40 MHz.

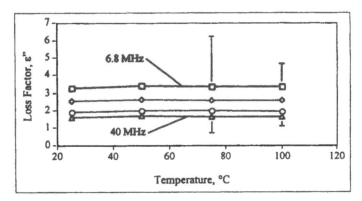

Figure 3.11 The temperature effect on the imaginary part of the relative permittivity, ε", at RF frequencies of 6.8, 13.6, 27.1, and 40 MHz.

square root of minus 1. The real part, ε', represents stored energy and the imaginary part, ε", or loss factor, represents dissipated energy. The means and standard deviations for relative permittivity values were calculated as a function of frequency at each temperature for an ensemble of samples to check repeatability. Measurements were made at 20, 50, 75, and 100°C.

3.4.5.2 Results

Figures 3.10 and 3.11 show the calculated real relative permittivity (ε') and loss factor (ε"), respectively, for Kirtland AFB surface sand as a function of temperature. Selected values at 6.8, 13.6, 27.1, and 40 MHz were plotted. The general trends were that (1) ε' decreased with increasing frequency, as expected; (2) ε' was nearly linear with temperature; and (3) ε" decreased with increasing frequency and was essentially independent of temperature. The mean values were plotted, and the bars represent 1 SD over the 3 batches in the ensemble (5 measurements for each batch at each temperature). The permittivity decreased monotonically with frequency in the figure, as was expected.

The four relations from the plots in Figure 3.10 can be approximately fitted to straight lines by the following equations:

$$\text{At 6.8 MHz: } \varepsilon' = 1.83 + 0.115 \text{ T (°C)} \tag{11}$$

$$\text{At 13.6 MHz: } \varepsilon' = 2.31 + 0.086 \text{ T (°C)} \tag{12}$$

$$\text{At 27.1 MHz: } \varepsilon' = 2.56 + 0.064 \text{ T (°C)} \tag{13}$$

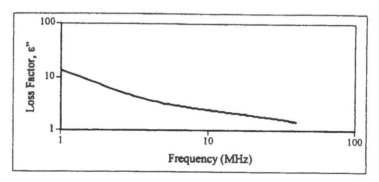

Figure 3.12 Plot of loss factor and RF frequency for the KAFB surface soil sample.

$$\text{At } 40 \text{ MHz: } \varepsilon' = 2.61 + 0.054 \text{ T (°C)} \qquad (14)$$

Temperature independence for the loss factor, ε'', was not expected *a priori*, but was easily understood in retrospect. When ε'' was measured, the soil was essentially dry at 0.8% dwb moisture. Measurements at low water content were appropriate to this treatment site since the target temperatures were more than 100°C. For that reason, nearly all of the soil moisture had to be vaporized to achieve the target range of 130 to 150°C. Therefore, the measured losses were due to solid state conduction and possibly dipole rotation rather than ionic conduction. Soils in the treatment zone were considerably wetter before RF heating, around 7 to 8% dwb. Hence, ionic conduction dominated the losses in the wet soils during the initial RF heating phase (see Section 3.2 for discussions on this topic). In wet soils, the losses are expected to be strongly temperature dependent on moisture content since the mobility of electrolytes in water increases with temperature rise at about 2% per °C. The relatively dry soil measured did not have significant ionic conduction since the carrier water was essentially gone; so the losses measured were essentially due to nonionic electrical conductivity and, perhaps, some dipole rotation. The measurements suggested that over this frequency range the losses were not temperature dependent. Again, the data points represented the means over the 3 batches (5 measurements for each batch at each temperature) and the bars 1 SD.

The loss factor decreased monotonically with frequency (Figure 3.12). The loss factor was essentially independent of temperature over this range and was approximately: (1) 3.4 at 6.8 MHz, (2) 2.5 at 13.6 MHz, (3) 1.95 at 27.1 MHz, and (4) 1.6 at 40 MHz. These four data points were fit over this range of frequencies by $\varepsilon'' = 7.52 \, \omega^{-0.416}$. The data suggested some contribution from electrical conductivity (it is not necessary to have ionic conduction to have measurable electrical conductivity). Recall from Section 3.2 that the loss has at least two parts: $\varepsilon'' = \varepsilon''_{pol} + \sigma/\omega\varepsilon_0$. A combination of conductive and polarization losses is expected to have a decreasing loss with increasing frequency (Figure 3.6). It is further expected that if dipole losses are significant, a transition from σ/ω line (on log–log axes) to a horizontal asymptote would be observed. In fact, this soil showed a similar trend (Figure 3.12), but the transition seemed to be to another σ/ω curve rather than a horizontal asymptote. If this was indeed the case, that would suggest that two conductive processes might be at work in parallel.

In both plots, the SD at 75°C is much larger than at any other temperature at all frequencies. The reason for the high SD of this particular measurement could not be ascertained.

These measurements were conducted on essentially dry soil samples that effectively simulated the electrical properties expected at high temperatures during RF heating. The loss factor of this soil was substantial at very low moisture content, most likely due to the distribution of **lossy** soil constituents like magnetite, hematite, kaolinite, illite, and smectite. The presence of these materials made it feasible to set target temperatures more than 100°C. The heating rate, especially during the initial phase of RF heating, will be strongly influenced by higher moisture content. An estimate of the strength of this influence can be obtained from the data on builder's sand described in Section

3.2. Assume that the wet Kirtland AFB soils have a moisture response slope similar to the builder's sand, where σ (S/m) $= 9.8 \times 10^{-5} + 3.18 \times 10^{-4}$ w (% moisture dwb). If so, the electrical conductivity and the loss factor at 27 MHz, where $\varepsilon'' = \sigma/\omega\varepsilon_0$, should be approximately linearly dependent on moisture. Then σ in the wet Kirtland AFB soil should be about 2 to 3 mS/m, (loss factor ε'' around 9.4) at 7% dwb moisture in the treatment zone before RF heating.

When dry, the Kirtland AFB soil was a complex heterogeneous mixture that contained about 1% dwb kaolinite, 3% dwb magnetite, and trace amounts of hematite. However, dry kaolinite powder was not a particularly strong absorber, as determined by measurements made at UT. The loss factor was measured to be about 0.2 to 0.3. Parkhomenko (1967) lists the electrical resistivity of hematite (Fe_2O_3) at either 0.35 to 0.7 Ω-cm or 220 Ω-cm. The differences were attributed to variations in sample impurities. These values correspond to electrical conductivities of 0.45 S/m or 143 to 286 S/m. At 27 MHz, a conductivity of 0.45 S/m is an effective loss factor of about 300. A conductivity of 200 S/m, near mid-range in the data reported, is equivalent to a loss factor of about 133. Thus, even if there was only a trace amount of hematite, some small contribution to the soil loss factor would be expected.

The magnetite concentration in the Kirtland AFB soil was 3%. Parkhomenko reports a resistivity for magnetite between about 0.036 and 57 Ω-cm, with the majority of measurements clustered around 0.15 to 0.2 Ω-cm. Choosing 0.2 Ω-cm as a representative and conservative value, the effective loss factor at 27 MHz would be about 3300. Thus, at 3% concentration, the magnetite would be expected to make a significant contribution to the measured loss factor.

3.4.6   Contaminant Types

The site contained a broad range of petroleum hydrocarbon compounds. The hydrocarbons at the site contained volatile GRO compounds, semivolatile DRO compounds, and nonvolatile motor-oil and mineral-oil range organic (MRO) compounds. These ranges are defined in Table 3.5. The major petroleum hydrocarbons at the site were the semivolatile DRO compounds and nonvolatile MRO compounds. The analyses focused on quantifying the concentrations of broad groups of compounds rather than concentrations of specific compounds since the measure of success of the demonstration would be based on the decrease in the semivolatile range compounds. Hence, one of the primary objectives of the RF-SVE demonstration was to measure the removal of semivolatile DRO compounds.

Table 3.5   Petroleum Product Ranges

Name	Acronym	Hydrocarbon Range	Usual Product Types
Gasoline range organic compounds	GRO	GRO $< C_{10} - C_{12}$	Light-end hydrocarbons
Diesel range organic compounds	DRO	$C_{10} - C_{12} <$ DRO $< C_{18} - C_{24}$	Diesel fuels, jet fuels, fuel oils
Motor-mineral-oil range organic compounds	MRO	$C_{18} - C_{24} <$ MRO	Heavy-end hydrocarbons, refinery sludge

3.4.7   Concentration Data

The demonstration's 132.0-m³ *sampled* volume was defined as the surface area enclosed by the lines joining the outer sample boreholes at the average depth of the boreholes. The defined areas for the pre- and post-demonstration boreholes are shown in Figures A1 and A2 (Appendix A). The pre- and post-demonstration borehole analyses for the TPH, DRO, and water components are listed in Tables A1, A2, and A4, respectively. Within the pre-demonstration *sampled* volume, the water content ranged from 1.6 to 13.1% dwb, with an average at 6.5%. The TPH concentrations ranged from 530 mg/kg to 29,900 mg/kg, with an average of 6739 mg/kg. The DRO concentrations ranged from below detection limit (BDL) to 5060 mg/kg, with an average of 920 mg/kg.

3.5 PROCEDURES AND FIELD IMPLEMENTATION

3.5.1   Modeling and Design of Soil Vapor Extraction System

Envirogen (Canton, MA) analyzed the data collected from the point air permeability testing performed by Brown & Root Environmental on September 14 and 18, 1996. The permeability data
were used to perform two- and three-dimensional airflow modeling that determined the optimal
extraction-well configuration and maximized the airflow from the extraction well through the heated
design volume. The optimum well configuration included the number of wells, the screen intervals,
and the location of the wells.

The data were used in Envirogen's proprietary two-dimensional airflow model (MDFIT) to
estimate the intrinsic permeability at each point tested. The results indicated a large variability both
laterally and vertically within the relatively small area. The three-dimensional airflow model
(AIR3D) was then calibrated using the MDFIT results and the collected field data from wells TT1C,
TT2A, and TT3C. The vacuum propagation indicated by the model matched the field data reasonably well given the low vacuums detected in the field.

Simulations for four cases were performed for a treatment zone from 3.4 to 6.4 m bgs. These
four cases are summarized below:

> CASE #1: A well installed in the center portion of the treatment zone with a screen at a depth of 4.7
> to 5.2 m bgs and an airflow rate of 175 slpm.
>
> CASE #2: A well installed in the center portion of the treatment zone with a screen at a depth of 4.7
> to 5.2 m bgs and an airflow rate of 156 slpm. An additional well installed with a screen above the
> treatment zone at 0 to 3.4 m bgs and another well installed with a screen below the zone at 6.4 to
> 7.3 m bgs. The airflow rates for the 2 wells were 164 slpm and 37 slpm, respectively.
>
> CASE #3: A well installed in the center portion of the treatment zone with a screen at a depth of 4.7
> to 5.2 m bgs with an airflow rate of 150 slpm. Additional wells installed with one having a screen
> at 0 to 3.4 m bgs and one having a screen at 6.4 to 7.3 m bgs with airflow rates of 331 and 91
> slpm, respectively.
>
> CASE #4: A well installed in the center portion of the treatment zone with a screen at a depth of 5.2
> to 5.8 m bgs with an airflow rate of 195 slpm. An additional 2 wells installed with one having a
> screen at 2.7 to 3.4 m bgs and another having a screen at 6.4 to 7.3 m bgs with airflow rates of
> 400 and 184 slpm, respectively.

One goal of the SVE was to maximize the flow of air through the treatment zone, i.e., radially,
to enter the central well screen at 4.7 to 5.2 m bgs. As illustrated in Figure 3.13, the results of
CASE #1 indicated that 50% or more of the airflow into the single well was derived from above
and below the treatment zone, potentially creating problems with respect to cooling the extracted
air. Considering that the well-screen length in CASE #1 was 0.6 m, greater problems would be
expected for a treatment zone well screened across the entire zone at 3.4 to 6.4 m bgs.

CASE #2 and CASE #3 improved the quantity of airflow derived from within the treatment
zone. The idealized flow pattern resulting from the use of three well screens is illustrated in Figure
3.14. The quantity of airflow to the central well screen derived from below the treatment zone was
decreased to zero by increasing the flow rate in the lower well screen (CASE #2 vs. CASE #3).
However, to further decrease the amount of airflow derived from above the treatment zone, it was
necessary to lower the depth of the treatment zone well from 4.7 to 5.2 m down to 5.2 to 5.8 m
bgs. It was also necessary to shorten the screen length of the well above the treatment zone from
3.4 m to 0.6 m (CASE #4). For CASE #4, about 22% of the airflow into the treatment zone well
would be added from above the treatment zone. No airflow was expected from below the treatment
zone at the edge of the 3.0 m perimeter. In addition, from the edge of a 1.8 m perimeter within
the 3.0 m area, only 1% of the airflow would be contributed from above and below the treatment
zone.

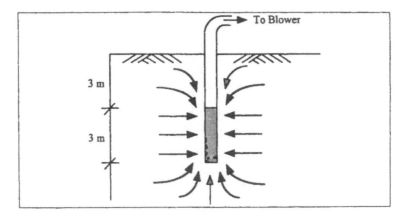

Figure 3.13 Idealized airflow field using a single, 3-m long well screen centered within the design volume.

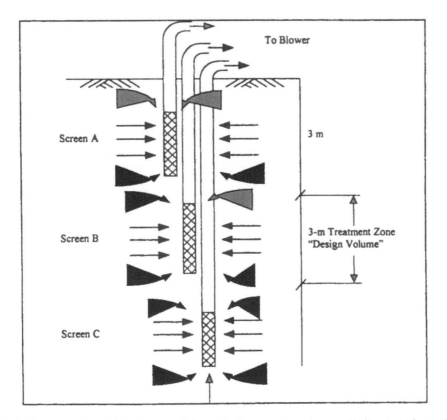

Figure 3.14 Idealized airflow field using three independent well screens with variable flow rates for each screen.
Note: Balancing the airflow rate using the three screens maximizes the radial flow into well screen B.

Based on the modeling results for the four cases, ENVIROGEN recommended installing three wells in the spatial center of the treatment volume. For a treatment zone ranging from 3.4 to 6.4 m bgs, the recommended installation was as follows:

- 1 well immediately above the treatment zone (screened 2.7 to 3.3 m bgs),
- 1 well immediately below the treatment zone (screened 6.4 to 7.3 m bgs), and
- 1 well in the treatment zone (screened 5.2 to 5.8 m bgs).

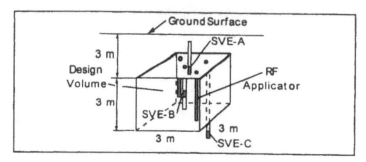

Figure 3.15 Design volume (3 m × 3 m × 3 m) for RF-SVE demonstration.

Due to the observed variability of the intrinsic permeability at the site and the modeling limitations associated with heterogeneity, the design and operating parameters for CASE #4 were recommended as a starting point for the demonstration's operational configuration. It was also recommended that the vertical vacuum gradients be measured at the perimeter of the treatment area and used as an aid for optimizing the final airflow rates from each well.

3.5.2 RF-SVE Demonstration System

The objective of this field demonstration was to determine and document the reduction in SVOC concentrations (specifically for DRO compounds) in a 3 m × 3 m × 3 m block of unsaturated-zone soil. To facilitate evaluation of SVOC removal, the RF-SVE treatment and monitoring instrumentation systems were focused on this specific *design* volume. The top surface of the *design* volume was at a depth of about 3 m bgs as shown in Figure 3.15.

The size of the *design* volume was selected to provide a volume that would be treatable given the limitations of the size of the RF heating system and vapor extraction system to be used in this project. The 3-m cube was centered spatially and vertically at the site to enclose the zone of expected highest contaminant concentrations as determined by the results of the analyses of soil samples obtained during the initial site investigation.

The well layout for the demonstration site is shown in Figure 3.16. There were 19 wells used for the RF-SVE operating system and pressure-temperature monitoring: 2 RF antenna wells, 3 SVE well screens, 8 P/T wells, and 6 PITT wells. The RF antennae were placed inside two 18-cm (7-inch) fiberglass well casings to a depth of 6.2 m (20 ft). The length of each antenna was 3.5 m (11.5 ft). The dipole antennae were remotely operated and computer controlled at an ISM frequency of about 27 MHz. This frequency provided the most efficient conversion of AC power to RF energy for the soil characteristics at the Kirtland AFB site. Power was alternated between each antenna so that only one antenna transmitted energy at any one time. RF power was switched automatically from 1 antenna to the other when the temperature of the vertical center of the antenna reached 175°C or the temperature differential between the 2 antennae exceeded 15°C.

As shown in Figure 3.17, the three 5-cm (2-in.) diameter fiberglass SVE wells were placed at varying depths to enable radial airflow through the *design* volume. The 3 SVE well depths and well screen lengths were 3.4 m and 0.6 m (SVE A); 6.1 m and 0.6 m (SVE B); and 7.8 m and 0.9 m (SVE C). SVE A and SVE B wells were installed into the same borehole. However, as shown in Figure 3.16, the SVE C well was offset laterally 1.1 m from the center of the *design* volume.

All pressure-temperature (P/T) and PITT wells were fabricated with 5-cm (2-in.) diameter fiberglass casings. At all the P/T locations, the depth of the temperature-monitoring wells was 6.7 m (22 ft). Most of the P/T locations had 3 pressure-monitoring wells at depths of 3.4 m (11 ft), 6.1 m (20 ft), and 7.6 m (25 ft). Each PITT location had 3 fiberglass casings at depths of 4.0 m (13 ft), 5.9 m (19 ft), and 7.0 m (23 ft). The temperature wells had caps attached to the bottom of the casing. The pressure and PITT wells had 0.6-m (2-ft) well screens at the bottom of the casing.

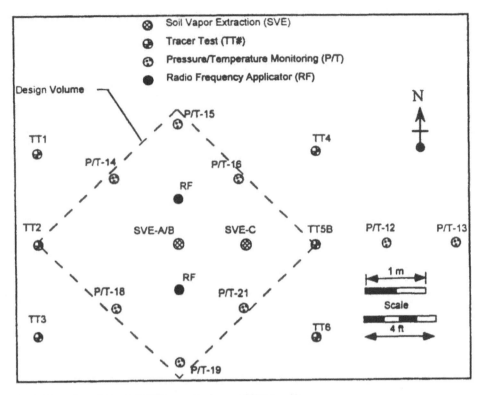

Figure 3.16 Plan view of the RF, SVE, monitoring, and PITT wells.

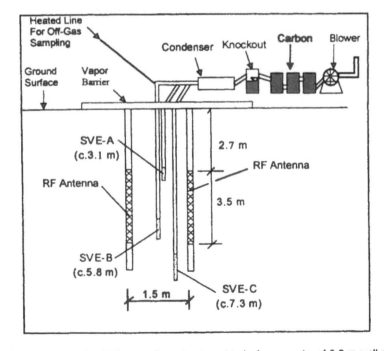

Figure 3.17 Schematic of the RF-SVE and off-gas treatment train (c. — center of 0.6-m well screen).

A vapor barrier covered the demonstration site to minimize the escape of hot DRO vapors and to hinder air from being drawn vertically from the ground surface down into the *design* volume. The barrier consisted of a 12.0-m × 12.8-m (about 40-ft × 42-ft) nylon-reinforced polyethylene sheet that was centered on the SVE B well. Openings for the aboveground extensions of all wells were sealed with nylon-reinforced polyethylene tape.

3.5.3 Operating Parameters

The following SVE system parameters were monitored during system operation:

- SVE Extraction Wells — temperature and pressure (vacuum) at a depth of 4.6 m (15 ft) and temperature at the wellhead
- Antenna Wells — temperature at a depth of 4.6 m (15 ft)
- P/T Monitoring Wells — temperatures at depths 1.5, 2.4, 3.4, 4.3, 5.2, 6.1 m (5, 8, 11, 14, 17, and 20 ft) and pressures (vacuum) at depths of about 3.0 m, 5.8 m, and 7.3 m (10 ft, 19 ft, and 24 ft)
- PITT Wells — temperatures at a depth of 4.6 m (15 ft)
- Differential pressures (pitot tube) between the wellhead and heat exchanger, between the heat exchanger and activated carbon drums, and between the activated carbon drums and blower
- Temperatures at the outlet of the rotameter and blower and pressures at the inlet and outlet of the blower
- Airflow rate coming from SVE B (and from other SVE wells when used as extraction wells)

The measured data for the demonstration are summarized in Appendix B.

During the first month of operation, the air–vapor flow rate was measured with a Dwyer anemometer (air velocity meter) as the mixture left the wellheads. For the last 2 months, the flow rate was measured with a rotameter attached to the knock-out drum outlet.

The initial operating conditions were based on the results from the testing of the SVE system before RF heating. During this unheated test, the system was operated at several airflow rates to draw ambient vapors through the demonstration volume. Soil vapor-phase pressures were allowed to equilibrate and were then measured at the P/T wells. The results from this test established the minimum airflow rate to induce measurable negative pressure (vacuum) readings within the *design* volume. As the volume was heated, the airflow rate was adjusted to maximize the removal rates of water and DRO components from the soil.

After extracting the water and organic vapors from the subsurface, the off-gas was drawn through a gas-phase treatment train that consisted of:

- A large copper-tube condenser cooled with ambient air to remove water and DRO vapors from the off-gas
- A 55-gal knock-out drum to collect the water and DRO condensates
- Three 55-gal activated-carbon drums in series to remove residual DRO vapor from the exhaust air
- A blower with an exhaust stack to draw the off-gas through the SVE system

In addition to serving as the final treatment unit for the exhaust air, the carbon drums also coalesced any water and DRO mist that escaped the condenser and knock-out drum.

3.5.4 Volumes Considered for Analysis

In this demonstration, the target treatment volume was called the *design* volume, which was roughly a 3.05-m (10-ft) cube containing 28.3 m³ (37.0 yd³) of soil at a depth beginning at 3.05 m.

In addition to the *design* volume, four additional volumes of soil were used to evaluate the data collected during the demonstration. These were the *sampled* volume, *extended design* volume, *heated* volume, and *tracer* volume. The centroids of all but the *tracer* volume nearly coincided with each other.

Initially, the *sampled* volume was defined by the project team to be an elliptical cylinder of soil that contained about 320 m³ (422 yd³) of soil. This volume included two outlying pre-demonstration boreholes and one outlying post-demonstration borehole (BH 12, 13, and 32 in Figures A1 and A2, Appendix A). It was determined that these three boreholes should not be included within the *sampled* volume because they adversely biased the calculations for the pre- and post-demonstration mass balances.

Parks (1949) listed calculation methods that can be used to estimate mass balances. These methods applied to boreholes with uniform spacing on rectangular coordinates; to boreholes with semiregular spacing on rectangular coordinates; and to boreholes with irregular spacing. The third category applied to the pre- and post-borehole patterns for the RF-SVE demonstration. This category included the area of influence method and the triangular grouping method. Because the borehole patterns were quite irregular, the triangular grouping method was used to calculate the pre- and post-soil volumes and mass balances for the demonstration.

Consequently, the *sampled* volume of soil was defined by the average ground-surface area (16.5 m²) that was enclosed by the lines joining the outer pre- or post-demonstration sample boreholes and their average borehole depth of 8.0 m (26 ft). The borehole layouts and calculations for estimating the average *sampled* volume and the TPH, DRO, and water masses are in Appendix A. The *sampled* volume measured approximately 4.6 m × 3.6 m × 8.0 m deep (15 ft × 12 ft × 26 ft). Thus, the averaged *sampled* volume contained 132 m³ (173 yd³) of soil.

A portion of the *sampled* volume included the *design* volume. The *extended design* volume (50 m³) had the same 3.05 m height and was at the same 3.05-m depth as the *design* volume. However, it had the same 4.6 m × 3.6 m ground-surface area as the *sampled* volume.

The *heated* volume (324 m³) was defined by the measured-temperature profile and the heat balance results when RF heating ceased on February 28, 1997. The NO-e-SYS computer program from Fortner Software calculated the profile and the heat balance. The program defined time-dependent average temperatures for seven volumes of soil within specific temperature ranges. Soil temperatures between 40 and 130°C defined the extent of the *heated* volume. This *heated* volume measured 8.78 m × 8.78 m × 4.21 m deep (28.8 ft × 28.8 ft × 13.8 ft). The centroids of the *design*, *heated*, and *extended design* volumes coincided at a depth of 4.57 m (15 ft).

The *tracer* volume (480 m³) was defined using the PITT. The PITT employed selectively partitioning tracers to locate and quantify the mass of contaminant in the subsurface. The volume of the subsurface swept by the tracers was called the *tracer* volume, and included the design, *extended design*, *sampled*, and *heated* volumes.

3.5.5 RF-SVE Modes of Operation

Two modes of RF-SVE operation, *countercurrent* and *cocurrent* flow of air and heat, were used in the demonstration. The *"five-spot"* mode was proposed, but it was not implemented. These three modes are described in the following sections.

From December 3, 1996 to January 17, 1997, the contractors operated the RF-SVE system in the countercurrent mode. On January 17, they began operating the RF-SVE system in the cocurrent mode. The project team concluded, however, that the cocurrent mode removed excessive amounts of cool air and water vapor from outside the *design* volume. The cool air decreased the peripheral P/T well temperatures. Furthermore, air did not flow into the SVE B well toward the peripheral wells as originally proposed. Therefore, on January 27, 1997, they returned to the countercurrent mode.

3.5.5.1 Countercurrent Mode of RF-SVE Operation

In the *countercurrent* mode of operation, the air and water–DRO vapors were extracted from the central SVE B well. The air flowed from outside the *design* volume toward the central well, SVE

B. Within 0.75 m (2.5 ft) of the SVE B well 2 RF antennae generated heat within the soil. The heat radiated outward toward the perimeter of the *design* volume *countercurrent* to the airflow. The heat increased the temperature of the soil and evaporated the water and DRO compounds into the air. As the air moves through the soil toward SVE B, it crossed zones of increasing temperatures that kept the water–DRO vapors in the air from condensing. Because the hot air and vapors flow toward the central SVE B well, there was no mechanism to remove the heat that was produced by the RF antennae in the central portion of the *design* volume. Accordingly, the temperatures of the RF antennae increased toward their maximum operating temperature of 175°C. Consequently, the RF antennae had to be cooled by a constant flow of ambient air into wells A1 and A2 or operated at lower power output.

3.5.5.2 Cocurrent Mode of SVE Operation

In the *cocurrent* mode, air entered the central SVE B well and moved through the *design* volume outward toward three or four extraction wells (P/T wells 14, 16, 18, and 21 in Figure 3.16). To draw air into the test volume of soil, the central wellhead was open to the atmosphere. The heat produced from the RF antennae was emitted outward through the soil toward the periphery of the *design* volume. The heat flow was *cocurrent* with the airflow. The heat increased the temperature of the soil and evaporated the water and DRO compounds into the air. As the air moved through the soil toward the P/T wells, it crossed zones of decreasing temperatures and the water and DRO vapor could condense and release heat to the cooler wet soil. The temperature zones could also prevent any moisture in the air being drawn from outside the *design* volume. When the airflow removed the water from the soil, the temperatures increased above 94°C, especially in zones within the *design* volume near the RF antennae. However, at 94°C, high-energy output from the antennae could produce a steam front, a zone of positive pressures, around the central SVE B well. This front could block the airflow into the central SVE well. Hence, a considerable amount of air and water vapor could be drawn from the soil outside the perimeter P/T wells. This "outside air", in turn, would decrease the temperature at the extraction wells and decrease the efficiency of water–DRO removal from the *design* volume.

3.5.5.3 Five-Spot Mode of RF-SVE Operation

After the demonstration was completed, an evaluation of RF-SVE results indicated that the *five-spot* mode of SVE operation could decrease the overheating of the RF antennae. It also could increase the amount of heat added to the soil while heating to 100°C. The *five-spot* mode is a modification of the *cocurrent* mode. In this mode the RF antennae send heat outward toward the perimeter of the *design* volume. A blower injects air into the central SVE B well to maintain airflow toward the extraction wells. A second blower removes the air and water–DRO vapors from four extraction wells that are equally spaced around the perimeter of the *design* volume. This mode decreases the amount of "outside air" flowing into the extraction wells and it can move the heat away from the two RF antenna wells.

3.6 DEMONSTRATION PROJECT CHRONOLOGY

Table 3.6 lists the sequence of field operations for the RF-SVE demonstration project. Table 3.7 summarizes operating parameters used during the five major periods for the RF-SVE phase of the project. The effects on the temperature at the centroid of the *design* volume and the water and DRO removal rates are also included.

Table 3.6 Sequence of Field Operations

Date	Operation
3/8/96	Drill & sample 2 boreholes for site evaluation
6/24–6/25/96	Start site characterization; drill & sample 4 boreholes and obtain bulk soil samples
9/13/96	Drill tracer wells FT14-07, 08, 09; perform *in situ* air permeability test
10/7–7/10/96	Drill boreholes FT14-10 through FT14-21 for soil sample recovery and RF-SVE system installation
10/11–10/28/96	Install SVE wells, casings; construct off-gas treatment system
10/29–10/31/96	Perform "SVE Only" demonstration
11/11–11/18/96	Complete pre-demonstration PITT
11/19–11/20/96	Complete pre-demonstration *in situ* respiration test
11/18–11/24/96	Begin RF system testing
11/25/96	Initial start-up of RF-SVE full-scale operation
11/25–11/26/96	Continue RF system testing
11/26/96	Stop RF system-generator failure
12/3/96	Begin RF-SVE full-scale operations
12/3/96–1/3/97	Set airflow from SVE-A, B, C (countercurrent mode)
1/3–1/9/97	Set airflow from SVE-B
1/9–1/17/97	Set airflow from SVE-B (using rotameter)
1/17–1/27/97	Set airflow from P/T wells 14,16,18, 21 (cocurrent mode)
1/27–2/28/97	Set airflow from SVE-B (countercurrent mode)
2/28/97	Stop RF heating (continue SVE operation)
3/17/97	Stop SVE (complete RF-SVE field demonstration)
3/17–3/20/97	Complete post-demonstration respiration tests
4/14–4/16/97	Complete post-demonstration soil sampling
5/2–5/16/97	Add water to site soil using an irrigation system
6/19–6/28/97	Complete post-demonstration PITT
7/7–7/10/97	Complete final closure work at the site

3.7 DEMONSTRATION RESULTS

The Appendices contain discussions, data, and results from the Kirtland AFB demonstration. The following sections discuss the demonstration data and results, conclusions, and recommendations. Tables 3.8, 3.9, and 3.10 summarize the measured and calculated data.

Table 3.8 divides the major periods listed in Table 3.7 according to changes in the extraction-well used for SVE and in the RF kilowatt output. Table 3.8 lists averaged values for each period for the following measured field data.

- Airflow rate, slpm, and the humidity of air leaving the SVE well, kg water per kg dry air
- The temperature (°C) of air within SVE B, SVE C; six peripheral P/T wells (14, 15, 16, 18, 19, and 21); and P/T wells 15 plus 16
- Total water and DRO product removed, liters

The temperature for each P/T well is an average of the readings at depths of 4.3 m and 5.2 m below ground surface. The temperature for each SVE well is the value measured at a depth of 4.6 m. Details of the measured data can be found in Appendix B.

3.7.1 Summary of Results

The RF-SVE demonstration began on December 3, 1996 and ended on February 28, 1997, for 87 days of operation. Applying the RF energy resulted in the center of the *design* volume at SVE B reaching a maximum temperature of 139°C, somewhat lower than the goal of 150°C. Temperatures at the 6-perimeter P/T wells of the *design* volume achieved an average maximum temperature of

Table 3.7   RF-SVE Operation Periods

Period (time, days)	Dates	Description
Low power RF tests (15 days)	11/18/96–12/3/96	1. RF unit setup 2. RF power below 2 kW 3. RF power intermittent but continuously operated for 1 day
RF and SVE shakedown (17 days)	12/3/96–12/20/96	1. Extract air from SVE A, B, and C wells 2. 12/3: start up at low RF power, <2 kW 3. 12/12–12/20: increase RF power from 2 kW–10 kW 4. Average RF power for the period at 4.8 kW 5. 12/3–12/12: decrease airflow from 470 slpm to 260 slpm 6. 12/12: decrease airflow to below 150 slpm 7. Heat soil around SVE B to 74°C 8. Low water and DRO removal rates
RF and SVE systems operational evaluation (28 days)	12/20/96–1/17/97	1. Extract air from SVE A, B, and C wells; SVE A and C wells; or SVE B well 2. Keep RF power at 10 kW until 1/9/97 3. Increase RF power from 10 kW to +14 kW after 1/9; average at 10.4 kW 4. Decrease airflow from 150 slpm to 50 slpm 5. Heat soil around SVE B to 90°C by 1/9/97 6. Heat soil around SVE B to 99°C by 1/17/97 7. Low water and DRO removal rates
RF-SVE operation (42 days)	1/17/97–2/28/97	1. Extract air from four P/T wells (14, 16, 18, 21) until 1/27/97 2. Extract air from SVE B well after 1/27/97 3. Maintain RF power between 13 kW and +15 kW; av at 13.2 kW 4. Increase airflow to 450 slpm on 1/17/97 5. Decrease airflow from 450 slpm to 250 slpm by 2/14/97 6. Heat soil around SVE B to +130°C by 1/27/97 7. High water and DRO removal rates
Soil mass cool down (17 days)	2/28/97–3/17/97	1. Extract air from SVE B well 2. Shut down RF power unit on 2/28/97 3. Maintain airflow at about 275 slpm 4. Cool soil around SVE B to 79°C 5. Decreasing water and DRO removal rates

107°C. P/T wells (15 and 16) attained an average maximum of 114°C. The temperatures at the perimeter of the *design* volume were below the goal of 130°C.

As expected, operation of *SVE-only* revealed that the constituents in the off-gas as analyzed through gas chromatography were the lighter GRO constituents rather than the DRO constituents. Analysis of the off-gas DRO condensate and the site soils during and after the RF-SVE demonstration indicated that chemicals as heavy as C_{21} were removed from the soil during RF-SVE operation. The off-gas monitoring was sporadic and of somewhat uncertain quality during the demonstration due to intermittent sampling interruptions and analytical equipment malfunctions.

RF heating surely did enhance the removal rate of DRO constituents. RF-SVE removed about 116 L (~100 kg) of DRO constituents as measured by off-gas condensate collection. The mass balances based on soil sampling showed that 116 kg of DRO constituents were removed from the *sampled* volume or about 48% of the initial DRO mass in that volume. Furthermore, the average DRO soil concentration decreased from 1020 mg/kg to 560 mg/kg. The *extended design* volume that included the *design* volume also showed a decrease in the average DRO concentration from 1520 mg/kg to 690 mg/kg.

These results, as discussed later, should be viewed in light of the nonoptimized operating conditions that were employed during the RF-SVE demonstration.

Table 3.8 Summary of Demonstration Results

Time Period	RF Heat Time (days)	Extraction Wells	Average RF Output (kW)	Airflow (slpm)	SVE B[a] (°C)	SVE C[c] (°C)	Six P/T[b] (°C)	P/T (15+16)[b] (°C)	Humidity of Air (kg/kg)	Water[c] Removed (L)	DRO Removed (L)
11/20/96 12/3/96	0[d]	A, B, C	0.00	nm	24–27	24–27	24–27	24–27	nm	nm	nm
12/3/96 12/12/96	9	A, B, C	2.11	348	27–46	27–33	27–38	27–38	0.020	104	0.4
12/12/96 12/20/96	17	A, B, C	7.88	245	46–74	33–49	38–62	38–64	0.074	196	0.8
12/20/96 1/3/97	31	A, B, C	9.91	88	74–89	49–57	62–79	64–79	0.302	571	2.8
1/3/97 1/9/97	37	B	9.65	59	89–94	57–nm	79–78	79–72	0.372	754	7.5
1/9/97 1/17/97	45	B	11.96	82	94–99	nm–75	78–nm	72–nm	0.390	1068	11.5
1/17/97 1/27/97	55	14, 16 18, 21	13.35	434	99–139	75–83	nm–82	nm–85	0.246	2655	23.8
1/27/97 2/14/97	73	B	13.40	327	139–131	83–116	85–99	85–106	0.270	5292	71.6
2/14/97 2/28/97	87	B	12.77	263	131–134	116–123	99–107	106–114	0.265	6961	104.0
2/28/97 3/17/97	0[d]	B	0.00	279	134–79	123–79	107–73	114–77	0.105	7821	115.9

Note: nm, not measured.

[a] The temperatures are the measured values at 4.6 m depth in the SVE well.
[b] The temperatures are averages of the measurements at 4.3 m and 5.2 m depths in the P/T wells.
[c] The data for the Water Removed are from the Water (Daily) column in Table B3, Appendix B.
[d] No RF heating from 11/20–12/3 or 13 days and from 2/28–3/17 or 18 days.

3.7.2   RF Energy Supplied

KAI's RF field data are included in Appendix D. From December 3, 1996 through February 28, 1997 (87 days), 21,000 kWh of energy were produced by the RF generator and transmitted to 1 of 2 subsurface antennae. During the 87 days, the antennae radiated into the soil 17,000 kWh or 61.1 million kiloJoules (kJ) as heat. This amount of heat equates to about 81% of the energy produced by the RF generator.

Properly designed antennae should have a conversion efficiency of approximately 93%. The difference was due, in part, to the difference between the actual soil moisture content at 6.8% dwb and the bulk-sample moisture content at about 1%. Tests on the bulk soil sample were used to design the RF antennae for the Kirtland demonstration. The lower moisture content of the sample affected KAI's design that decreased the conversion efficiency of RF energy to heat from 93% down to 81%.

According to KAI, the vacuum-tube RF generator that was used at the Kirtland demonstration had a previously measured conversion efficiency of AC power to RF energy of 60%. The data in Appendix D show that the recorded conversion efficiency of AC power to RF energy was 44%. However, KAI estimated that about 6.5% of the AC power was used to heat the generator trailer. Therefore, the actual conversion efficiency of AC power supplied to the RF generator was 47%.

For the Kirtland demonstration, the overall efficiency of AC power to RF heat was about 38% (efficiency = 0.81 × 47). If the conversion efficiency of RF energy to heat was 93% as originally expected, the AC power to RF heat conversion efficiency would be about 43%. During the demonstration, KAI's RF equipment experienced a 6% down time.

KAI's newer solid-state RF units have a conversion efficiency of AC power to RF energy of about 72%. Using the antenna conversion efficiency of 93%, KAI now expects a 67% conversion efficiency of AC power to heat instead of 43%.

3.7.3   Heat Utilization

As shown in Tables 3.7 and 3.8, the output of the RF generator increased from 2.1 kW to 10 kW during the shakedown and system evaluation period, December 3, 1996 to January 3, 1997. During this 31-day period, the average generator output was 7.1 kW. From January 17 to February 28, the RF-SVE operation period, the output was maintained between 13 and 15 kW. For this 42-day period, the average RF output was 13.2 kW. Higher RF output could not be attained because the RF-SVE system was operating in a *countercurrent* mode. Heat was building at each antenna causing its temperature to approach 175°C, which was the maximum allowed for safe operation of the antenna and adjacent casing.

Table 3.9 is a summary of the heat balances for each period defined in Table 3.8. The details for the calculated heat balances can be found in Appendix C. Table 3.9 contains the following data:

- The amount of RF energy added to the soil as heat
- The energy used to heat the dry soil
- The energy used to heat the residual soil moisture (water)
- The heat added to the dry air
- The heat required to vaporize water into the dry air
- An estimate of the heat loss to the surroundings

The vaporized water in the air includes the water condensed in the air-vapor treatment system and an estimate of the water vapor in the blower exhaust air.

The heat loss to surroundings was calculated by computer program routines (or macros) that used the temperature profile at the end of each period and heat conduction across the six boundary surfaces of a constant 324-m³ *heated* volume. The profiles were developed from the temperature data in Table B1, Appendix B, using the NO-e-SYS data management program. The codes for the macros are listed in Appendix C.

Table 3.9 Heat Balance Summary for RF-SVE Demonstration at Kirtland AFB

Period Ending	12/12/96	12/20/96	1/3/97	1/9/97	1/17/97	1/27/97	2/14/97	2/28/97	Total to 2/28/97
RF Heat Input, kJ									
Total	1,485,000	3,918,000	9,116,000	4,375,000	6,139,000	8,095,000	16,215,000	11,774,000	61,118,000
Total operating days	9	17	31	37	45	55	73	87	87
Heat utilization, kJ									
Dry soil	923,000	1,346,000	3,521,000	−266,000	2,232,000	2,080,000	4,984,000	2,225,000	17,044,000
Residual soil water	320,000	463,000	1,185,000	−89,000	736,000	655,000	1,437,000	606,000	5,313,000
SVE airflow	87,000	48,000	64,000	49,000	79,000	501,000	1,256,000	839,000	2,923,000
Condensed water vapor	254,000	232,000	954,000	497,000	770,000	4,015,000	6,912,000	4,387,000	18,021,000
Water (g) in exhaust air	67,000	13,000	11,000	10,000	16,000	365,000	728,000	229,000	1,439,000
Loss to surroundings	617,000	804,000	2,240,000	1,005,000	1,261,000	1,470,000	3,837,000	3,155,000	14,389,000
Total utilized, kJ	2,268,000	2,906,000	7,975,000	1,206,000	5,094,000	9,138,000	19,278,000	11,441,000	59,130,000
% of input	153%	74%	87%	28%	83%	112%	118%	97%	97%

Note: Temperature averages and energy balance for total temperature volume. Volume = 8.78 m × 8.78 m × 4.21 m deep (28.8 ft × 28.8 ft × 13.8 ft) = 324.1 m³.

Cartesian coordinate system centered at the middle of the design volume. X(Depth): −3.05 m to +1.52 m; Y(North): −4.57 m to +4.57 m; Z(East): −4.57 m to +4.57 m. Calculations in SI units.

The heat loss calculation used an average heat conductivity for wet soil and the temperature drop across a 0.366-m (1.2-ft) layer at the boundary surfaces. The estimated heat loss to the surroundings (Q_S) was calculated using the following heat-conductance formula:

$$Q_S = kA(dT/dx)dt$$

where

Q_S = heat loss to the surroundings (kJ)
k = thermal conductivity of the soil (kW/m-°C)
A = area of boundary surface (m²)
dT = rate of change of temperature at the boundary surface
dx (or dy or dz) = 0.366 m (1.2 ft)
dt = time period of heat = (days)(86,400 s/day)

As shown in Appendix C, the values for dT/dx, dT/dy, and dT/dz are calculated using the NO-e-SYS computer program from Fortner Software.

Table 3.9 shows that for the initial 31-day period, the RF antennae added about 14.5 million kJ of heat to the soil with the following calculated distribution:

- Heating dry soil mass: 5.8 million kJ (0.40 kJ/kJ$_{input}$)
- Heating residual soil moisture: 2.0 million kJ (0.14 kJ/kJ$_{input}$)
- Vaporizing soil moisture into the SVE Air: 1.5 million kJ (0.10 kJ/kJ$_{input}$)
- Heating SVE Air: 0.2 million kJ (0.01 kJ/kJ$_{input}$)
- Heat loss to the surroundings (calculated): 3.7 million kJ (0.25 kJ/kJ$_{input}$)

The heat lost to the surroundings by difference was 5.0 million kJ (0.35 kJ/kJ$_{input}$) or [14.5 million kJ – (5.8 + 2.0 + 1.5 + 0.2) million kJ].

During the initial 31 days, the average airflow decreased from 350 to 60 slpm. Since the RF energy input was slowly increased from 2.1 to 9.6 kW, the wet-soil temperatures slowly increased to 94°C. Consequently, about 0.35 kJ per kJ$_{input}$ of the total heat input was lost to the soil outside the *heated* volume. About 0.54 kJ/ kJ$_{input}$ went into heating the wet soil. Since the airflow rate decreased as the RF energy increased, the air and contained water vapor consumed only 0.12 kJ/ kJ$_{input}$.

The table also shows that for the final 42-day period of RF operation, about 36.1 million kJ of heat was added to the soil by the RF antennae. The calculated distribution of heat was as follows:

- Heating dry soil mass: 9.3 million kJ (0.26 kJ/kJ$_{input}$)
- Heating residual soil moisture: 2.7 million kJ (0.08 kJ/kJ$_{input}$)
- Vaporizing soil moisture into the SVE Air: 16.6 million kJ (0.46 kJ/kJ$_{input}$)
- Heating SVE Air: 2.6 million kJ (0.07 kJ/kJ$_{input}$)
- Heat loss to the surroundings (calculated): 8.5 million kJ (0.23 kJ/kJ$_{input}$)

The heat lost to the surroundings by difference was 4.9 million kJ (0.14 kJ/kJ$_{input}$).

During this period, the average airflow varied between 265 to 435 slpm. The RF energy input was high, 13.2 kW, and the soil dried out as the temperatures increased toward the desired 150°C. Hence, the heat lost to the soil outside the *heated* volume was about 0.14 kJ/kJ$_{input}$. About 0.33 kJ/kJ$_{input}$ went into heating the wet soil. Although the airflow was increased, only 0.07 kJ /kJ$_{input}$ was added to the dry SVE exhaust air. However, the water vapor in the exhaust air required 0.46 kJ/kJ$_{input}$. Consequently, 0.53 kJ/kJ$_{input}$ was removed in the wet air leaving the central well SVE B. Operating in the "five-spot" mode could have added more heat to the wet soil.

For the total 87 days of RF-SVE operation, about 61.1 million kJ of heat were added to the soil by the RF antennae. The distribution of heat was as follows:

- Heating dry soil mass: 17.1 million kJ (0.28 kJ/kJ$_{input}$)
- Heating residual soil moisture: 5.3 million kJ (0.09 kJ/kJ$_{input}$)
- Vaporizing soil moisture into SVE Air: 19.4 million kJ (0.32 kJ/kJ$_{input}$)
- Heating SVE Air: 2.9 million kJ (0.05 kJ/kJ$_{input}$)
- Heat loss to the surroundings (calculated): 14.4 million kJ (0.24 kJ/kJ$_{input}$)

The heat loss to the surroundings by difference was 16.4 million kJ (0.27 kJ/kJ$_{input}$).

The *heated* volume contained 324 m^3 of soil with a surface area of 302 m^2 (or 0.932 m^2/m^3). As discussed in this section, the estimated heat loss is a function of the area of the surface surrounding the *heated* volume. This surface should be defined by the area per unit volume of soil (m^2/m^3). Therefore, when the soil was slowly heated during the initial 31 days, the estimated heat loss was about 0.35 kJ/kJ$_{input}$ per (m^2/m^3) (535 kJ/m^2/day). When the soil was heated at the maximum rate, the final 42 days, the estimated heat loss was lower at about 0.14 kJ/kJ$_{input}$ per (m^2/m^3) (385 kJ/m^2/day). The average value for the 87 days of RF heating was about 0.27 kJ/kJ$_{input}$ per (m^2/m^3) (at 625 kJ/m^2/day).

Table 3.10 gives the volume and mass of soil for each period that is heated to one of seven temperature ranges. The temperature ranges are below 40°C, 40 to 60°C, 60 to 80°C, 80 to 94°C, 94 to 110°C, 110 to 130°C, and above 130°C. The table also lists the average temperature for each volume and mass. The volumes and their average temperatures were calculated using the computer program NO-e-SYS.

When the RF generator operated at an output of 13.2 kW and at an airflow rate of 190 slpm, idealized heat balance calculations showed that the 28-m^3 *design* volume could be heated to 94°C in about 10 days (Case 1 in Appendix E). However, the average RF generator output was only 7.1 kW during the initial 31 days of actual RF heating. The lower RF output would be expected to increase the heating time to 19 days (Case 2 in Appendix E). The January 3 data in Table 3.10 showed that the RF antennae heated at least 90 m^3 of soil beyond the 28-m^3 *design* volume. Because of the heat loss beyond the *design* volume, the time needed to heat the soil to 94°C could exceed the estimated time of 19 days and approach the 31 days of heating.

3.7.4 Test Volume Temperature

Figure 3.18 shows time plots for the soil temperatures within the P/T and SVE wells and for the amounts of water and DRO constituents removed by RF-SVE. The air and soil temperatures for each well remained below 94°C until the water had vaporized from the soil surrounding each well. Once the water was removed from this portion of the *design* volume, the temperatures increased toward the desired 150°C.

Table 3.8 and Figure 3.18 show that on January 17 (45 days of RF heating), the air in SVE B well reached a temperature of 99°C after 1070 L of water were removed. The data in Table 3.10 show that only about 2 m^3 of soil was above 94°C with an average temperature of about 95°C. By January 27 (55 days), SVE B reached a temperature of 133°C after the RF-SVE systems removed 2655 L of water. The volume above 94°C increased to about 7 m^3 with an average temperature of about 102°C. Table 3.8 shows that the average temperatures in all wells exceeded 94°C on February 14 (73 days) after the removal of 5292 L of water. The volume above 94°C increased to 26 m^3 with an average temperature of 106°C. When RF heating ended on February 28, the average temperatures in most of the SVE and P/T wells were above 110°C. The volume above 94°C increased to 36 m^3 with an average temperature of 108°C. Of this volume, 14 m^3 attained an average temperature of 118°C. This volume was close to one-half the *design* volume. However, 6961 L of water were removed from the heated soil. This amount was considerably higher than the original 3500-L estimate for the *design* volume. Therefore, considerable water was removed from beyond the *design* volume. The 100% increase in the mass of extracted water

Table 3.10 Heated Volume Data

Ending Date	12/3/96	12/12/96	12/20/96	1/3/97	1/9/97	1/17/96	1/27/97	2/14/97	2/28/97	3/12/97	3/18/97
Below 40°C											
Av temp (°C)	28	28	28	28	28	30	34	36	37	34	34
Vol (m³)	324	308	259	207	205	178	145	66	37	75	119
Mass (kg)	576,300	548,000	460,500	368,200	365,200	316,500	257,300	118,200	66,300	133,600	212,400
40–60°C											
Av temp (°C)		42	48	49	49	49	48	48	49	49	48
Vol (m³)		16	56	66	68	86	111	141	153	140	149
Mass (kg)		28,300	99,300	116,700	120,700	153,100	196,700	251,500	271,800	248,100	265,100
60–80°C											
Av temp (°C)			63	69	69	69	69	69	69	69	68
Vol (m³)			9	37	40	41	47	64	68	67	50
Mass (kg)			16,500	66,400	71,400	72,400	83,800	113,100	121,300	118,500	88,400
80–94°C											
Av temp (°C)				85	84	87	86	86	87	87	83
Vol (m³)				14	11	17	15	27	30	26	6
Mass (kg)				25,000	19,000	30,300	26,200	48,000	52,600	46,400	10,400
94–110°C											
Av temp (°C)						95	100	101	101	101	0
Vol (m³)						2	6	18	21	15	0
Mass (kg)						4,000	10,500	31,900	37,700	25,800	0
110–130°C											
Av temp (°C)							115	116	118	113	0
Vol (m³)							1	8	14	2	0
Mass (kg)							1,800	13,700	25,500	4,000	0
Above 130°C											
Av temp (°C)							131	130	131	0	0
Vol (m³)							0.01	0.01	0.61	0	0
Mass (kg)							10	10	1,100	0	0
Totals											
Av temp (°C)	28	30	32	40	39	44	48	57	62	55	47
Vol (m³)	324	324	324	324	324	324	324	324	324	324	324
Mass (kg)	576,300	576,300	576,300	576,300	576,300	576,300	576,310	576,410	576,300	576,400	576,300

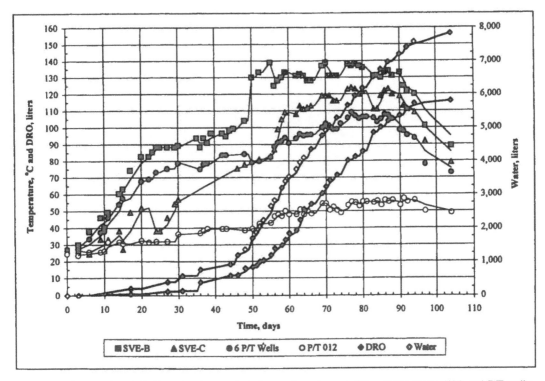

Figure 3.18 Time plots of DRO and water products in condensate and of temperatures in SVE and P/T wells.

increased the heat requirement. At a constant 13.1 kW energy input, this greater heat requirement increased the time to attain 94°C

Figures 3.19, 3.20, and 3.21 show the temperature profiles within the *heated* volume during and at the end of RF heating. The profiles were produced while estimating the heated volumes and average temperatures in Table 3.10 using the computer program NO-e-SYS. Figures 3.22, 3.23, and 3.24 show the temperature contours on vertical and horizontal slices through the *heated* volume at the end of 87 days of RF heating..

3.7.5 Water and DRO Contaminant Removal

The project team originally estimated that the 28.3 m³ *design* volume contained about 3500 L of water and 130 L of DRO contaminants. The project schedule called for 6 weeks (42 days) of RF-SVE operation to remove the water and contaminants. One of the sub-objectives was to heat the wet soil to 94°C and remove 3500 L of water and some of the contaminants within the first 3 weeks of RF-SVE operation. Therefore, the expected rate of water removal would have to be about 170 L per day. The remaining 3 weeks were to be used to heat the soil to 150°C and remove the remaining DRO contaminants in the *design* volume.

The RF-SVE system removed about 600 L of water and less than 5 L of DRO contaminants in the 31 days between December 3, 1996 and January 3, 1997 (Table 3.8). The average water-removal rate was about 18 Lpd. This rate was much lower than the expected 170 Lpd. During this 31-day period, the average airflow rate and RF antenna output were 205 slpm (7.2 scfm) and 7.1 kW, respectively. Since the RF output was low, the soil heating rate was also too low. Moreover, considerable heat was lost to the surroundings beyond the *heated* volume (~35%). On January 17, the project team decided to increase the airflow and RF output of the antennae. From

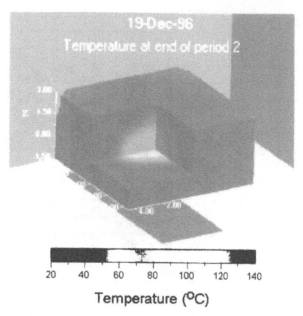

Figure 3.19 Temperature profile within the heated volume after 17 days of RF-SVE operation (December 20, 1996).

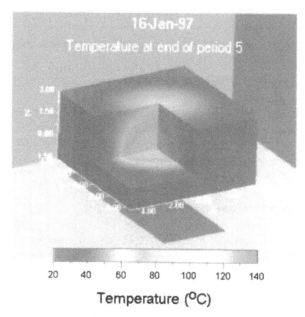

Figure 3.20 Temperature profile within the heated volume after 45 days of RF-SVE operation (January 17, 1997).

January 17 to February 28, 1997 (42 days), the RF-SVE system removed an additional 5893 L of water and about 93 L of DRO contaminants. Hence, the water-removal rate was about 140 Lpd. This rate was closer to the expected rate of 170 Lpd. For the 42 days, the average airflow rate was about 330 slpm (~11.5 scfm), and the average humidity of the air was 0.26 kg water per kg dry air.

From January 27 through February 28 (32 days), the RF-SVE system removed 80 liters (~70 kg) of DRO contaminant. During this period, the airflow rate averaged about 300 slpm (10.5 scfm); and the rate of DRO removal using the SVE-B well was about 2.2 kg/day (or 0.0043 kg/kg dry

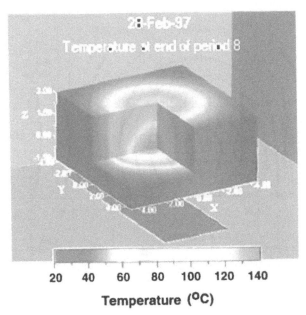

Figure 3.21 Temperature profile within the heated volume after 87 Days of RF-SVE operation (February 28, 1997).

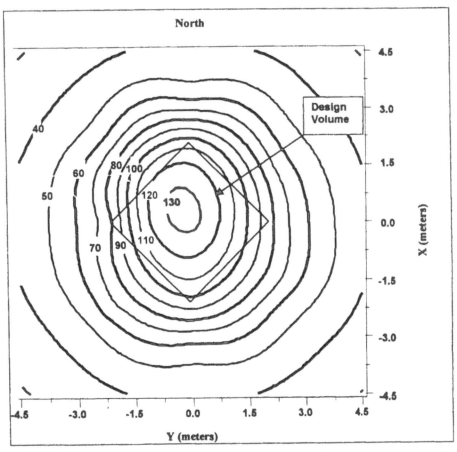

Figure 3.22 Temperature contours on a horizontal plane through Z = 0 (centroid of the heated volume) after 87 days of RF-SVE.

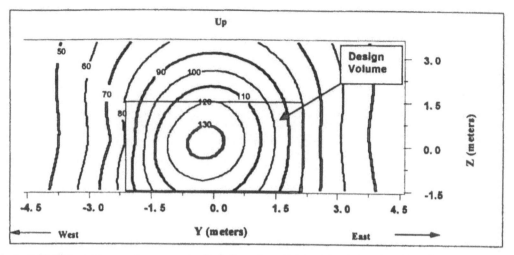

Figure 3.23 Temperature contours on a vertical plane through X = 0 (centroid of the heated volume) after 87 days of RF-SVE.

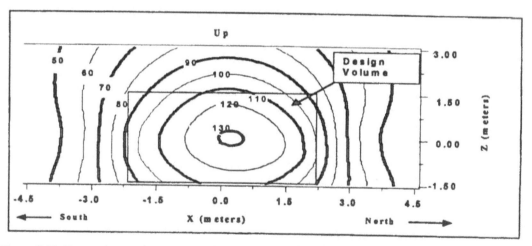

Figure 3.24 Temperature contours on a vertical plane through Y = 0 (centroid of the heated volume) after 87 days of RF-SVE.

air). As shown in Figures 3.1 and 3.2, this value was well below the saturation limits of most DRO compounds. For example, at 135°C (~410 K), nonodecane ($C_{19}H_{40}$) has a saturation limit of 0.020 kg/kg dry air and hexadecane ($C_{16}H_{34}$) has a limit of 0.080 kg/kg dry air.

3.7.6 Mass Balances

The balances for the pre- and post-demonstration TPH, DRO, and water masses are calculated in Appendix A for both the 132-m³ *sampled* and the 50-m³ *extended design* volumes. The calculated results are listed in Table 3.11.

Because the 28-m³ *design* volume did not contain physical barriers, the RF-SVE was considered an open system. Consequently, the RF-SVE process was able to have an impact on the soil outside the *design* volume. At the end of the demonstration, about 116 L (about 100 kg) of DRO contaminants were collected as off-gas condensate. The RF-SVE treatment removed 116 kg DRO from the *sampled* volume (~48% decrease in DRO mass) and 70 kg from the *extended design* volume

Table 3.11 Demonstration DRO and TPH Mass Balances

Volume	DRO (kg)	DRO (mg/kg)	TPH (kg)	TPH (mg/kg)	H_2O (kg)	H_2O (g/100 g)
Presampled	244	1,020	1,700	7,070	16,320	6.8
Postsampled	128	560	1,400	6,070	8,790	3.8
Amount removed	116	NA	300	NA	7,530	NA
Preextended design	133	1,520			6,110	7.0
Postextended design	63	690			2,760	3.0
Amount removed	70	NA			3,350	NA

(~53% decrease). Therefore, at least one third of the DRO condensate came from outside the *extended design* volume. Thus, the decrease in DRO mass resulting from RF-SVE essentially was assessed to volumes larger than the original *design* volume.

The average DRO concentrations of the post-*sampled* and *extended design* volumes were, respectively, 560 and 690 mg/kg. These low values indicated that further treatment could have resulted in a decrease in the DRO removal rate.

The data also showed that 300 kg of TPH was removed from the 132-m³ *sampled* volume. This value exceeded the 116-kg DRO value by 184 kg. This difference represented the potential loss in MRO compounds. Furthermore, the calculated DRO mass exceeded the amount of DRO condensate by about 16 kg. The four possible causes for these differences were as follows:

- Errors introduced by soil sampling, handling, holding, subsampling, and analyses
- Movement of DRO and TPH compounds out of the *sampled* volume
- Hydrous pyrolysis oxidation of DRO and TPH compounds
- Biooxidation of DRO and TPH compounds

As shown in Table 3.8, 7800 kg (L) of water were condensed from the SVE air during the 104-day demonstration. An additional 680 kg (L) was lost in the exhaust air leaving the blower. This estimated value was calculated using the daily volume of SVE air and the saturated humidity of air at the exhaust air temperature (Appendix C, Table C8). Therefore, the total water removed was 8480 kg. The water-balance data showed that the *sampled* volume had a net loss of 7530 kg. The 50-m³ *extended design* volume lost 3350 kg of water. The difference between the total water removed and *sampled*-volume net loss was about 950 kg (about 13% excess). In addition to the sampling-analytical errors, another probable cause for this difference was the removal of water from outside the 132-m³ *sampled* volume by the SVE air flowing through warm soil. On February 28, the amount of wet soil heated above 40°C was about 290 m³ (Table 3.9).

3.7.7 DRO Condensate Analyses

As illustrated in Figure 3.18, the removal rate of the DRO components increased significantly when the temperatures in the SVE B and SVE C exceeded 94°C. The slope of the curve for the cumulative DRO volume increased significantly after 55 days of RF heating indicating an overall increase in the DRO removal rate. Furthermore, the slope of the curve had not declined significantly by the end of 87 days of RF heating. However, not all of the increase in the removal rates was caused by increases in soil temperature. The volumetric flow rate through the system was also increased during the period when temperature was increased.

DRO condensate samples from each collection drum were also analyzed to assess changes in the free-product composition over time. Four marker chemicals, tridecane, pentadecane, heptadecane, and nonadecane, were analyzed to determine their total masses in the DRO condensate. The total mass for each chemical is listed in Table 3.12.

Table 3.12 Total Contaminant Released from Soil Based on DRO Condensate Analyses

Compound	Mass of Contaminant in DRO Condensate (kg)
Tridecane (C_{13})	1.7
Pentadecane (C_{15})	2.3
Heptadecane (C_{17})	0.7
Nonadecane (C_{19})	0.2
DRO condensate after 87 days	~90

As expected, the rates of removal of DRO compounds from the soil were not constant throughout the duration of the project. Generally, the removal rates increased as temperatures increased, gas flow rates increased, and the soil temperature exceeded 94°C. The removal rates for each compound before the soil temperatures reached 94°C were less than the rates after the temperature exceeded 94°C (see Table 3.13).

Table 3.13 Comparison of Mass Removal Rates Prior to and After the Boiling Point of Water was Reached

Soil temperature	Average Mass Removal Rates (g/day)			
	Tridecane	Pentadecane	Heptadecane	Nonadecane
<94°C	10	13	4	0.5
>94°C	24	36	17	4

A comparison of the DRO contaminant data in Table 3.12 was made to the analysis of the typical fuel oil in Table 3.14. Using the analyses for the n-paraffins, the hypothetical amounts of the marker chemicals were estimated and compared to their condensate concentrations:

- Tridecane (C_{13})   = 0.9 kg (condensate = 1.7 kg)
- Pentadecane (C_{15})   = 1.0 kg (condensate = 2.3 kg)
- Heptadecane (C_{17})   = 0.6 kg (condensate = 0.7 kg)
- Nonadecane (C_{19})   = 0.3 kg (condensate = 0.2 kg)

Except for nonadecane (C_{19}), the quantities of the marker chemicals in the condensate were comparable to the amount in a typical fuel oil.

The fractions for each marker chemical relative to the amount of DRO condensate were also determined. The results are presented in Figure 3.25. The fractions for each compound were relatively constant after 55 days of RF heating. Mass removal rates for each marker chemical were also assessed as a function of time and soil temperature. Results are plotted in Figure 3.26 through Figure 3.29.

The most volatile marker chemical, tridecane, represented a larger fraction of the DRO condensate at lower temperatures, with initial contributions ranging from 3 to 4.5% (Figure 3.26A). This value decreased to about 1.5% at the termination of the project. Although the fraction of tridecane in the condensate was largest during the first half of the project, less than 1% of the total tridecane removal occurred during the first 10 days of the project (Figure 3.26B). The maximum removal rate of tridecane was 4 g/h on days 62 and 82.

Pentadecane represented approximately 1% of the DRO condensate at the beginning of the project (Figure 3.27A). Less than 1% of the total pentadecane was removed in the off-gas after 30 days of RF heating (Figure 3.27B). Once the soil temperature in the center of the *design* volume reached the 94°C, the pentadecane contribution was relatively constant at 2.5% of total condensate. Removal rates increased rapidly after 30 days. The maximum removal rates of pentadecane were approximately 6.3 g/h on day 62 and 5.8 g/h on day 82.

Table 3.14 Physical and Chemical Characteristics
of No. 2 Fuel Oil

Characteristic or Component	
API gravity (20°C, oAPI)	31.6
Density (20°C, g/L)	0.868
Saturates, wt%	61.8
n-paraffins	8.07
$C_{10} + C_{11}$	1.26
C_{12}	0.84
C_{13}	0.96
C_{14}	1.03
C_{15}	1.13
C_{16}	1.05
C_{17}	0.65
C_{18}	0.55
C_{19}	0.33
C_{20}	0.18
C_{21}	0.09
Iso-paraffins	22.3
1-ring cycloparaffins	17.5
2-ring cycloparaffins	9.4
3-ring cycloparaffins	4.5
Aromatics, wt%	38.2

(Brown & Root Environmental, private communication)

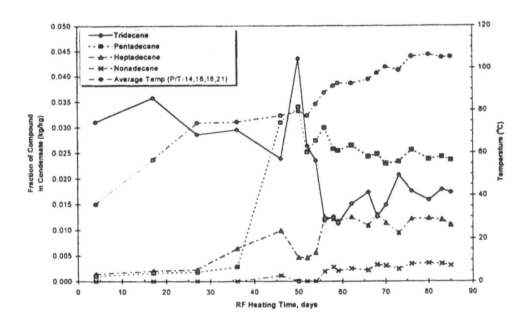

Figure 3.25 Fraction of individual DRO constituents in condensate.

Initially, the percentages of DRO contributed by heptadecane and nonadecane were less than 0.2% and 0, respectively (Figures 3.28A and 3.29A). Less than 1% of the total heptadecane and nonadecane masses was removed prior to day 30 and day 45, respectively (Figures 3.28B and 3.29B). Once the soil temperature reached 94°C, the relative contributions of heptadecane and nonadecane in DRO condensate were approximately constant at 1% and 0.3%, respectively. As

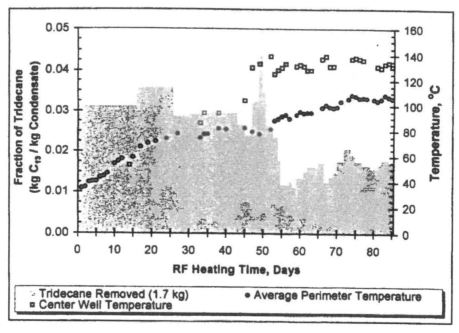

(A)

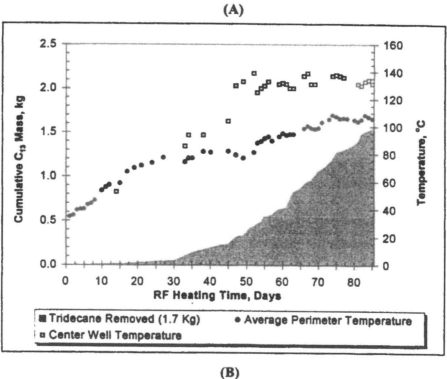

(B)

Figure 3.26 C_{13} in condensate. (A) As fraction of condensate; (B) cumulative mass collected in condensate.

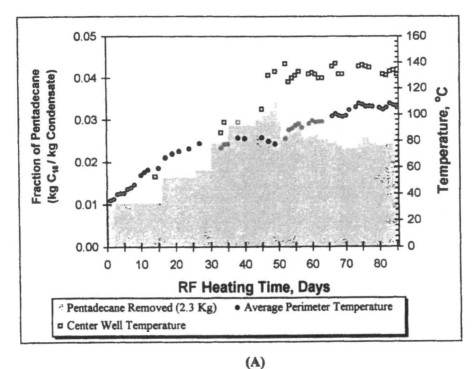

(A)

(B)

Figure 3.27 C_{15} in condensate. (A) As fraction of condensate; (B) cumulative mass collected in condensate.

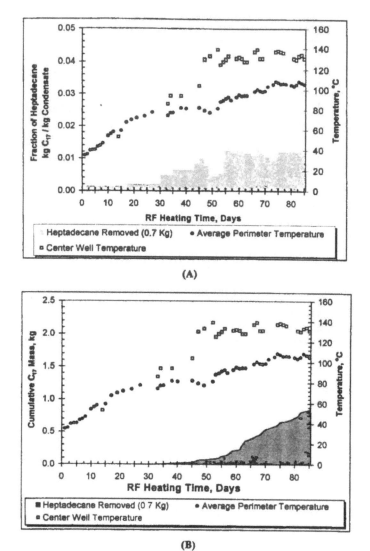

Figure 3.28 C_{17} in condensate. (A) As fraction of condensate; (B) cumulative mass collected in condensate.

with tridecane and pentadecane, the maximum removal rates for heptadecane and nonadecane occurred on days 62 and 82. On these days, the removal rate for heptadecane was 3 g/h, and the removal rate for nonadecane ranged from 0.6 to 0.8 g/h.

3.7.8   Partitioning Interwell Tracer Tests

3.7.8.1   Background on Tracer Tests

The PITT has been used since the 1970s for characterizing oil in reservoirs. Recently, the PITT has been adapted for detection of nonaqueous phase liquids (NAPL) in the subsurface. This method was first applied to the detection of NAPL contamination in the saturated zone and has been demonstrated to be effective at both laboratory and field scales. The PITT technology has also been applied to the detection of NAPL in the vadose zone.

A PITT involves several steps. First, a hydraulic flow field is set up between injection and extraction wells that bracket the zone of possible contamination. A mix of tracers is then introduced

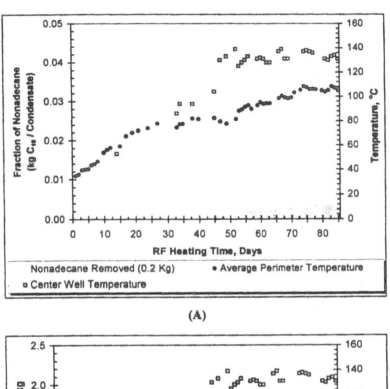

(A)

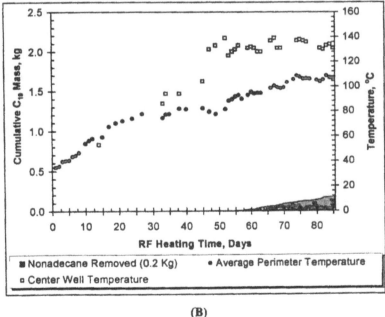

(B)

Figure 3.29 C_{19} in condensate. (A) as fraction of condensate; (B) cumulative mass collected in condensate.

in the injection well and detected at the extraction site. The injected mixture includes a nonpartitioning tracer and at least one partitioning tracer. The nonpartitioning tracer exists solely in the
mobile phase; so, its travel time is dependent only on the flow rate and swept volume. The
partitioning tracer is distributed between the mobile and resident immobile phases as it moves
along the stream paths and, therefore, lags behind relative to the fraction of time spent in the
immobile phases. A comparison of the partitioning and nonpartitioning tracer response curves can
thus provide quantitative information about immobile phases present in the subsurface. Method of
moments analyses are used to delineate the NAPL saturation from the tracer response curves.

Appendix G describes detailed procedures for both preliminary laboratory partitioning tests and field demonstration. Fundamental considerations and calculation techniques are included.

3.7.8.2 Purpose of the PITT Study

The PITT was used to help characterize the hydrocarbon contamination beneath a fire-training pit at Kirtland Air Force Base in Albuquerque, NM. The second purpose of the PITT was to evaluate the remedial effectiveness of RF-SVE.

The PITT was intended to provide information about the total petroleum hydrocarbons (TPH) and the DRO existing in the subsurface before and after the RF-SVE demonstration. After the site was selected and soil samples were analyzed, UT determined that the removal of DRO was the most important factor concerning the remediation effort. Consequently, thermodynamic modeling was employed concurrent with the PITT results to help delineate the removal of the TPH and DRO fractions.

3.7.8.3 Laboratory Studies

The purpose of the laboratory studies was to gather information about the thermodynamic interaction of the tracers and the site hydrocarbons. Specifically, this included two tasks:

1. Determine the equilibrium partition coefficients, K_N, between each tracer and the hydrocarbons.
2. Develop a thermodynamic model to predict how the composition of the hydrocarbon mixture affects these partition coefficients.

Laboratory Results — Three laboratory tests (KIRT11, KIRT12, and KIRT14) were completed to measure the equilibrium partition coefficients, K_N, between five tracers and the TPH hydrocarbons. The partitioning tracers were selected from the perfluorocarbon ($C_X F_{2X}$) family of compounds. Methane (CH_4) was used as the nonpartitioning tracer.

For the laboratory tests, two composite bulk samples of soil were prepared using soil samples from six pre-demonstration boreholes. The soil samples were from the depth interval of 3 to 4 m bgs. The first composite contained 10,800 mg TPH per kg of soil. The second contained 13,500 mg TPH per kg. The first sample was used in the KIRT11 test that was carried out in a 30-cm long, glass chromatography column. The second sample was used in the KIRT12 and KIRT14 tests. These tests were carried out in a 30-cm long, stainless-steel column.

Figure 3.30 shows the tracer response curves for the experiment (KIRT12), which are typical of all three experiments. The partition coefficient values (K_N) for a given tracer compound were consistent among the 3 laboratory experiments, varying 9.8% at most. For example, K_N values for $C_7 F_{14}$ were 8.3, 8.8, and 9.4 for the 3 tests. This range of variation is typical for this type of test.

Compositional Analysis — Because of the compositional complexity of the TPH, a relationship between the tracer partition coefficients and the TPH composition was required. Hydrocarbon mixtures are suitable for thermodynamic analysis using the Peng-Robinson equation of state (EOS). Useful routines of the compositional simulator UTCOMP were arranged in a single program that included this EOS. The EOS requires the PVT properties of the components as input; so this information was needed for both the tracers and the TPH. The thermodynamic properties of the tracers were available from the literature, but the TPH properties were unknown. Therefore, a true boiling point (TBP) analysis of the TPH was completed. The TBP analysis was used with carbon-number correlations to develop a pseudofractional representation of the site hydrocarbons for input into the EOS.

The EOS was "tuned" to match the tracer gas-TPH partition coefficients determined in the laboratory experiments described earlier. The tuned EOS was then used as a predictive tool to

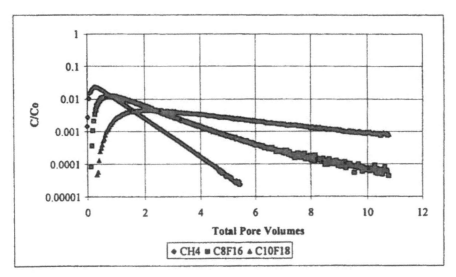

Figure 3.30 Tracer response curves, experiment KIRT12.

determine tracer gas-TPH partition coefficients for different mixture compositions. This was necessary due to the change in TPH composition that was caused by the RF-SVE remediation of the
site soils.

3.7.8.4   PITT Field Tests

Field Tracer Test Design — The well layout called for three well pair groupings at three depths
per grouping as shown in Figure 3.31, plan view. Figure 3.32 shows the various depths of each
well grouping. Designated TT-numbers are consistent with those set out in Section 3.5.2. The letter
designations in Figure 3.31 indicate the different depths for each numbered well. Computer simulations with this layout were run to help determine the PITT operational parameters. The purpose
of these simulations was to design a PITT likely to be effective given variable subsurface parameters,
since the exact soil permeability (k) distribution and TPH saturation (S_N) distribution were unknown.
UTCHEM, the three-dimensional UT multicomponent multiphase flow and transport simulator,
was used for the simulations. The concentration breakthrough curves were analyzed with the method
of moments. The flow rate and mass of tracers injected were the two main characteristics that
needed to be determined with the simulations. The flow rate was adjusted to achieve a mean
residence time of one day for the conservative tracer, methane. After the flow rate was set, the mass
of tracer injected was adjusted according to the concentration envelope. This envelope was defined
as the two curves limiting the concentrations of the nine wells at any given time-step.

   Over 50 simulations were completed and the results suggested that the following test characteristics were appropriate for the design volume:

   - A total airflow rate of 23.85 m³/day (6300 ft³/day) evenly distributed over the 9 screens
   - A total mass of 900 g of the nonpartitioning tracer, methane
   - A total mass of 200 g for each of the partitioning tracers, the perfluorocarbon series

The simulations indicated that the PITT should last 7 days to increase the accuracy of the
results. The expected pressure drops were between 690 Pa and 1380 Pa. These were the design
parameters indicated by the simulations and were followed whenever possible in the actual test.
The actual PITT test parameters are described in Appendix G1.

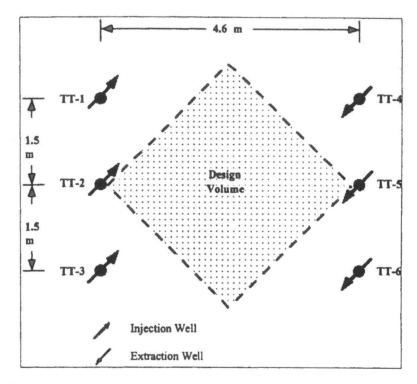

Figure 3.31 Plan view of PITT well layout.

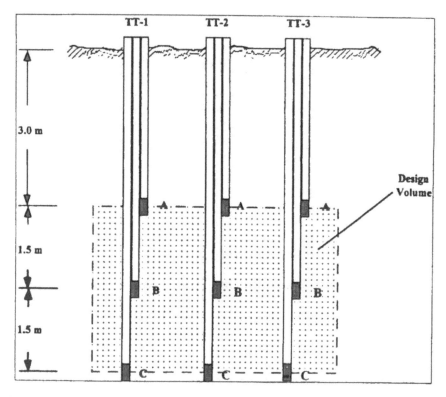

Figure 3.32 Profile view of PITT well layout.

Field Test Results —

Overall Remedial Performance

PITT1 was completed before the RF-SVE demonstration began at Kirtland AFB. PITT2 was completed after the demonstration. However, before PITT2 was completed, the site was irrigated with water to bring the soil's moisture content to about 7% dwb. The heating process had dried out the soil to the point where the tracer compounds might sorb to the soil. The data from PITT1 and PITT2 yielded an estimate of the amount of TPH and DRO hydrocarbons removed by the RF-SVE process (Table 3.15).

Table 3.15 TPH and DRO Mass Removal Calculated by Method of Moments

		Mole Weight$_{av}$	ρ(g/mL)	Mole Fraction	Volume (L)	Mass (kg)	% Removal by Mass
PITT1	TPH	366	0.905	1.000	2,280	2,100	—
	DRO	253	0.843	0.320	490	410	—
PITT2	TPH	403	0.914	1.000	1,990	1,800	—
	DRO	253	0.843	0.150	180	150	—
Decrease	TPH					300	13
	DRO					260	63

Note: These values should be considered with a variability of approximately 19%.

The values in Table 3.15 were determined based on pseudofractional modeling of the site TPH (Appendix G1). The TBP analysis was used to define fractions of the TPH by carbon number, including the DRO fraction (C_{10} to C_{21}). The EOS was used to correlate the tracer–air–TPH K values based on the composition of the site hydrocarbon, then the previously outlined iterative procedure was used with the field data and thermodynamic data to converge to the values given in Table 3.15. Appendices G2 and G3 explain the methods used to obtain these results.

Remedial Performance by Inverse Modeling

Appendix G3 provides an explanation of the inverse modeling process. Table 3.16 shows the inverse modeling results for hydrocarbon mass determination in the tracer volume. The decrease in TPH using this method was 290 kg, which was about 10 kg lower than that determined through the method of moments analysis. Given the range of uncertainty for both methods, the reductions determined from the two methods can be considered to be in good agreement.

Table 3.16 Pre- and Post-PITT TPH Mass Calculated by Inverse Modeling

Test	Mass (kg)
PITT1	2470
PITT2	2180
Decrease	290

3.7.8.5 PITT Conclusions

Laboratory Experiments
* Partition coefficients can be reliably determined for hydrocarbon mixtures derived from field samples, with less than 10% variation among the experiments.
* Column testing was adequate for both screening candidate tracer compounds and quantifying their interaction with the site hydrocarbon mixture.
* For the Kirtland AFB site, the perfluorocarbon compounds were suitable for use as gas-hydrocarbon partitioning tracers.

Field Test
- The *tracer* volume was much larger, 480 m³, than the *sampled* volume (132 m³) and the *design* volume (28.3 m³).
- The PITT provided information about the amount and composition of the TPH in the tracer volume before and after remediation.
- The PITT results indicated that the remedial effort decreased the TPH by 300 kg and DRO by 260 kg.
- The decrease in TPH from the PITT was similar to the 300 kg decrease that was calculated using the borehole analyses for the *sampled* volume in Section 3.7.6.
- The PITT results showed that 2100 kg of TPH and 410 kg of DRO range hydrocarbons existed in the tracer volume before remediation and 1800 kg of TPH and 150 kg DRO hydrocarbons existed after remediation.

3.7.9   Thermal and Biological Oxidation Rates

3.7.9.1   Respirometry Tests

Pre- and post-respirometry tests were part of the Kirtland demonstration. These tests measured the oxidation rate of organic compounds as hexadecane ($C_{16}H_{34}$) equivalent by the formula:

$$C_{16}H_{34} + 24.5O_2 \rightarrow 16CO_2 + 17H_2O \tag{R1}$$

where the formula weight of $C_{16}H_{34}$ is 226.5 g/g-mole. In these respiration tests, helium was not added to monitor oxygen loss by other mechanisms such as inorganic oxidation, adsorption, or outward migration. Furthermore, background tests were not conducted outside the contaminated zone.

The pre-demonstration respirometry tests were completed in 3 wells while the temperature of the soil was about 24°C. Assuming no other oxygen uptake mechanisms, hexadecane oxidation rates from 0.7 to 10 mg hexadecane per kg/day (with an average of 5.1 mg/kg/day) were inferred.

In 12 post-demonstration respirometry tests, a range of oxidation rates from 0.5 to 1.4 mg/kg-day was indicated. In these tests, the soil temperatures increased from 30 to 68°C. Figure 3.33 is a plot of the rate (moles hexadecane per kg/s) and the inverse temperature (1/K). The average rate for the 12 tests was about 0.9 mg hexadecane per kg/day with an average temperature of 55°C. Using the least squares equation, the rate increases from about 0.6 mg/kg/day at 30°C to 1.1 mg/kg/day at 70°C.

Two inferred causes for the pre- and post-demonstration respiration results are as follows:

- Thermal oxidation of DRO and TPH compounds, for example, hydrous pyrolysis oxidation (HPO) and
- Biooxidation of DRO and TPH compounds

3.7.9.2   Thermal Oxidation

Investigators at Lawrence Livermore National Laboratories (LLNL) have completed considerable laboratory and field work on thermal oxidation processes, primarily HPO (Knauss, K.G. et al., March 1997; Newmark, R.L. and Aines, R.D., March 1997; Leif, R.N. et al., November 15, 1997; Leif, R.N. et al., November 25, 1998; and Knauss, K.G. et al., May 1998). HPO is an aqueous phase reaction between organic compounds and oxygen that readily occurs at temperatures above 90°C. In this reaction, the dissolved organic compounds react with dissolved oxygen to form carbon dioxide and water (and other mineral compounds). For example, the reaction between trichloroethene (TCE) and oxygen in the aqueous (aq) and gaseous (g) phases are as follows:

$$2C_2HCl_{3\,(aq)} + 3O_{2\,(aq)} + 2H_2O \rightarrow 4CO_{2\,(aq)} + 6H^{+2} + 6Cl^{-1} \tag{R2}$$

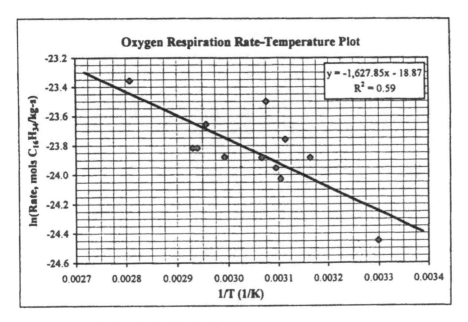

Figure 3.33 Oxygen respiration rate–temperature plot.

$$2C_2HCl_{3\,(g)} + 3O_{2\,(g)} + 2H_2O \rightarrow 4CO_{2\,(g)} + 6HCl_{(g)} \tag{R3}$$

Most of the laboratory and field studies have been directed toward the application of HPO to saturated soil because liquid water is necessary for the reaction to take place. Apparently, the gaseous phase reaction does not readily occur. However, HPO can occur in unsaturated porous soil if sufficient liquid water is present in the capillaries.

The referenced field process that was directed toward a pool of DNAPL at a site in California involved the following steps:

- Steam was injected to heat the soil and water to 100°C. The investigators estimated that this requires ~238 kg steam per m³ of "saturated" soil.
- While at 100°C, steam and air were injected into the soil. The steam–air mixture displaced the groundwater within several meters of the injection well. The organic compounds dissolved into the displaced hot water.
- When the steam condensed, oxygen in the air dissolved into the liquid water. There were 100 ppm O_2 (100 mL O_2 per m³) in the steam–air mixture. During condensation, the displaced water returned and mixed with the oxygenated water. About 1/15 of the mixed water mass consisted of the condensed steam. However, a good portion of the remaining nitrogen-rich air remained in the saturated-zone soil.
- After the HPO reaction was nearly complete, more steam and air were added to repeat the displacement–condensation–reaction process. Sufficient steam must be always added to make-up for the heat loss to the surroundings.

Knauss (1998) measured the rate of reaction for R2 in laboratory experiments. Figure 3.34 shows the rate–temperature plot for the initial oxidation rate of TCE (dC_o/dt). The rate of TCE oxidation (r_{TCE}) was dependent upon the concentration of TCE in the aqueous phase (Figure 3.35). From this plot, the following rate equation was proposed:

$$r_{TCE} = dC/dt = kC^{0.85} \tag{15}$$

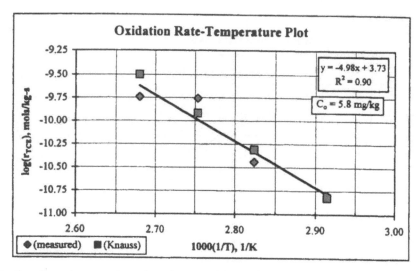

Figure 3.34 The effect of temperature (T) on the rate of TCE oxidation (r_{TCE}).

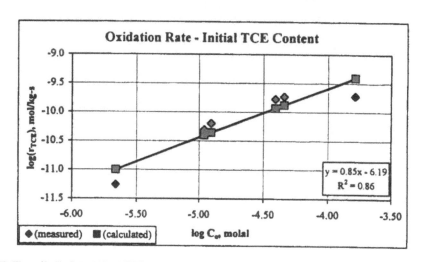

Figure 3.35 The effect of the initial TCE concentration (C_o) on the rate of TCE oxidation (r_{TCE}).

where k is the equilibrium constant at a particular temperature. For example, at 100°C, k equals 1.50×10^{-6} mol/kg/s. Literature data show that the maximum solubility of TCE in water at 25°C is 1000 mg/kg. The LLNL investigators measured the solubility of TCE at increasing temperatures. Their results are plotted in Figure 3.36. The plot shows that at 25°C the solubility of TCE in water is about 1500 mg/kg (11.0 mmol). At 100°C, the solubility increases to 2,900 mg/kg (or 22.0 mmol). Using the oxidation rate formula and the values for k and C_o at 100°C, the maximum rate of reaction would be 8.80×10^{-8} mol/kg/s (999 mg/kg/day). In an earlier report (Knauss, 1997), the investigators reported an initial reaction rate of 9.82×10^{-9} mols/kg/s (112 mg/kg/day). At this rate, the initial TCE concentration should be 860 mg/kg (C_o = 6.50 mmol).

LLNL investigators measured similar reaction rates for naphthalene ($C_{10}H_8$) and pentachlorophenol (C_6HCl_5O). For example, they measured an oxidation rate of 1.4×10^{-11} mol/kg/s for naphthalene at 114°C. Subsequently, they measured the rate using a low initial concentration of 2.3 mg/kg water (0.0180 mmol). However, solubility measurements showed that at 80°C, the naphthalene has a concentration of about 300 mg/kg water (2.34 mmol). Therefore, one would expect a marked increase in the reaction rate if, as with TCE, the rates were concentration dependent.

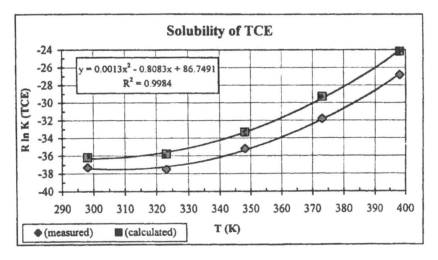

Figure 3.36 The temperature effect of TCE solubility in water.

As the data for TCE indicate, the oxidation process is quite dependent upon both an elevated temperature (~100°C) and a high solubility of the contaminant in the aqueous phase. Furthermore, the mass-transfer rate of oxygen from air into the aqueous phase must be adequate to keep a high rate of reaction.

When RF-SVE is applied to remove SVOCs from a wet unsaturated soil, similar steps to the HPO process may take place. For example, a soil with a porosity of 0.34 and moisture content of 10% dwb will be 52% saturated by volume with water. The contaminants could dissolve in the hot aqueous phase. Simultaneously, oxygen would be absorbed from the air, and the oxidation reaction would take place. However, during SVE the airflow rate must be kept low to minimize the removal of water from the soil, but high enough to supply the oxygen for the oxidation reaction.

3.7.9.3 Biooxidation

Another possible cause for the pre- and post-demonstration respiration was biooxidation of the DRO and TPH compounds. However, the heavier DRO and TPH compounds are more difficult to biodegrade than the light ones. At the optimum temperatures of 25 to 35°C, mesophilic bacteria can oxidize organic compounds at the rates listed for the pre-demonstration respiration tests. At the optimum temperatures of 60 and 70°C, thermophilic bacteria could oxidize organic compounds. At 100°C, no biooxidation should occur.

Bacteria can only oxidize organic compounds that are dissolved in the aqueous phase. Bacteria cannot directly oxidize bulk organic compounds. Furthermore, the water content of the soil must be sufficient to maintain metabolic activity of the bacteria. The bacteria live in water or at water interfaces and oxidize dissolved organic compounds. During the RF-SVE demonstration, the water within the site continuously decreased and the soil temperature continuously increased. If the water content of the soil is decreased enough, the bacteria will enter a nonmetabolizing state. Furthermore, above 50°C, the mesophilic bacteria should become dormant or die.

Thermophilic bacteria also need stable conditions of water content, nutrient availability, and temperature to increase in number and thrive at temperatures above 50°C. Therefore, applying RF-SVE to remove SVOCs from wet soil, one would expect to see a marked decrease in the biooxidation reactions. Bearing in mind the decreasing soil moisture, similar results should also occur for thermal oxidation reactions. Although both oxidation mechanisms probably occurred at the Kirtland AFB site during the RF-SVE demonstration, they were of little consequence in the removal of DRO (and TPH) compounds; hence, the low-average post-demonstration respiration tests results.

3.8 CONCLUSIONS

The following conclusions were drawn by the project team regarding the future use of RF-SVE for full-scale remediation.

3.8.1   RF Heating of Soil

- A better understanding of the interactions between RF heating and the SVE system (e.g., flow rates and flow field) is required to optimize RF-SVE system design with respect to the rate of soil heating, moisture removal, and ultimately chemical removal. The suboptimal SVE system for the demonstration was probably responsible for the lower soil temperatures than the 130 to 150°C goals. In addition, the direction of airflow had an adverse effect by keeping the temperatures in the *design* volume below the targeted temperatures. The countercurrent airflow toward the SVE B well and the RF heat sources removed a considerable amount of the RF energy in humid air. Furthermore, build-up of heat in the soil around the antennae prevented the attainment of their maximum power output of up to 25 kW. The build-up required constant removal of heat from the soil and well casings using ambient air from a blower. Moving the airflow away from the heat sources toward the periphery of the *design* volume would move heat away from the antennae and add heat to cooler soil by water–vapor condensation from the air.

3.8.2   Chemical Removal from Soil

- The goals of determining and documenting the removal of chemicals, specifically DRO compounds, from the site soil and achieving a significant reduction in chemical concentrations were met. However, SVE system flexibility could have permitted the RF-SVE system to achieve and sustain higher soil temperatures and increased chemical removal.
- The project goals did not include the removal of all petroleum hydrocarbons from the soil. The goals only targeted ranges of DRO or SVOCs. One can conclude from the results of this demonstration that specific constituents can be removed from soils using this technology. Therefore, constituents with similar chemical and physical characteristics to GRO and DRO compounds can be assumed removable using the RF-SVE technology.
- For most MRO compounds, the vapor pressures at 140 to 150°C are too low for effective removal by RF-SVE. However, organic compounds that have similar vapor–pressure–temperature curves to the DRO can be removed by heating toward 150°C.
- The decrease in TPH from the PITT tests was similar to the decrease that was calculated using the borehole analyses. However, the *tracer* volume was about four times as large as the *sampled* volume. Therefore, most of the decrease in TPH probably occurred in the *sampled* volume.
- Before and during RF-SVE treatment, thermal oxidation and biooxidation played a minor role in decreasing the TPH (and DRO) content within the *sampled* volume. As suggested by the prerespiration test results, biooxidation probably occurred during the pre-RF-SVE period. During RF heating, some thermal oxidation also could have occurred as the temperatures of the wet soil approached 100°C. The increasing soil temperature should have decreased the biooxidation of the organic compounds. However, both oxidation mechanisms required sufficient soil moisture to be effective. As shown by post-demonstration respiration rates, both mechanisms were adversely affected by removal of water from the *sampled* volume.

3.9 DESIGN RECOMMENDATIONS

3.9.1   Design Considerations

The following demonstration observations should be considered in the design of the full-scale RF-SVE treatment systems covered in Chapter 4:

- For the Kirtland demonstration, the AC power to RF-energy conversion efficiency for the vacuum-tube generator was ~47%. The RF antenna for the Kirtland demonstration had a RF-energy to heat conversion efficiency of ~81%. For commercial development, KAI expects to use a solid-state generator that will have an efficiency of 72%. A properly designed RF antenna should have a RF-energy to heat conversion efficiency of ~93%.
- RF heating produced temperatures of about 140°C at the center of 28-m^3 *design* volume and from 110 to 120°C at the periphery. These temperatures were below the goal of 150°C at the center and 130°C at the periphery.
- The open treatment site allowed the RF heat to go beyond the boundaries of the *design* volume as shown by the temperature profiles within the *heated* volume. In fact, the 324-m^3 *heated* volume achieved an average temperature of 62°C and lost about 25% of the RF energy to its surroundings after 87 days of RF-SVE treatment.
- Low RF heating rates increased the heat loss to the surroundings from the *heated* volume. For example, the heat loss at a RF generator output of 7.1 kW was ~35% as compared to ~15% heat loss at an output of 13.2 kW.
- The airflow rate and temperature should be measured using the near dry air entering the blower and not the humid air leaving the SVE extraction wells. The near dry air is produced after the water and SVOCs are removed in a condenser. The humidity (kg water vapor per kg dry air), the SVOC content, and temperature of the air–vapor mixture leaving the extraction wells should also be measured.
- For the initial 31 days of RF heating, the airflow rate decreased from 350 slpm down to 60 slpm. The average flow rate per unit of the *heated* volume was about 0.6 slpm per m^3. For the last 42 days of RF heating, the airflow rate per unit volume was increased to 1.0 slpm per m^3.
- For the Kirtland demonstration, the *heated* volume had a surface area to volume ratio of 0.932 m^2/m^3. Larger treated volumes can have lower ratios. Lower ratios decrease the heat loss to the surroundings.
- For the 87 days of RF heating, the minimum volumetric heat loss to the surroundings per unit of total heat input was 0.288 kJ/kJ$_{input}$ per (m^2/m^3). For the final 42 days of RF heating, the minimum heat loss was 0.140 kJ/kJ$_{input}$ per (m^2/m^3). For a larger treatment volume, one could expect a lower heat loss to the surroundings.
- Heat lost in the dry air leaving the SVE B well was less than 5% of the total RF heat added to the soil. The water vapor mixed with the air contained ~85% of the total heat in the off-gas from the SVE B well. The open site and the airflow toward the SVE B well removed considerable water from beyond the *design* volume. The total water removed was 8480 kg; the total in the *design* volume was 3500 kg. The additional water consumed a major portion of the total RF heat (~18%) that could have been used to heat the wet and dry soils within the *design* volume.
- The RF heat removed about ~7530 kg of water from the *sampled* volume. The off-gas treatment produced 8480 kg of water condensate. Therefore, 950 kg of water was brought in from outside the *sampled* volume.
- The humidity of dry air can depend upon the dry airflow rate. For example, the humidity was ~0.40 kg water per kg dry air at a flow rate of ~90 slpm (~3 scfm) and 0.25 kg/kg at a flow of ~425 slpm (~15 scfm). For the final 42 days of RF heating, the average humidity was 0.262 kg/kg dry air at an average airflow rate of 330 slpm (~12 scfm). In both cases the RF energy input (heat input) to the soil was between 12 and 13 kW. At constant heat input, a nearly constant rate of water vaporization will occur. Moreover, the humid volumetric flow leaving the extraction wells consists of the liters of water vapor plus the liters of dry air. Hence, increasing the mass flow rate of air (and dry volumetric flow rate) decreases the humidity. If the airflow rates are measured at the extraction wellheads, the flow rates must be adjusted to maintain constant flow rates of dry air by accounting for the water vapor in the air.
- RF-SVE treatment removed 116 kg of DRO (~48% of the total DRO mass) from the *sampled* volume. The final average DRO content in the soil was ~500 mg/kg. If the project team had continued the RF-SVE operation, the rate of DRO removal probably would have decreased.
- For the last 41 days of RF heating, the amount of DRO removed per unit of dry air was 0.0041 kg/kg. As shown in Appendix F, this value is well below the calculated saturation limits (kg/kg dry air) for C_{12} through C_{21} organic compounds at 135°C.

- For the sampled volume, the TPH decreased by ~300 kg (~19% of the total TPH mass). Sampling and analysis error, oxidation processes, and migration out of the *sampled* volume are the major reasons for the decrease in TPH. However, the average oxidation rate in the *heated* volume after RF-SVE treatment was less than 1 mg hexadecane per day/kg of soil. At this rate, one can only account for a fraction of the TPH loss (less than 10%).

3.9.2 RF-SVE Operation

The following conclusions are drawn concerning RF-SVE operation:

- Heat flow and airflow should operate in the cocurrent mode:
  - Air injection into central well screen and air extraction from three or four peripheral well screens appears to be desirable.
  - RF antennae should be located near central well screen.
- The SVE design should be flexible to allow for the operation of the SVE system in both cocurrent and countercurrent modes.
- The plan design of SVE injection well screens should be on a square (four well screens) or hexagonal pattern (three well screens).
- Phases of operation:
  - Maximize RF antenna energy output by heating the soil from the center outward toward the periphery of the treatment zone.
  - Heat wet soil to boiling point of water (94°C at Kirtland):
    - Use a low dry airflow (1.0 slpm/m³ of soil) to minimize heat loss in the extracted air. The objective of this phase is to rapidly increase the temperature of wet soil to 94°C. Vaporization of soil moisture will consume a large portion of the RF energy. Minimizing the vaporization of moisture could increase the heating rate of the wet soil.
    - Remove some light-end organic compounds.
    - Minimize humidity of dry air (at or below 0.20 kg/kg).
  - Remove water and light-end organic compounds while the soil is at the boiling point of water (94°C)
    - Optimize airflow to maximize the evaporation of liquid water (2.5 slpm/m³ of soil).
    - Optimize humidity of dry air (between 0.20 and 0.40 kg/kg).
  - Heat the dry soil toward 150°C
    - Remove heavy-end organic compounds.
    - Maintain an optimized airflow to maximize the removal of the organic compounds.
  - Achieve final DRO content of less than 500 mg/kg, or other cleanup goal, if appropriate.

CHAPTER 4

Engineering Design

This engineering design of a typical RF-SVE system was based on a hypothetical site with conditions similar to those exhibited at the Kirtland demonstration site. The objective was to develop engineering design concepts for a full-scale remediation project for discussion of the magnitude of costs and expected performance. In addition, modifications to the design that could decrease costs and anticipated time for remediation will be discussed.

To facilitate the future reference and retrieval of general information as it relates to other technologies, standard terminology and general information are presented below as suggested in the *FRTR Guide*. This format, as used in reporting the demonstration site results, provides a concise and consistent presentation of the hypothetical model site chosen and the RF heating technology in general.

Site Characteristics:
  Media Treated
    Soil (*in situ*)
  Contaminants Treated
    Semivolatile organic compounds (SVOCs)
Treatment Systems:
  Primary Treatment Technology
    Thermally enhanced recovery with soil vapor extraction
  Supplementary Treatment Technology
    Post-treatment (air)–catalytic oxidation unit

4.1 HYPOTHETICAL SITE CONDITIONS

Results from the demonstration were used to project the specifications for the design of the full-scale RF-SVE system. Accordingly, the soils of the two hypothetical sites were assumed to have similar physical and chemical properties to the Kirtland site. To establish the parameters for these hypothetical sites, some assumptions were made to complete the design.

Two different sized hypothetical sites, Site 1 and Site 2, were used to examine base cases for development of full-scale RF-SVE system designs. These two cases attempted to provide representative examples of the expected performance of the RF-SVE technology when utilized to target the removal of water and SVOCs, such as the DRO compounds. These same goals were also the targets for the Kirtland demonstration. The two cases illustrated the cost-effectiveness of the RF-SVE technology for large and relatively small sites, Case 1 and Case 2, respectively.

The following conditions were assumed to exist at both sites:

- The contaminated soil was 5.79 m (19 ft) thick, ranging from the surface soil to 5.79 m bgs.
- Average contaminant concentration was 8000 mg/kg of SVOCs with a maximum value of 10,000 mg/kg.

- The sites received about 61 cm (24 in.) of precipitation per year.
- The air temperature ranged from 15 to 30°C (yearly average at 20°C).
- Contaminants of concern were any SVOCs that have similar vapor–pressure–temperature relationships as the DRO compounds.
- Soil was weathered granite (similar to the Kirtland AFB site). The soil was classified as a sand and gravel mixture with some silt and clay.
- Average soil bulk dry density was approximately 1778 kg/m³ (111 lb/ft³).
- Average soil moisture was 10 kg of water per 100 kg of dry soil or 10% dwb.
- Average soil porosity was 34%.
- The depth to groundwater was at least 6 m (~20 ft) below the maximum depth of the target treatment zone.
- The initial soil temperature in the treatment zone was generally constant at 15°C.
- The cleanup criterion for the SVOCs contaminants in the soil was less than 500 mg/kg in soil.

Using similar conditions as the demonstration site provided a sound basis for the design parameters and equipment that was selected for the two base cases.

The two sites and their basic RF-SVE system layouts are shown in Appendix H, Drawings S1 and S2. For the two cases, Table 4.1 lists the main operational sequences that were discussed in Section 3.9.2, RF-SVE Operation. Further descriptions for Site 1 and Site 2 are provided in Table 4.2.

Table 4.1 Case 1 and 2 Operational Sequences

1. Heat the wet soil from 15°C to the boiling point of water (100°C)
2. Maintain 100°C temperature and remove water and light-end SVOCs to produce "dry" soil areas
3. Heat "dry" soil from 100 to 131.5°C and remove heavy-end SVOCs
4. Cool down heated soil volume with SVE only

Table 4.2 Hypothetical Site Dimensions and Well Layouts

	Site 1	Site 2
Area shape	Rectangular	Hexagonal
Length	66.0 m	18.05 m sides
Width	65.0 m	Not applicable
Depth	5.79 m	5.79 m
Total area	4,290 m²	847 m²
Total volume	24,843 m³	4,906 m³
Number of RF wells	36	7
Number of extraction wells	42	12
Number of central air vent wells	36	7

Since the site conditions and characteristics were the same for Sites 1 and 2, only the areas, and therefore volumes, of soil treated differed at each site. The number of RF wells and SVE wells at the two sites was different. However, the well spacing or well density was identical for both designs. Each layout used the same 6.096-m (20-ft) radius of influence for the RF energy that was emitted from each central antenna. The area of influence for each antenna was 116.7 m² (1260 ft²). At each site, multiple areas of influence were arranged in a hexagonal pattern. For Site 1, the pattern formed a rectangular area. For Site 2, the pattern formed a hexagonal area.

For both cases, the SVE systems were assumed to be operating continuously. The air intake wells were placed adjacent to each RF well as a precaution to prevent the overheating of the RF antenna. The air intake wells were not connected to SVE vacuum (blowers) as the SVE extraction wells, and therefore, were considered passive. However, if insufficient airflows were measured at these passive wells, blowers were to be used to inject air into the wells. Each RF well was heated with one 20-kW mobile RF heating unit. For the Kirtland demonstration, 2 RF antennae in their corresponding RF wells were supplied RF energy from the same 25-kW RF generator unit.

4.2 TREATMENT SYSTEMS

For both cases, the RF-SVE treatment system consisted of the following major components: (1) mobile RF heating units, (2) soil vapor extraction (SVE) system, and (3) air-vapor (off-gas) treatment system. KAI's 20-kW mobile RF unit was used for both design cases. The 20-kW RF heating units were dispersed throughout the site near the RF antennae wells they were designed to heat. The SVE-system and air-vapor, off-gas, treatment equipment were necessary for the removal and control of the volatilized contaminants from the subsurface. The SVE and air-vapor treatment equipment and components were placed within a concrete equipment compound in the southwest corner of the site. Drawing P1 in Appendix H provides a simplified process flow diagram for the RF-SVE system. Drawing P2 in Appendix H provides a piping and instrumentation diagram for the SVE and off-gas treatment systems.

4.2.1 Treatment Compound

A concrete compound for treatment equipment was located in the southwest corner of the site. The compound measured 6.10 m (20 ft) by 6.10 m (20 ft) and it included a small containment berm around the periphery. The compound held all of the remediation equipment utilized to perform SVE and air-vapor treatment including the electrical control panels.

4.2.2 Mobile RF Heating Units

KAI's commercial RF heating units used in the hypothetical cases were designed to be mobile, to require minimal supervision, and to be operationally effective, reliable, and rugged. The RF units provided up to 10 kW or up to 20 kW of RF power using single-phase AC power. The RF units operated at discrete ISM frequencies that ranged from 13.56 MHz up to 40.68 MHz. Each RF heating unit consisted of the following components:

- Solid-state RF generator
- Couplers for measuring forward and reflected RF power
- Matching network, or tuner, to assist in soil matching
- Computer for continuous monitoring of performance of the RF system from a remote site
- Temperature monitoring system
- One or more interactive antennae

Most of the built-in test hardware that existed in the original KAI 25-kW research RF unit was not necessary for the commercial unit. Instead, three hand-held instruments, the high-frequency standing-wave ratio (HF/SWR) analyzer, cable analyzer, and field meter, were substituted for the network analyzer, time-domain reflectometer, and magnetic field probe. The HF/SWR analyzer was used to check the antenna and verify its ability to efficiently radiate RF power into the soil. The cable analyzer checked the transmission line. If a fault existed, it pinpointed location for repair. The field meter insured that any background electromagnetic energy was below the acceptable OSHA limit.

4.2.2.1 Control and Diagnostic Computer

The commercial RF heating systems were designed by KAI. A rack-mounted computer monitored and controlled the RF process and equipment using the Lab View® software from National Instrument. This software also provided a remote operation and monitoring capability. The RF system computer was accessed and controlled remotely from an off-site computer via a telephone line and modem. The computer was also used to store data on the performance of the RF system and soil temperatures as well as other data such as the airflow rates and vacuum pressure levels. The data were either downloaded by connecting to the control computer directly or by sending the data at periodic intervals to a remote computer system. Because the computer's control software was accessed remotely, it was used at any time to shut down the RFH system or to adjust the output RF power level.

4.2.2.2 Radio Frequency Generator and Tuner

The commercial unit contained modularized RF generators comprised of multiple solid-state power modules. Each generator was programmed to deliver up to 20 kW of RF power while operating at frequencies of 6.78, 13.56, 27.12, or 40.68 MHz. To monitor the performance of the RF generator, the RF heating system continuously measured both forward power delivered to and the reverse power reflected from the RF antenna. A built-in matching network, or tuner, located between the RF generator and the antennae protected the generator from excessive reflected power by maintaining a fixed impedance of close to 50 Ω. If, however, the reflected power exceeded a set value, the control computer shut down the RF generator before any damage occurred. Before starting up RF heating, an air-cooled dummy load was used to evaluate the performance of the generator.

A shielded, rigid, coaxial transmission line was used to deliver RF power from the generator to either of two external ports located on the outside of the equipment trailer. These ports can be used to consecutively feed power to one or two antennae using a RF switch. To allow the transmission of power, the antennae and transmission line were pressurized with nitrogen gas.

4.2.2.3 RF Antenna

To calculate the antenna length for the most efficient energy delivery in a particular soil or resonant, NEC 4, developed by Lawrence Livermore National Laboratory, was again employed. The antenna designed for use in the hypothetical site soil operated at 20 kW and at a frequency of 13.56 MHz. This antenna was approximately 6 m (20 ft) long with a diameter of 88.9 mm (3.5 in.).

The 6-m RF antenna for both design cases heated a volume of soil with a radius of influence (roi) of about 6 m and a thickness of about 6 m. The assumed roi was based on KAI's experience and previous pilot-scale data. Operating at a frequency of 27.12 MHz, a 3-m antenna could heat a volume of soil with a roi of about 3 m and a thickness of about 3 m. Therefore, each commercial RF unit had phase control for a multiple antennae array that permitted flexibility in the control of the heating pattern.

Drawings S1 and S2 in Appendix H show the RF-well layouts for each site. Case 1 used 36 RF wells and 36 antennae, while for the smaller site in Case 2, 7 RF wells and 7 antennae were used.

4.2.2.4 Fiber Optic Thermometer

During RF heating, a fiber optic thermometer measured the temperature of the soil contacting the antenna casing. The data were used by the control and diagnostic computer to control operation . of the RF generator, to optimize the RF system's output, and to achieve the desired soil-heating pattern.

4.2.3 Soil Vapor Extraction System

The SVE system can be used to increase the effective distribution of heat by advection. The SVE system consisted of a series of SVE wells connected to an air-water (moisture) separator and vacuum blower. The blower exhaust gas was fed to a catalytic oxidizer unit to convert the organic vapors into carbon dioxide and water.

The SVE wells were screened in the vadose zone across the area of contamination to maximize the airflow through the subsurface. The configuration and spacing between the wells were typically determined from SVE pilot tests. Based on the data from the SVE tests, a properly sized blower was used to apply a vacuum for removing contaminants volatilized through RF heating.

After the off-gas left the extraction wells, the water vapor and some organic vapors were near or at saturation as the gas temperature decreased. Condensing moisture in the off-gas stream had a detrimental effect on vapor treatment equipment, affecting performance and energy requirements. Hence, the moisture was removed from the vapor stream using an air-water separator before entering the blower and catalytic oxidizer. The air leaving the air-water separator was reheated in a heat exchanger with hot gases from the catalytic oxidizer before entering the blowers of the SVE system. Condensate collected in the air-water separator was treated and disposed.

The drawings S1 and S2 in Appendix H show the SVE-well layouts for Cases 1 and 2, respectively. The SVE system design for Case 1 used 42 SVE-extraction wells approximately 6.10 m (20 ft) in depth, each with an assumed roi of 6.10 m (20 ft). The area of influence for each well was 117 m^2 (1260 ft^2). The volume of influence for each extraction well was 712 m^3 (25,130 ft^3). The system design for Case 2 used 12 SVE-extraction wells with the same 6.10 m roi and 6.10 m depth. Air-vent wells were installed adjacent to the RF wells for each of the sites (36 wells for Case 1 and 7 for Case 2). The SVE wells were screened from approximately 1.5 m to 6 m (5 to 20 ft).

The Kirtland demonstration used one extraction well. For the three stages of RF-SVE operation, the recommended off-gas flows in slpm per m^3 of soil and the corresponding full-scale airflow rates for each extraction well were as follows:

- Heat wet soil from 15°C to the boiling point of water (100°C)
  - Off-gas Flow ~0.6 slpm/m^3
  - Case 1 Full-scale airflow = ~350 slpm (~13 scfm)
  - Case 2 Full-scale airflow = ~250 slpm (~9 scfm)
- Remove water and light-end organic SVOCs contaminants while at 100°C
  - Off-gas Flow ~1.0 slpm/m^3
  - Case 1 Full-scale airflow = ~600 slpm (~21 scfm)
  - Case 2 Full-scale airflow = ~400 slpm (~15 scfm)
- Heat dry soil from 100°C to 132°C
  - Off-gas Flow ~1.0 slpm/m^3
  - Case 1 Full-scale airflow = ~600 slpm (~21 scfm)
  - Case 2 Full-scale airflow = ~400 slpm (~15 scfm)

The above airflow rates are recommended values. The optimum airflow is not only a function of the volume of soil, but is also dependent upon the optimum water and SVOC vapor concentrations that are needed to remove these components from the soil. The airflow is also dependent upon the amount of heat available to vaporize the water from the wet soil. All of these relationships will be used in Section 4.4 to establish actual off-gas flow rates for each operational phase.

Soil Vapor Extraction Blowers

For Case 1, three ROOTS positive-displacement blowers (Model URAI 56) were utilized for operation of the SVE system. Each blower was powered with a 14.9-kW (20 hp) TEFC motor. Each of these blowers was designed to deliver 12,750 slpm (450 scfm) at 47.4 kPa (14 in. of Hg) vacuum. For Case 2, one ROOTS Model URAI 56 positive-displacement blower powered with a 14.9-kW (20 hp) TEFC motor was utilized. This blower was designed to deliver 12,750 slpm (450 scfm) at 47.4 kPa (14 in. of Hg) vacuum.

The SVE blowers for both systems were equipped with an inlet filter, high-temperature switch on the outlet, inlet low-pressure switch, outlet temperature gauge, inlet flame arrestor, magnehelic differential-pressure gauge, and an outlet silencer. The inlet and outlet of each blower unit were attached by manifolds to a common header. Check valves and ball valves were provided on the inlet and outlet manifolds to allow the operation of a single unit or all units.

Moisture Separator Unit

A moisture separator unit was provided on the common inlet header of the SVE unit to provide for the removal of any entrained water in the air stream. The separator unit consisted of a main condensate tank, a secondary collection tank, a water transfer pump, and associated valves, switches, and gauges to properly operate the unit. The main condensate tank was equipped with an auto-relief valve, an auto-dilution valve with an ambient air filter, a vacuum gauge, high-level switch, and a solenoid valve. The solenoid valve was used to facilitate the equalization of pressure between the main moisture separator and secondary transfer tank to allow the gravity flow of water from the main to secondary tank. The secondary collection tank was equipped with a high-level switch, a low-level switch, and a vacuum gauge. This tank also used (1) a 0.74-kW (1 hp) Myers Centri-Thrift Model 100MM transfer pump, (2) a pressure gauge on the pump discharge, and (3) a ISTA Model 1720 totaling cold water meter. The meter measured the amount of water collected for treatment and discharge.

4.2.4   Vapor Treatment System

The RF heating increased the rate of SVOCs removal using SVE. The selection of the type of off-gas treatment equipment depended upon many factors such as:

- The type of SVOC contaminant (for example, petroleum hydrocarbons, chlorinated hydrocarbons)
- Rate of removal from soil
- Concentration of SVOCs in the extracted air
- Available power

Some applications may not require any vapor treatment or may need only an adsorption system like activated carbon. Minimizing vapor treatment cost while achieving air discharge requirements was the driving requirement. The design developed for Case 1 used two catalytic oxidation units that were equipped with air hot-gas heat exchangers. Each unit was rated at 21,240 slpm (750 scfm), for off-gas treatment. The Case 2 design required only 1 catalytic oxidation unit rated at 14,200 slpm (500 scfm).

4.2.5   Operations Trailer

An operations trailer was selected to serve as an office space for on-site system personnel during RF-SVE operations.

4.2.6   System Piping

Drawings S1 and S2 in Appendix H show the piping layouts. The piping for the SVE and off-gas treatment system was fabricated from Schedule 40 threaded carbon steel. The system piping consisted of a main header routed along the south portion of the site and secondary lateral piping leading to the extraction wells. The main header consisted of 7.62-cm (3-in.) pipes. The piping from the main header to the wells consisted of 5.08-cm (2-in.) pipes.

4.2.7 Instrumentation and Controls

Wellhead Instrumentation and Controls

SVE and vapor-treatment operations were monitored using a system of pressure-indicating gauges, temperature-indicating gauges, and flow monitors. Test ports were also installed at the secondary lateral piping leading to individual well locations near the main header piping. Annubar-type flow monitors installed in the lateral piping were used to measure off-gas flow rates from each extraction well.

4.2.8 Utilities

Electricity

The electrical distribution services for the two hypothetical cases consisted of the main service transformer, equipment electric service, and trailer electric service.

The main service transformer in a weatherproofed enclosure consisted of a 12,470 V to 480/277 V, 3-Phase, 4-wire, 1000-KVA unit for Case 1 or 300 KVA unit for Case 2. The service transformer was assumed to be fed by the local utility company from a primary transformer located adjacent to the new service transformer. The service transformer was fed by 12,470 V, 3-phase, 3-wire electrical service. A power meter was installed at the new service transformer to measure the total electrical power requirements for the project.

The service transformer fed the main disconnect panel (MDP) installed adjacent to the equipment control panel. The MDP provided the disconnect switches for all electrical circuits for the equipment and trailer. All remediation equipment was controlled and connected through this control panel with the exception of the catalytic oxidizer which was fed through a separate disconnect panel. This panel was connected to a 480 to 208/120 V, 30-KVA transformer in a NEMA 3R enclosure.

A separate service transformer was installed at the trailer location to supply electrical service. A 480 to 240/120 V, single-phase, 3-wire, 25-KVA electrical transformer (NEMA 3R enclosure) was installed to serve electrical services for the trailer.

Telephone

Three telephone lines were available at the operations trailer for use during the project. One line was used to operate the remote monitoring system (modem) installed in the main equipment control panel. The second line was used for the fax machine in the operations trailer. The third line served as the main telephone line for the operations trailer.

Propane Gas

A 3.79-m^3 (1000-gal) propane storage tank was provided and placed adjacent to the equipment compound to serve as make-up fuel for the catalytic oxidation unit. A 1-cm (3/8-in.) copper feed line was installed underground from the propane storage tank to the auxiliary fuel supply inlet on the catalytic oxidation unit.

4.3 PROCESS EQUIPMENT PERFORMANCE FACTORS AND SPECIFICATIONS

Based on the size and conditions at Case 1 and 2 sites, Table 4.3 lists the major equipment required for the RF-SVE treatment system. References to catalog cut sheets for these equipment items can be found in Appendix I.

Table 4.3 RF-SVE Major Equipment Lists, Cases 1 & 2

Components	Specifications/Information	Qty (Case 1)	Qty (Case 2)
RF generator	KAI mobile 20-kW trailer	36	7
Control and diagnostic computer	KAI (included in KAI's trailer)	36	7
RF antenna	KAI (includes transmission lines from RF well to generator)	36	7
Fiber optic thermometer	KAI supplied equipment	36	7
SVE blowers/control equipment	14.9 kW (20 hp); rated @ 12,750 slpm (450 scfm) @ 48 kPa (14 in. Hg)	3	1
Air-water separator	With 0.75-kW (1-hp) transfer pump	1	1
Catalytic oxidation units	21,240 slpm (750 scfm) capacity	2	
	14,150 slpm (500 scfm) capacity		1

The equipment and component designs for the two case studies were based on continuous operation of all SVE wells and simultaneous heating with all RF antennae over the entire soil volume. The continuous operation of the RF-SVE system must take into account the process down time. The RF unit at the Kirtland demonstration had a down time of about 6%, and the heat balance calculations were based on actual operating time to remove water and SVOCs. Therefore, the total-remediation time must account for this lost 6% operating time (see Chapter 5).

4.4 HEAT BALANCES

The results of the Kirtland AFB RF-SVE demonstration showed that a single 3-m (10-ft) antenna at a depth of 3 m into the soil can achieve a heating radius of 3 m (10 ft) or larger. Heating the soil within this radius to a temperature between 120 and 140°C required about 2 months of operation at 13 to 14 kW total RF output. Using longer, multiple antennae, the heating radius was extended through the superposition of heating patterns and the interaction of the electromagnetic fields of the antennae. The RF antenna heated the soil uniformly by first radiating energy into the soil and then allowing continued radiation, thermal conduction, and convection by wet airflow to smooth out the temperature pattern. For the hypothetical base cases, a 6-m (20-ft) heating radius for each antenna was assumed. Uniform heating by the antenna was achieved by first heating the soil locally around the antenna to approximately 130 to 140°C. Controlling the power input into the antenna and the direction, amount, and humidity of the wet airflow expanded this high temperature zone. Heat was distributed when wet airflows from the central air vent well near the antenna toward the extraction wells. As the wet air moved through the cooler soil, some of the water vapor in the air condensed, adding heat to the cool wet soil. Over time, the soil temperatures within the heated radius increased through convection, conduction, and RF radiation to between 130 and 140°C.

The heat calculations for Case 1 and Case 2 were based on using KAI's new solid-state RF generator that has a conversion efficiency for AC power to RF energy of about 72%. For a properly designed antenna, the conversion efficiency of RF energy to heat in the soil is about 93%. This efficiency factor accounts for the losses within RF transmission lines from the RF mobile unit to the RF antenna in the soil. The overall efficiency of AC power to heat should be about 67%. It should be noted that the vacuum-tube RF generator at the Kirtland demonstration operated at an overall efficiency of about 38%.

In the base design cases, a heat balance for each operational phase was calculated through an iterative process. Appendix J contains the detailed calculation sheets listing the SVE and RF design parameters that were used to calculate the heat balance for base design Cases 1 and 2. Case 3 in Appendix J is an alternative mode of operation that will be discussed in Chapter 5. In the base case calculations, the heat added by the RF antennae was balanced with the heat loss to the surroundings and the heat required to increase the temperature of the wet soil and off-gases and to volatilize the water. Since the value was small, the amount of heat needed to vaporize the SVOCs was not included in the heat balance calculation. For each phase, the SVE airflow rate and humidity were adjusted according to operational parameters developed in the Kirtland demonstration. The rate and total amount of contaminating organic compounds removed from the soil were established through relationships between soil temperature, airflow rate, and expected saturation limits in air of the vaporized organic compounds.

4.4.1 Heat Balance for Case 1

For the Case 1 design study, the RF-SVE system heated the soil to 132°C and removed the water and organic contaminants. After treatment, the concentration of the SVOCs was below the target level of 500 mg/kg. The total heat (RF energy) requirement to heat the soil to 132°C assumed that heat was required to vaporize the soil moisture at 100°C and decrease the soil moisture to about 1% dwb. The removal of water produced a dry soil. After the moisture was decreased, the soil was then heated to 132°C. The heat requirement also included the heat added to the air that flowed through the soil toward the SVE wells. The heat content calculations include the following:

- Heat requirement to increase the temperature of the "wet" soil from ambient (15°C) to the boiling point of water or 100°C, assuming the initial moisture content of wet soil was 10% dwb.
- Heat requirement to remove water vapor while the system operated at 100°C, assuming the final moisture content of dry soil was 1%.
- Heat requirement to increase the temperature of the dry soil from 100 to 132°C.
- Heat requirements to increase the temperature of the SVE air from 20 to 100°C and then from 20 to 132°C.
- Heat loss to the surroundings while increasing the temperature of the wet soil from 15 to 100°C, while operating at 100°C, and while increasing the temperature of the dry soil from 100 to 132°C.

The heat losses to the surroundings and in the SVE vapor-air for Cases 1 and 2 were based on the following demonstration data from Section 3.7.3:

- The average heat loss for the 87-day demonstration was 0.29 kJ/kJ$_{input}$ per (m^2/m^3).
- The minimum heat loss for the last 42 days of the demonstration was 0.14 kJ/kJ$_{input}$ per (m^2/m^3).
- The heat loss surface area surrounding the treatment volume was lower for Cases 1 and 2 than for the demonstration:
  - Area per unit volume for the demonstration was 0.093 m^2/m^3
  - Area per unit volume for Case 1 was 0.41 m^2/m^3
  - Area per unit volume for Case 2 was 0.30 m^2/m^3
- Using the average heat-loss and area per unit volume from the demonstration and the area per unit volume for each case, the heat loss was 0.120 kJ/kJ$_{input}$ for Case 1 and 0.090 kJ/kJ$_{input}$ for Case 2.
- When wet soil was heated to 100°C, the amount of water removed should be less than 25% of the total mass in the wet soil.
- Before removing the water at 100°C, the airflow and humidity were adjusted to within the following demonstration guidelines:
  - The humidity approached 0.26 kg of water per kg of air. This humidity was the average value for the final 42 days of RF-SVE demonstration.
  - Airflow rate per unit volume was between 0.6 and 1.0 slpm per m^3.
  - While removing the water and SVOCs at 100°C, the airflow rate was adjusted to optimize the removal rates of SVOCs and water.

From the heat content calculations, the total energy required from the RF system was calculated.

For Cases 1 and 2, the airflow for each SVE extraction well was set at 300 slpm (~11 scfm) while heating the wet soil from 15 to 100°C. The total airflow from the 42 SVE wells was 12.6 m³/min (~450 scfm). The humidity was set according to the guidelines above. However, during this initial phase, the airflow was adjusted to keep the humidity at a value that would remove less than 25% of the total soil moisture. When the water and SVOCs were removed at 100°C, the airflow and humidity were increased but still within the above guidelines. For this treatment period, the airflow was set at 750 slpm (~26 scfm) with a humidity of 0.36 kg/kg of dry air. The total airflow for the 42 SVE wells was 31.5 m³/min (~1,110 scfm).

Table 4.4 summarizes the heat and energy quantities for Case 1 that were calculated to remove the water and SVOC contaminants and to raise the soil temperature to the specified 132°C at Site 1. Table 4.5 divides the values in Table 4.4 into three RF-SVE phases:

- Heating wet soil from 15 to 100°C and removing a small portion of water and light-end SVOCs
- Operating at 100°C and removing the remaining water and light-end SVOCs
- Heating dry soil from 100 to 132°C and removing heavy-end SVOCs

Each phase occurred sequentially as the temperature increased from the center well toward the periphery of the antenna's radius (volume) of influence.

Using the heat required in kWh for the soil and SVE system in Table 4.5 and the total energy supplied by the RF system, the number of days that the system must operate in each phase can be calculated. The RF energy supplied to the soil was the product of the number of RF wells and the heat energy provided by each RF antenna well:

$$(36 \text{ RF antennae})(20 \text{ kW/antennae}) = 720 \text{ kW}$$

The conversion efficiency of RF energy to heat energy was assumed to be 92.5%; therefore, the total heat output of the RF antennae was

$$(0.925)(720 \text{ kW}) = 666 \text{ kW}$$

The total heat needed to increase the wet-soil temperature from 15 to 100°C was 2,044,000 kWh (Table 4.4 or 4.5). The number of days required for heating the wet soil to 100°C was

$$((2,044,200 \text{ kWh})/(666 \text{ kW}))/(24 \text{ h/day}) = 128 \text{ days}$$

Operating at 100°C, the total heat energy required to remove the water and SVOCs from the soil was 2,763,600 kWh (Table 4.4 or 4.5). Therefore, the number of days required for removing water and SVOCs at 100°C was

$$((2,764,600 \text{ kWh})/(666 \text{ kW}))/(24 \text{ h/day}) = 173 \text{ days}$$

The total heat needed to increase the dry soil temperature from 100 to 132°C was 558,500 kWh (Table 4.4 or 4.5). The number of days required for heating the dry soil to 132°C was

$$((558,500 \text{ kW-h})/(666 \text{ kW}))/(24 \text{ h/day}) = 35 \text{ days}$$

The total number of days of operation for the RF-SVE system to remove the water and SVOCs and to heat the soil to 132°C was

$$128 \text{ days} + 173 \text{ days} + 35 \text{ days} = 336 \text{ days}$$

Table 4.4   Total Heat & Energy Requirements for Case 1

Operating Phase	Heat Content (10^6 kJ)	Energy (kWh)
Heat wet soil from 15–100°C	5,220	1,448,600
Heat to vaporize water at 100°C	8,870	2,491,400
Heat Dry soil from 100–132°C	1,520	423,600
Heat SVE air to 100–132°C	1,170	325,400
Heat loss to surroundings	2,440	678,300
Total heat requirement from RF generator	19,320	5,367,300

Table 4.5   Heat & Energy Requirements by Operating Phase, Case 1

Heating wet soil from 15–100°C	Heat Content (106 kJ)	Energy (kWh)
Heat wet soil to 100°C	5,220	1,448,600
Vaporize water into the SVE air	1,260	350,500
Heat SVE air from 20–100°C	230	63,200
Heat loss to surroundings	650	181,900
Total heat requirement from RF generator	7,360	2,044,200
Operating at 100°C to remove water SVOCs		
Vaporize water into SVE air	7,700	2,140,900
Heat SVE air from 20–100°C	770	212,700
Heat loss to surroundings	1,480	410,000
Total heat requirement from RF generator	9,950	2,764,600
Heating dry soil from 100–132°C		
Heat dry soil to 132°C	1,530	423,600
Heat SVE air from 20–132°C	170	48,500
Heat loss to surroundings	310	86,400
Total heat requirement from RF generator	2,010	558,500

4.4.2   Heat Balance for Case 2

Similar heat balance calculations were performed for Case 2, in which the size of the site and the number of RF heating wells and antennae applied to the soil were the variables.

For Case 2, Site 2 was utilized to compare the results and effectiveness of the RF-SVE technology for a smaller site remediation, one similar to a site spill or small refueling station or area. The soil was heated to the same 132°C temperature as in Case 1 and, therefore, the assumed final SVOCs concentration was expected to be below the 500 mg/kg concentration.

The design assumptions identified in Case 1 were nearly identical for Case 2 since the soil and RF-SVE system was operated to the same 132°C heating temperature. The total heat (RF energy) requirement to heat the soil to 132°C assumed that heat was required to vaporize the soil moisture at 100°C and decrease the soil moisture to about 1% dwb. After the moisture was decreased, the soil was then heated to 132°C producing a dry soil. The heat requirement also included the heat added to the air that flowed through the soil toward the SVE wells.

From the heat content calculations, the total heat and energy required from the RF system were calculated. The total energy requirement also included the heat loss to the surroundings of 0.090 kJ/kJ$_{input}$.

While heating the wet soil from 15 to 100°C, the airflow for each SVE extraction well was again set at 300 slpm (~11 scfm). The total airflow for the 12 SVE wells was 3.6 m³/min (~125 scfm). The humidity was set according to the guidelines in Section 4.4.1. However, during this initial phase, the airflow was adjusted to keep the humidity at a value that would remove less than

Table 4.6   Total Heat & Energy Requirements for Case 2

Operating Phase	Heat Content (10⁶ kJ)	Energy (kWh)
Heat wet soil from 15–100°C	1,030	286,100
Heat to vaporize water at 100°C	1,770	492,000
Heat dry soil from 100–132°C	300	83,600
Heat SVE air to 100 or 132°C	280	76,400
Heat loss to surroundings	340	92,,800
Total heat requirement from RF generator	3,720	1,030,900

Table 4.7   Heat & Energy Requirements by Operating Phase, Case 2

Heating Wet Soil from 15°C to 100°C	Heat Content (10⁶ kJ)	Energy (kWh)
Heat wet soil to 100°C	1,030	286,100
Vaporize water into the SVE air	410	112,700
Heat SVE air from 20–100°C	70	20,300
Heat loss to surroundings	100	28,300
Total heat requirement from RF generator	1,610	447,400
Operating at 100°C to Remove Water-SVOCs		
Vaporize water into SVE air	1,370	379,400
Heat SVE air from 20–100°C	150	42,100
Heat loss to surroundings	190	51,700
Total heat requirement from RF generator	1,710	473,200
Heating Dry Soil from 100°C to 132°C		
Heat dry soil to 132°C	300	83,600
Heat SVE air from 20–132°C	50	13,900
Heat loss to surroundings	50	12,800
Total heat requirement from RF generator	400	110,300

25% of the total moisture in the soil. When the water and SVOCs were removed at 100°C, the airflow and humidity were increased but still within the guidelines. The airflow was set at 600 slpm (~21 scfm) with a humidity of 0.317 kg/kg of dry air. The total airflow for the 12 SVE wells was 7.20 m³/min (~255 scfm).

Table 4.6 summarizes the heat and energy quantities for Case 2 that have been calculated to remove the water and SVOCs and to raise the soil temperature to the specified 132°C at Site 2. Table 4.7 divides the values in Table 4.6 into 3 RF-SVE phases: heating wet soil from 15 to 100°C; operating at 100°C to remove water and SVOCs; and heating dry soil from 100 to 132°C. Each phase occurred consecutively as the temperature increased from the center well toward the periphery of the antenna's radius (volume) of influence.

As calculated for Case 1, the RF energy supplied by the system was the product of the number of RF wells and the heat energy applied by each RF antenna. For Site 2, this was

$$(7 \text{ RF antennae})(20 \text{ kW/antenna})(0.925 \text{ efficiency}) = 130 \text{ kW}$$

The total heat energy needed to heat the wet soil from ambient to 100°C was 447,400 kWh (Tables 4.6 or 4.7). The time needed to heat the soil to 100°C was 144 days. Operating at 100°C, the total heat energy needed to remove water and SVOCs was 473,200 kWh. The corresponding time required to remove the water and SVOCs was 153 days. The total heat energy needed to heat the dry soil from 100 to 132°C was 110,300 kWh with a corresponding heating time of 36 days. The total time to heat the soil to 132°C and remove the water and SVOCs for Case 2 was 333 days.

The AC power requirements for the RF generator were calculated assuming a 72% efficiency for the conversion of AC power to RF signal energy. Therefore, the overall conversion efficiency of AC power to heat energy at the antenna was

$$\text{Conversion Efficiency} = (92.5\%)(72\%) = 66.6\%.$$

This efficiency value was used when the total AC power required and the energy costs to produce the calculated heat energy were calculated for each case.

4.5 CONTAMINANT MASS REMOVAL

The mass of contaminant that was removed by RF-SVE can only be estimated in both hypothetical case studies. The SVOC contaminant mass balance in Appendix J includes the initial mass, the mass that was removed, and the mass remaining after treatment. The mass balance was calculated using the following design information:

- Initial average: SVOC concentration at 8000 mg/kg (0.008 kg/kg) of dry soil with the highest concentrations at 10,000 mg/kg
- Expected final average: SVOC concentration at 500 mg/kg of soil
- Average dry soil density at 1780 kg/m³ (111 lb/ft³)
- Volume of soil for Site 1 at 24,800 m³
- Volume of soil for Site 2 at 4910 m³

4.5.1 Mass Removal — Case 1

The initial mass of SVOC contaminant for Site 1 was calculated to be

$$(0.008 \text{ kg/kg})(24,800 \text{ m}^3)(1,780 \text{ kg/m}^3) = 353,400 \text{ kg}$$

The mass of SVOC contaminant remaining after treatment was

$$(0.0005 \text{ kg/kg})(24,800 \text{ m}^3)(1,780 \text{ kg/m}^3) = 22,100 \text{ kg}$$

Hence, the total mass of SVOC contaminant that was removed by RF-SVE was estimated to be

$$(353,400 \text{ kg} - 22,100 \text{ kg}) = 331,300 \text{ kg}$$

Since pilot testing was not conducted, the expected rate of contaminant removal was difficult to estimate. However, results from the mass and heat balance calculations were used to estimate the SVOC removal rates during two major operating phases:

- Heating the wet soil from ambient to 100°C
- Removing water and SVOCs at 100°C and removing SVOCs while heating the dry soil to 132°C

The latter phase is the combination of the last two phases in Table 4.5 (and Table 4.7).

To simplify the initial estimation of the SVOC removal rate (to be adjusted later), the fraction of SVOCs removed during each phase was assumed similar to the fraction of water removed. Accordingly, the mass of soil moisture removed during the two operating phases was used to estimate the SVOC removal rates. Finally, using the mass of dry air for each phase, the kilograms of SVOC vapor per kilogram of dry air were compared to the vapor–pressure–temperature values in Appendix F. At a particular temperature, the calculated value should be lower than the Appendix F values.

The mass of contaminated soil in Site 1 was 44,172,000 kg. The total mass of soil moisture that must be removed by RF-SVE was 3,975,400 kg. When the soil was heated from ambient to 100°C, the amount of moisture removed was calculated to be 559,200 kg. Therefore, the average SVOC removal rate for this phase was calculated as follows:

$$(331,300 \text{ kg})(559,200 \text{ kg})/(3,975,400 \text{ kg}) = 46,600 \text{ kg}$$

$$(46,600 \text{ kg})/(128 \text{ days}) = 364 \text{ kg/day } (8.7 \text{ kg/day per well})$$

The amount of SVOCs removed during the second phase was the mass remaining after the initial 128 days. The remaining water was also removed during this phase. Therefore, the average SVOC removal rate for the final 208 days of operation (173 days + 35 days) was as follows:

$$(331,300 \text{ kg}) - (364 \text{ kg/day})(128 \text{ days}) = 284,700 \text{ kg}$$

$$(284,700 \text{ kg})/(208 \text{ days}) = 1,370 \text{ kg/day } (32.6 \text{ kg/day per well})$$

The rate of SVOCs removal was higher, as expected, during the final operating phase of RF-SVE treatment. It should also be noted, although not illustrated here, that the removal rates were not constant, especially in the earlier stages of heating. The rates of SVOC contaminant removal for each phase were presented as average removal rates.

Finally, as validation of the above calculations for both phases, the mass of SVOCs per kg of dry air was compared to the vapor–pressure–temperature values in Appendix F. For the initial 128-day period, the average mass of SVOCs per kg of air was as follows:

$$\text{Mass of SVOCs} = 46,600 \text{ kg}$$

$$\text{Mass of Dry Air} = 2,796,000 \text{ kg}$$

$$(46,600 \text{ kg})/(2,796,000 \text{ kg}) = 0.0167 \text{ kg SVOCs/kg dry air}$$

For the final 220-day period, the average mass of SVOCs per kg of air was as follows:

$$\text{Mass of SVOCs} = 284,700 \text{ kg}$$

$$\text{Mass of Dry Air} = (9,447,500 \text{ kg} + 1,529,100 \text{ kg})$$

$$\text{Mass of Dry Air} = 10,976,600 \text{ kg}$$

$$(284,700 \text{ kg})/(10,976,600 \text{ kg}) = 0.0259 \text{ kg SVOCs/kg dry air}$$

The data in Appendix F show that up to 100°C, C_{12} to C_{15} organic compounds can be removed by the SVE air. Plots of the data are shown in Figures 3.1, 3.2, and 3.3. The data also show that the C_{19} to C_{20} organic compounds can be removed at temperatures up to 132°C. Since the calculated values for SVOC compounds were close to the values in Figure 3.2, the rate of SVOCs removal (kg/day) could have been lower than the calculated values for each operating phase. Additional treatment time might have been necessary to remove the SVOC compounds. However, the only heat requirement from the RF antennae included the heat in the SVE air and the heat loss to the surroundings. This heat value was estimated using the results for heating the dry soil from 100 to 132°C or about ~2500 kWh/day. Nine 20-kW RF units would be needed to maintain this heat input.

4.5.2 Mass Removal — Case 2

The contaminant mass balance and removal was similarly calculated for Case 2. Due to the smaller volume of soil within Site 2, the amount of moisture and, consequently, the amount of SVOCs to be removed, were also less than with Case 1. Thus, the total mass of SVOC contaminant to be removed from Site 2 during RF-SVE treatment was estimated to be 65,423 kg.

For the two operating phases of the RF-SVE treatment for Site 2, values for the SVOC mass removal rates in kilograms per day and the kilograms of SVOC vapor per kg of dry air were as follows:

- Heating the wet soil from 15 to 100°C (144 days):
  - Average removal rate = 8.7 kg/day
  - Average dry air content = 0.0167 kg/kg
- Removing water and SVOCs at 100°C and removing SVOCs while heating the dry soil to 132°C (189 days):
  - Average removal rate = 22.3 kg/day
  - Average dry air content = 0.0215 kg/kg

Similarly, for Case 2, the data in Appendix F show that up to 100°C, C_{12} to C_{15} organic compounds can be removed by the SVE air. The data also show that the C_{19} to C_{20} organic compounds can be removed at temperatures up to 132°C. Since the calculated values for SVOC compounds were close to the appendix values, the rate of SVOC removal (kg/day) could have been lower than the calculated values for each phase. Additional treatment time might be necessary to remove the SVOC compounds. However, the only heat requirement from the RF antennae included the heat in the SVE air and the heat loss to the surroundings. This heat value was estimated using the results for heating the dry soil from 100 to 132°C or about ~750 kWh per day. Two 20-kW RF units would be needed to maintain this heat input.

CHAPTER 5

Costs and Economic Analysis

This chapter presents cost estimates for the two hypothetical cases developed in Chapter 4. An economic analysis of the remediation costs for full-scale application of the RF-SVE technology is also presented. The analysis discusses factors that influence costs and how these costs could be decreased.

The project costs shown in the two preliminary cost estimates below reflect only the cost to design and perform the remediation of the site. Typical project conceptual design, site assessment and characterization, and laboratory analysis costs are **not** included in the cost estimates. It was assumed that the RF-SVE technology had already been determined to be the preferred treatment methodology for these hypothetical sites. Therefore, these characterization costs were not included so as to focus on the costs to design and implement this technology. Additional characterization and laboratory costs are a part of most remediation projects regardless of the technology chosen.

The cost estimates in Tables 5.1 and 5.2 are considered order-of-magnitude level, which is generally considered accurate to within –30% or +50%.

5.1 STANDARDIZED COST BREAKDOWN

Table 5.1 for Case 1 and Table 5.2 for Case 2 list the summary of costs for these studies categorized as "before", "during", and "after treatment" according to the second-level work breakdown structure (WBS) as defined in the *FRTR Guide*. More detail of the cost breakdown of the summary estimates of the hypothetical proposed designs is located in Appendix K.

5.1.1 Significant Cost Items

The total capital cost to install the RF-SVE system for the full-scale hypothetical Site 1, Case 1, was estimated to be $761,800. This, however, did not include the most significant cost factor, the lease costs of the RF trailer units. One RF unit rents for $6,000/month. Multiplied by the 36 units required for the calculated 336-day duration of treatment gave a total of $2,592,000 for approximately 11 months for operations plus 1 month for equipment setup. The RF unit costs were not included as an up-front capital cost since the RF units were rented and, therefore, were represented as an operation and maintenance (O&M) cost. The mobile RF units' lease was the single most significant cost factor for the full-scale remediation project. This lease cost drove up the total O&M cost for SVE and RF operation to approximately $4,352,900 of the total remediation cost of $4,528,900 (Table 5.1). The electrical power requirement for the RF units was the second most significant cost factor of the O&M costs. Confirmation sampling ($57,900) to be performed after remediation was the most significant "after treatment" cost. The "RF engineering design" was not

Table 5.1   Summary of RF-SVE Full-Scale Remediation Costs, Case 1 (Larger Site)

Cost Element	Unit Cost ($)	No. of Units	Cost ($)
Before treatment costs			
Mobilization and preparatory work – mobilization of equipment, personnel, and material	20,000	Lump sum	20,000
Site work — clearing, fencing, concrete slab	41,000	Lump sum	41,000
Engineering design of SVE	17,400	Lump sum	17,400
Total "before treatment" cost with 15% markup			87,500
During treatment costs			
SVE system equipment and installation (1 month)	417,600	Lump sum	417,600
SVE operating costs (11 months)	220,000	Lump sum	220,000
RF system equipment and installation (1 month)	150,300	Lump sum	150,300
RF operating cost (11 months)	3,380,800	Lump sum	3,380,900
Total "during treatment" cost with 15% markup			4,352,900
After treatment costs			
Decontamination & decommissioning	7,000	Lump sum	7,000
Disposal	7,500	Lump sum	7,500
Site restoration	5,000	Lump sum	5,000
Demobilization	4,000	Lump sum	4,000
Confirmation sampling	57,800	Lump sum	57,900
As-built drawings/final closure reports	3,600	Lump sum	3,600
Total "after treatment" cost with 15% markup			88,500
Total remediation cost			4,528,900

Table 5.2   Summary of RF-SVE Full-Scale Remediation Costs, Case 2 (Smaller Site)

Cost Element	Unit Cost ($)	No. of Units	Cost ($)
Before treatment costs			
Mobilization and preparatory work – mobilization of equipment, personnel, and material	12,000	Lump sum	12,000
Site work — clearing, fencing, concrete slab	26,900	Lump sum	26,900
Engineering design of SVE	14,000	Lump sum	14,000
Total "before treatment" cost with 15% markup			58,600
During treatment costs			
SVE system equipment and installation (1 month)	146,700	Lump sum	146,700
SVE operating costs (11 months)	201,800	Lump sum	201,800
RF system equipment and installation (1 month)	79,200	Lump sum	79,200
RF operating cost (11 months)	807,000	Lump sum	806,700
Total "during treatment" cost with 15% markup			1,297,200
After treatment costs			
Decontamination & decommissioning	2,500	Lump sum	2,500
Disposal	5,000	Lump sum	5,000
Site restoration	5,000	Lump sum	5,000
Demobilization	3,000	Lump sum	3,000
Confirmation sampling	35,500	Lump sum	35,500
As-built drawings/final closure reports	2,400	Lump sum	2,400
Total "after treatment" cost with 15% markup			55,800
Total remediation cost			1,411,600

considered a pre-treatment engineering cost. It was considered a "during treatment" cost since final design and fine-tuning of the RF system were performed during installation and setup of the mobile RF units and during start-up testing.

The significant cost items for Case 1 were also applicable to the smaller hypothetical site in Case 2. However, since much of the engineering, SVE and off-gas system installation, and operating costs were comparable to the larger site, O&M oversight, RF lease, and electricity costs were proportionately higher. The economic factors are further discussed in Section 5.2.

5.1.2 Unit Costs — Case 1 and Case 2

Assuming the total quantity of soil media to be treated in Site 1 was 24,843 m³ (32,493 yd³ or 48,700 ton), the unit remediation cost was about \$182/m³ (\$139/yd³ or \$92/ton). The unit remediation cost for Site 2, which contained 4903 m³ (6374 yd³ or 9610 ton) of soil, was approximately \$288/m³ (\$221/yd³ or \$147/ton). Detailed information and economic factors are contained in Section 5.2.

5.1.3 Project Management

Project management costs for these cost estimates were assumed to occur only during the project design and execution. The subcontractors who will perform the fieldwork have been selected. Therefore, no procurement or subcontracting strategy-planning costs has been included.

Remedial Action Work Plan

The development of the remediation work plan is typically categorized with project management duties. This plan is often submitted to regulatory agencies for approval before the start of any remedial activities. Many times this work plan will include the proposed contracting strategy, a general description of methodology, and general sequence in which the remediation activities will be conducted. The factors to be considered when predicting costs associated with preparing these work plans included

- Level of governmental involvement
- Amount of emphasis placed on the environmental and monetary risks associated with the remediation
- Expected timeframe for remediation to be completed

Treatability Studies

Evaluating the results of treatability studies helps confirm contaminant properties and distribution, energy requirements, and contaminant behavior. The cost estimate assumed that pilot testing was completed, the results were evaluated, and RF-SVE technology was selected to remediate the hypothetical sites.

Soil Sampling/Site Assessments

Extensive soil sampling and analysis to assess the nature and extent of the contaminants, as well as the collection of geotechnical soil borings, are imperative to determine the feasibility of using any remediation technology. Due to variability between sites, characterization was not included in the proposed cost to perform the remediation using RF-SVE.

5.1.4 Health and Safety Considerations

In addition to the typical health and safety concerns that are considered for most typical construction and remediation work, the RF heating technology mandates specific health and safety issues. Adequate RF transmission line and wellhead grounding, as well as RF shielding, are important for safe operation. The monitoring of electromagnetic field power densities must be performed by trained personnel to assure potential exposures are within safe limits. The cost to write a health and safety plan regarding RF safety, as well as provide a safety professional trained in RF safety, was budgeted into the cost estimates.

As with any heating technology, health and safety precautions should be of special concern if underground tanks or aboveground containers are in the proximity due to the potential ruptures induced by increased temperatures.

Some heavy lifting and hoisting are required up to 7.6 m (25 ft) in the air for the placement of the antenna in the RF well casing. Therefore, hard hat compliance and eye protection is required. Air monitoring is also required at the exhaust stack in most applications to ensure air quality standards are being maintained at the local, state, and/or federal agencies.

5.1.5   Utility Requirements and Considerations

The electrical interface hardware within the RF trailers allows the RF trailer units to operate with typical, single phase, 240 V, 60 Hz, 125-A electrical power with neutral service. However, with the implementation of several RF trailers operating at one time, 36 for Case 1 and 7 for Case 2, additional electrical service was required. A separate transformer at 1000 KVA was priced into the estimate since a large electrical demand (900 kW for RF trailers alone) is required at all times during RF-SVE operations. For Case 2, a 300-KVA transformer was priced into the estimate.

Telephone service was required since the RF heating system was interfaced through a modem system that enabled access and control via telephone service. The telephone line can also be used for telephone calls during site visits.

5.1.6   Start-Up, Modifications, and Repairs

The RF heating systems can be operated on a continuous basis. The operational aspects of the system are based on procedural algorithms that are manipulated via a digital computer interface. The computer system is interfaced to the RF generator so that specific parameters can be controlled to enhance the effectiveness of the RF energy delivery to the wells. Other peripheral components that support operational requirements to the RF generator are simple selector switches for each component.

The RF heating systems are solid-state circuitry that should require minimal maintenance. The most important task is replacement of air filters to protect against overheating due to clogged filters. The filters collect particles in the air that are being delivered to electronic components for cooling purposes.

Potential maintenance actions would be instigated with the occurrence of reduced efficiency of the RF system. In the case of reduced efficiency, a computer-driven diagnostic program is executed to isolate the failed component.

5.2 ECONOMIC ANALYSIS FACTORS

The following factors might influence typical project economics for the RF-SVE technology.

5.2.1   Process Design Assumptions/Basis for Costs

The hypothetical design assumptions were generally conservative. The components and their operational configurations were based on favorable site conditions for the RF-SVE technology. The RF mobile units were selected for the full-scale hypothetical designs based on the RF equipment known to be available for use at this time. These currently available RF units have been utilized only on small treatment or pilot-scale sites. For larger, full-scale remediation sites, higher output RF units might prove to be more cost effective (see Section 5.2.3).

The component quantities and sizing were conservatively designed due to the assumption of the continuous operation of both the RF heating units and the SVE system. Greater efficiencies could possibly be realized by decreasing the airflow from each of the SVE wells. Decreasing the SVE airflow through the blowers during early stages of heating and allowing the RF energy to heat the soil to 100°C might accelerate the heating time. Therefore, cost could be decreased by a reduction in RF-SVE operating time and energy consumption.

The number of RF heating units and the method and timing of RF heating to heat larger areas of soil might also affect the remediation performance and cost. The two design cases assume a simultaneous heating of all soil areas at once. If heat losses from the surroundings and SVE airflow can be minimized, RF energy could be applied sequentially to the RF wells starting in the center and heating outward towards the perimeter. Sequential heating could reduce the number of mobile RF units needed at the site.

5.2.2 Alternative SVE-HPO Cost Study

As an alternative cost study, the RF-SVE system would heat the soil within the site to 100°C and oxidize the SVOC contaminants by hydrous pyrolysis oxidation (HPO) (Appendix J, Case 3). The temperature of the soil would remain at 100°C during the oxidation phase. Since soil moisture is essential to the HPO process, the amount of water removed by the SVE air has to be minimized by using low airflows. However, sufficient airflow would be used to assist the heating of the soil to 100°C and to provide oxygen for the HPO reaction at 100°C.

In Case 3, the area and volume of the treated soil were the same as for Case 2 (Site 2), 4910 m^3 (6375 yd^3 or 9610 ton). The main RF-SVE design and operating parameters were similar to values used in Case 2. The site layout for Case 3 was the same as for Case 2 (Appendix H, Drawing 2). However, for Case 3, the treatment of the soil was carried out in the following two periods:

- Period 1: Heating the wet soil from 15 to 100°C.
- Period 2: Operating at 100°C to oxidize the SVOC contaminants by HPO reactions.

The oxidation reaction has to be included in the mass and heat balance calculations for Case 3. The oxidation reaction produced considerable heat and some water and carbon dioxide. The heat of reaction decreased the heat requirement from the RF units. To complete the mass and heat balances for Case 3 in Appendix J, the following reaction for pentadecane ($C_{15}H_{32}$) was assumed as typical for the oxidation of SVOCs:

$$C_{15}H_{32} + 23O_2 \rightarrow 15CO_2 + 16H_2O \qquad (R4)$$

The calculated heat of reaction for the oxidation of $C_{15}H_{32}$ was 47,600 kJ/kg alkane, Appendix J, Table J2. To calculate the net amount of heat produced by R4 and the mass and heat balances for both periods, the following assumptions were used:

- The SVOCs removed by SVE in Period 1 were 16,400 kg or 25% dwb of the total mass.
- The SVOCs removed by oxidation and SVE in Period 2 were 49,100 kg or 75% dwb of the total mass.
- The SVOCs removed by oxidation in Period 2 were 32,700 kg or 50% dwb of the total mass.
- During Period 2, the oxidation reaction consumed 113,200 kg of oxygen in 2.32 million kg of SVE air, or about 21% dwb of the total oxygen in air.
- The oxidation of SVOC produced 44,200 kg of liquid water and 101,700 kg of gaseous carbon dioxide.
- During Period 2, the net heat of reaction for the organic alkanes offset the heat requirements for the water vapor and SVE air at 100°C and the heat loss to the surroundings.

Table 5.3   Design, Heat, & Energy Requirements for SVE-HPO, Case 3

System Design Parameters	Site 2	
Treatment area shape	Hexagonal	
Length	18.05 m sides	
Depth of contamination	5.79 m	
Total ground surface area	847 m²	
Total volume of soil	4,910 m³ (6,375 yd³)	
Total mass of soil	9,610 tons	
Number of RF wells	7	
Number of peripheral extraction wells	12	
Number of central air vent wells	7	

Operating Parameters	Period 1	Period 2
Number of RF units	4	1 (standby)
RF energy from each antenna, kW	20	0
Number of days for each period	179	186
SVOC removal rate, mg/day/kg of soil	~10	~30

Heating Wet Soil from 15–100°C	Heat Content (10⁶ kJ)	Energy (kWh)
Heat wet soil from 15–100°C	1,030	286,100
Vaporize water into the SVE air	50	14,000
Heat SVE air from 20–100°C	10	2,500
Heat loss to surroundings	60	15,500
Total heat requirement heating to 100°C	1,150	318,100

Operating at 100°C to Oxidize SVOCs		
Vaporize water into SVE air at 100°C	1,175	326,500
Heat SVE air from 20–100°C	180	50,500
Heat loss to surroundings	185	51,700
Net heat from oxidation reaction R4	−1,540	−428,700
Total heat requirement operating at 100°C	0	0
Total Heat Requirement from RF Generator	1,140	318,000

With these assumptions, the net heat produced by R4 during Period 2 was estimated to be about 1540 million kJ.

For Case 2, the total treatment time was 333 days. To compare the results of Case 3 to Case 2, the total treatment time for Periods 1 and 2 was set at 365 days. During Period 1, the SVE airflow from each extraction well was kept low at 30 slpm (~1 scfm) to minimize the removal of soil moisture by the air and to maximize the heating rate of the soil. To decrease lease costs, the number of 20-kW RF units was decreased from 7 in Case 2 to 4 in Case 3. During Period 2, the airflow was increased to about 600 slpm (21 scfm) per well. For Period 2, RF heating was not necessary. The number of 20-kW RF units was decreased from 4 to 1 standby unit. Without the heat of reaction for R4, 2 RF units each operating at about 12 kW were needed to maintain the soil temperature at 100°C and to provide heat for the wet SVE air and loss to the surroundings. The heat requirements for Case 3 were calculated using similar methods as for Case 2. Table 5.3 summarizes for Case 3 the calculated results in Appendix J.

The calculated data in Case 3 showed that the soil's water content decreased by ~65% dwb or from 10% down to ~4%. The HPO reaction had to occur in the aqueous phase within the wet soil (Section 3.7.9.2). If the soil's moisture content decreases, the rate of oxidation will decrease. However, one could maintain the soil moisture content by irrigating the site with water. The added water consumes some of the heat produced by the oxidation reaction when its temperature increases to 100°C.

Table 5.4 Summary of RF-SVE Full Scale Remediation Costs for SVE-HPO, Case 3

Cost Element	Unit Cost ($)	No. of Units	Cost ($)
Before treatment costs			
Mobilization and preparatory work – mobilization of equipment, personnel, and material	12,000	Lump sum	12,000
Site work — clearing, fencing, concrete slab	26,900	Lump sum	26,900
Engineering design of SVE only	14,000	Lump sum	14,000
Total "before treatment" cost with 15% markup			58,600
During treatment cost			
SVE system equipment and installation (1 month)	138,700	Lump sum	138,700
SVE operating costs (12 months)	206,100	Lump sum	206,100
RF system equipment and installation (1 month)	79,200	Lump sum	79,200
RF operating cost (12 months)	425,400	Lump sum	435,300
Total "during treatment" cost with 15% markup			910,300
After treatment cost			
Decontamination & decommissioning	2,500	Lump sum	2,500
Disposal	5,000	Lump sum	5,000
Site restoration	5,000	Lump sum	5,000
Demobilization	3,000	Lump sum	3,000
Confirmation sampling	35,500	Lump sum	35,500
As-built drawings/final closure reports	2,400	Lump sum	2,400
Total "after treatment" cost with 15% markup			55,800
Total remediation cost			1,024,700

The SVOC rate of removal by SVE for both periods (~10 mg/day/kg dry soil) seemed reasonable when compared to the mass removal rates in Figure 3.3. Finally, after examining the figures in Section 3.7.9.2, the calculated oxidation rate of ~20 mg/day/kg dry soil for Period 2 was also reasonable.

The summary of full-scale remediation costs for Case 3 is given in Table 5.4. In comparing Case 3 to Case 2, the AC electrical power requirement was decreased from 1,431,800 to 441,900 kWh. This energy reduction decreased the treatment unit cost by $19/m^3 ($14/yd^3 or $10/ton). The lease cost for the RF units was also decreased from $504,000 to $204,000. This decrease equated to cost savings of $61/m^3 ($47/yd^3 or $31/ton). The total cost savings for these 2 items were $80/m^3 ($61/yd^3 or $41/ton).

5.2.3 Capital and O&M Costs

As mentioned in Section 5.2.1, larger output RF units would decrease both capital and operating costs. For example, if an increased output RF generator, a 50-kW unit, was utilized for Case 2, the number of RF units needed to provide the 240 kW could be decreased to 5. For Case 3, two units would be needed for Period 1 and one unit for Period 2.

Capitalization of the RF units was a more cost-effective option. The previous cost estimates assumed that the RF units were leased. For Cases 1 and 2, the estimated lease cost for a 12-month duration was $72,000 for each 20-kW RF unit. For Case 3, the estimated cost for 13 months averaged $78,000 for each unit. However, KAI's new 10-kW trailer unit has an estimated capital cost of $45,000. Using the 6/10 factor for equipment costs, the estimated cost for a 20-kW unit was about $70,000.

The annualized capital-recovery cost without salvage was calculated using the following formula:

$$A/P_{i,n} = i(1 + i)^n/[(1 + i)^n - 1]$$

where A was the annual capital-recovery payment; P was the capital cost; i was the interest rate; and n was the number of years to write off the capital expenditure. If an interest rate of 7% and a

5-year write-off period were used, the annual payment for an expenditure of $70,000 was $18,800, which included a 10% contingency factor.

For Case 1 with 36 20-kW RF units, the lease cost was $2,592,000. The capital-recovery payment for the project was $777,000, which included a 15% markup. Hence, the treatment cost decreased to $109/m³ ($84/yd³ or $56/ton). For Case 2 with seven 20-kW RF units, the lease cost was $504,000 and the capital payment for the project was $151,000. Hence, the treatment cost decreased to $216/m³ ($166/yd³ or $110/ton). For Case 3, four 20-kW RF units were required for the initial 179 days and 1 standby unit for the remaining 186 days. The project was completed in 13 months. The lease cost for the RF units was $204,000. Using the above formula, the total project payment for 4 units was $86,000. Therefore, the unit treatment rate decreased to $185/m³ ($142/yd³ or $94/ton).

Additional capital savings can be realized by pulsed operation of the SVE system. Decreasing the airflow at different stages of heating can produce the same effect. By targeting different areas with pulsed SVE in a sequential manner, the capacity of the SVE and air-vapor treatment systems could be decreased. For the hypothetical design Case 2, continuous operation of the SVE system required SVE and off-gas treatment systems having a capacity of 7200 slpm (~250 scfm) for 12 extraction wells at 600 slpm (~21 scfm) each. The total SVE airflow required one blower with a capacity of 12,750 slpm, 14.9 kW (~450 scfm, 20 hp) at a cost of $15,000. One catalytic oxidizer was also required with a capacity of 14,150 slpm (~500 scfm) and a unit cost of $40,000. The blower and oxidizer were moderate capital cost items. For a pulsed operation, the off-gas treatment capacity could be decreased as much as 50%; however, a corresponding increase in humidity must also occur with decreased airflow. Therefore, a smaller catalytic oxidizer, one with a capacity of 7200 slpm (~250 scfm), could save about $5,000. A smaller blower, one with a capacity of 7200 slpm (~250 scfm), 7.5 kW (10 hp), could save an additional $3,000. Operational cost would also decrease primarily as a result of reduced energy costs, both electrical and propane or natural gas fuel, as well as maintenance and routine monitoring costs.

Some up-front costs were decreased since the RF heating units were enclosed, mobile, self-contained, and were brought on site with relatively low labor intensity. Therefore, the installation cost for equipment in the SVE compound was decreased. However, the high lease cost of the RF trailers offset these costs. Cost for drilling the RF wells and placing the transmission lines and well antennae was similar to conventional methodologies for *in situ* on-site remediation.

Some labor cost savings were realized for O&M since the trailer units were remotely operated; however, the SVE system and vapor abatement systems required the typical operational monitoring and maintenance expected with these systems.

5.2.4 Scale-Up Requirements and Limitations

The designs presented in this monograph considered a relatively large, but realistic, remediation area. Because of the high lease cost of the RF trailer units, shortening the remediation time frames or decreasing the number of RF units required would result in the greatest cost savings.

When scaling up or down project designs based on the site parameters, it is imperative to maximize the efficiency of the RF units to minimize costs. To maximize the efficiency, the radius of influence, number of antennae per RF unit, proper tuner frequency, etc. must be considered. However, the RF-SVE technology appears to be less cost-effective for small volume projects. According to previous pilot-scale studies, the volume of contaminated soil should not be less than 350 m³ (455 yd³). The primary reasons for this limitation are the high mobilization and capital costs associated with installation of the SVE and vapor abatement systems and the high O&M costs associated with the RF trailer units. The economics of scale come into effect for larger systems that can justify the remediation costs per unit volume of soil (dollars/m³).

5.2.5 RF-SVE Project Costs Compared to Similar Technologies

Similar technologies will be defined as processes that use heat to enhance the *in situ* treatment of contaminated soil. The major competing technologies are steam injection-SVE and soil vitrification. The *in situ* vitrification process is still in the developmental stage and is best suited for removal of inorganic compounds. Results of pilot studies have shown that the unit cost for vitrification is more expensive than RF-SVE. The reliability of *in situ* vitrification technology has plagued the application of the technology. The lack of reliability can easily increase the O&M costs.

Steam injection used in conjunction with SVE is an emerging *in situ* treatment technology for contaminated soil. This technology targets hydrocarbon compounds that have favorable vapor pressures for SVE at 100°C. RF-SVE targets contaminants in soil that have favorable vapor pressures for SVE at temperatures between 100 and 150°C. The order-of-magnitude unit cost for steam injection-SVE has been reported to range from \$105 to \$250/m³ (\$80 to 200/yd³) (Balshaw-Biddle et al., 1998). The monograph authors have estimated the unit costs for the process using hydraulic fracturing to loosen tight soil and steam injection to remove light-end SVOCs. The estimated costs were \$144/m³ (\$110/yd³ or \$74/ton). Finally, the unit cost for enhanced SVE using 6-phase heating was also estimated to be about 112/m³ (\$86/yd³ or \$57/ton) (DOE, 1995).

The unit costs for the process of hydraulic fracturing with steam injection were for a hypothetical project that treated 24,000 m³ (32,000 yd³) of soil over a 560-day period. The hypothetical project for 6-phase soil heating treated 600,000 m³ (785,000 yd³) of soil over a 5-year period. The maximum treatment temperature for these processes was less than 100°C. The low operating temperature would limit their application toward VOCs and light-end SVOCs.

5.2.6 RF-SVE Project Costs Compared to Alternative Technologies

Alternative technologies will be defined as *in situ* treatment processes that do not use heat to enhance the removal of contaminants from soil. These *in situ* treatment technologies can be grouped into two categories: biological treatment and physical/chemical treatment.

The two biological treatment technologies that were reviewed are *in situ* biooxidation and *in situ* bioventing. In comparison to RF-SVE, both of these technologies are relatively inexpensive, ranging from \$20 to \$70/m³ (Fluor Daniel GTI, 1998); however, the treatment time for both of these technologies generally ranges from 5 to 30 years. This long treatment period may not be acceptable from a regulatory standpoint. Furthermore, the application of these technologies to soil might not achieve low cleanup levels (<500 or <1000 mg/kg), especially at high SVOC concentrations of, for example, 10,000 to 15,000 mg/kg.

In situ physical and chemical treatments include soil flushing, SVE, and solidification-stabilization technologies. SVE alone will obviously be less expensive than RF-SVE; however, it should not be compared to RF-SVE treatment. SVE enhanced with RF heating is required to remove SVOC compounds from soil. Treatment with SVE alone cannot remove the SVOCs.

Soil flushing technology is in the developmental phase. However, expected costs have been estimated to be generally higher and with a broader range from \$300/m³ to \$850/m³ (Lowe, 1998). The main reason for the variability in costs is the uncertainty of the technology at full scale.

Stabilization-solidification is generally less expensive, from \$60/m³ to \$350/m³, depending on the treatment target levels and off-site disposal location (Fluor Daniel GTI, 1998). However, its effectiveness is limited for contaminated soils that contain VOCs and SVOCs. Solidification-stabilization is more effective for inorganic contamination.

CHAPTER 6

Potential Applications

6.1 PERFORMANCE PARAMETERS

The expected RF-SVE treatment performance for the typical site, as well as the parameters that affect the treatment cost and/or performance, are discussed in this chapter. Treatment technology performance is typically measured in terms of the percentage of contaminants removed or determining whether or not the target concentration level was obtained. However, establishing performance levels for *in situ* treatment technologies such as RF-SVE is often challenging because it is difficult to accurately measure the level and extent of contamination within the site. Also, there is limited knowledge on removal rates of particular contaminants so that a targeted final concentration can be predicted. These parameters are summarized qualitatively in this chapter based on quantitative presentations in the previous chapters.

6.1.1 Maximizing Treatment Rate

The key to optimizing the RF-SVE system design is based on a clear understanding of the following relationships:

- The rate of soil heating
- The individual rates of moisture and contaminant removal
- Removal of contaminants through the SVE system
- Thermal oxidation and biooxidation reactions for organic compounds

SVE systems can be used to increase the effective distribution of heat by advection. Modeling can be developed that allows simulation of this process. Interpretation of these results can optimize full-scale system configurations. Further benefits can be realized if a system operates with minimal heat losses from the ground surface, boundary surfaces, and heated exhaust air. A properly designed system would allow retention of RF-generated heat in the contaminated soil and injection of preheated air into the ground. Hence, a well-designed system can decrease RF energy requirements by "containing" and "recycling" heat. This system could be simulated by treating soils beneath concrete or asphalt slab or other impermeable surface covers.

Maximizing the treatment rate for the RF-SVE, as with most remediation systems, can be attained through consideration of both design and operational optimization.

Design Optimization

During the design of the RF-SVE system, treatability testing should be performed to determine the performance of the SVE wells and to collect data to determine the heating properties of the soil.

The design of the heating pattern for a particular remediation application depends on the operating frequency of RF units; power level; antenna length, position, and orientation; the number of antennae; and the soil properties. The matching network is primarily responsible for allowing the RF generator to provide full, maximized power to the antenna. Therefore, the combination of the optimally designed tuner and antenna helps insure the maximum transfer of RF energy to the soil.

Operation Optimization

Pulse operation or reduced airflow through the SVE system can minimize the energy requirements without significantly affecting the treatment rate. Decreasing the airflow from the SVE blower and allowing the RF heating rate to accelerate in the early phase, up to 100°C, decreases the energy consumption and loss of heat from the subsurface. The monitoring of the consumption of oxygen will help determine the pulse frequency of the SVE blower.

The use of larger (if available) RF units to apply heat to several RF wells may also optimize operations by decreasing treatment times, utilizing smaller SVE blowers and capacities of the off-gas treatment system. The understanding of thermal and biooxidation processes could increase the SVOC removal rate and therefore further optimize the operational efficiency.

6.1.2   Site Characteristics Affecting Cost or Performance

Assessing the characteristics of the site is critical in determining if the remedial technology selected is applicable. The following subsections and Section 6.2.1 discuss these characteristics and their impact on the cost or performance.

Lithology

The lithology of the site is an important characteristic that may affect the cost and performance of remediation. Higher clay content and small particle size affect the permeability of the soils and decrease the radius of influence of the vapor extraction wells.

Topography

A relatively flat and accessible site lessens the burden of delivery and setup on the site, especially for multiple RF-generator trailers.

Surface Cover

If the area to be treated is covered by concrete or asphalt, there may be a reduction in the loss of heat to the atmosphere and infiltration of water into the soil. When the surface is sealed, short-circuiting of the SVE air from the surface is minimized, thus increasing the radius of influence of the airflow.

Moisture Content

The airflow is proportional to the air-filled porosity. As the moisture content increases, air-filled porosity decreases and, therefore, the airflow decreases. Any residual moisture in the soil also requires additional heat to reach the target temperature. The moisture content also affects the electrical properties of the soil. As the moisture content decreases, the ion mobility decreases, and the heating mechanism will change from primarily ionic conductivity to predominantly dielectric loss. If, however, the soil moisture content is too low, the RF heating capabilities of the soil will decrease.

Hydrogeology

If water-table fluctuations submerge the impacted soil for part of the year, then the amount of time the SVE system is operative will be decreased. Submerged RF antennae result in a loss of heat to the aquifer.

Precipitation

The moisture content will significantly increase following rainfall events, resulting in the need for additional heat to reach or maintain the target temperature. However, an impermeable membrane over the ground surface of the heated volume can decrease the flow of a rainfall into the treated soil. Moisture that is vaporized is collected in the air-water separator and must be properly treated and disposed.

Microbial Activity

Microbial degradation is not significant for removal of heavier-end SVOC contaminants. If the goal of the system is to promote *in situ* biodegradation of contaminant using RF-SVE at temperatures from ambient to about 35°C, microbial activity is an important characteristic for assessing the effect on cost and performance. Bioactivity also can be enhanced at the fringe of the RF-SVE design area where the soil temperatures are in the favorable range. Low soil moisture and high temperature adversely affect microbial activity. At the high temperatures of the proposed RF-SVE operation (>90°C), most biological activity becomes dormant.

Contaminant Type

If two contaminants with similar final concentration goals are present, only the less volatile contaminant needs to be considered in the design of the RF-SVE system. The more volatile contaminant will also be removed. If the concentration goals are different, it might be necessary to consider each contaminant in the design. In the case of high concentrations of chlorinated hydrocarbons, more expensive chemical-resistant piping and an acid scrubber will be required for treatment of the removed vapors.

Contaminant Concentration

Initial contaminant concentrations in the airflow leaving the extraction wells will be quite high. However, the concentrations will rapidly decrease to a fairly low, but constant, level. The initial high concentration is caused by the removal of saturated vapor from the pore space. Because the concentration rapidly decreases, it is not economical to design the vapor treatment system for the expected maximum concentration, but rather for the average over the lifetime of the project.

6.1.3   Site Matrix Characteristics and Potential Effects on Treatment Cost or Performance

Table 6.1 lists the matrix characteristics that can have potential effects on the remediation cost and performance of RF-SVE treatment.

Table 6.1   Matrix Characteristics Affecting Treatment Cost or Performance

Matrix Characteristics	Potential Effects on Cost or Performance
Contaminant: DRO compounds and SVOCs	Contaminants with a lower vapor pressure will require additional time for effective volatilization; contaminant concentration will affect the determination of the target temperature and operation strategy
Soil classification	Clay content, mineralogy, and particle size distribution affects airflow through the soil and has an impact on its heating properties
pH	The pH of the soil can have an impact on the solubility of the contaminant and oxidation processes
Thermal and biooxidation process activity	Thermal and biooxidation could contribute to the removal of SVOCs; at high operating temperatures, 100–131°C, bioactivity is dormant, and thermal oxidation processes can contribute removal of SVOCs; low soil moisture content and increasing soil temperatures can adversely affect bioactivity; low moisture content can also adversely affect thermal oxidation of SVOCs; RF energy could also have an effect on the development of biomass
Air permeability	Air permeability affects the zone of influence of the extraction wells through the advection of heat and, therefore, affects the number of wells needed for the remediation effort and cost of operating the extraction wells
Effective porosity	Porosity affects the driving force for transferring the contaminants into the air-filled space, and removal by the SVE system
Moisture content	The moisture content of the soil will affect the airflow rates during operation of the SVE system; moisture will reduce the permeability of the soil and increase the energy requirements for heating the soil to the target temperature; when water is volatilized and removed by SVE, the condensate must be treated and disposed; moisture content is also critical in thermal and biooxidation processes
Soil electric properties	An understanding of the soil electric properties is important in the design of the RF antenna system and can allow anticipation of the changes in the heating mechanisms

Table 6.2   Operating Parameters: Measurement Procedures and Potential Effects on Treatment Cost or Performance

Operating Parameter	Measurement Procedures	Potential Effects on Cost or Performance
Contaminant concentration in the vapors	Measured using field instruments, air bag, or adsorbent tube samples analyzed at an analytical laboratory	The contaminant concentration in the vapor stream reflects the rate of contaminant removal from the subsurface
Airflow rate	Measured at the SVE blower using an anemometer or other airflow-measuring device	Airflow rate affects the rate of volatilization of water and contaminants during the operation of the SVE system and the introduction of oxygen created by the vacuum induced in the subsurface
Operating pressure or vacuum	Measured using a pressure or vacuum gauge	Operating pressure or vacuum affects the rate of volatilization of contaminants when transferred from the adsorbed to the vapor phase
Oxygen concentration	Measured in the airflow of the SVE system and in vapor monitoring points	The oxygen concentration can be important for all oxidation processes that are temperature dependent
Oxidation capacity	Measured by conducting *in situ* respirometry tests	Oxidation capacity will affect thermal and biooxidation of the contaminants
Temperature	Measured within the soil and in the airflow of the SVE system	The temperature will affect the volatilization of the contaminant and oxidation processes

6.1.4   Operating Parameters: Measurement Procedures and Potential Effects on Treatment Cost or Performance

Table 6.2 provides a list of the measurement procedures for the different operating parameters that are used during RF-SVE. The table also explains how these parameters affect the cost and performance.

6.1.5 Regulatory and Compliance Factors

Regulatory requirements or compliance issues are the general group of factors that arguably could have the most impact on treatment cost or performance. These impacts are highly dependent upon the remedial objectives, including remediation time. The requirements for proving or verifying performance or compliance with the remedial objectives must also be considered.

6.2 POTENTIAL APPLICATIONS

The following section discusses site conditions and characteristics that may be suitable for RF-SVE applications.

6.2.1 Site Characteristics Suitable for RF-SVE

Although the RF heating/SVE technology can be used in virtually any soil, the methodology could pose problems in soils containing cobbles because thermal "shadows" could develop behind the cobbles. Along with this reasoning, extremely heterogeneous soils or areas with large obstructions buried in the soil could also pose problems in terms of nonuniform airflow.

RF heating could become ineffective in extremely dry soils since some water is needed for RF heating to occur. The soil must have sufficient moisture so that free charges (for example, dissolved ions) can migrate under the influence of the applied electric field. Soil mineralogy can also be an important factor in RF heating. The presence of iron minerals, such as hematite and magnetite, will produce heat although the soil moisture content is low. Moist permeable soils have an increased ability to absorb RF energy. Conversely, very high moisture levels can limit the effectiveness of RF heating by shrinking the radiated heating pattern and increasing the amount of energy required to heat and remove moisture where high heating applications (>100°C) are needed. Consequently, the application of RF heating in the saturated zone, although feasible, may not be practical due to the amount of energy required to heat migrating groundwater. RF heating is most suitable for use in the vadose zone or straddling the capillary fringe.

Permeable soils have increased ability to absorb RF energy and provide for rapid remediation through enhanced SVE, especially for high temperature RF application. RF heating may also be effective for tight clay soils. RF heating can be used to generate steam *in situ* within the tight soils, thereby fracturing the soil so that contaminants can be subsequently extracted.

As with any heating technology, the proximity of containers or tanks that could rupture may be of concern. If tanks and drums are present, the site should be evaluated to determine if the RF-SVE technology should be employed. Metal tanks and drums can also absorb considerable RF energy.

The RF heating system can be applied to source or hot spots or expanded to encompass an entire contaminant plume. The RF generator units are mobile and compact, resulting in minimal site preparatory work and disruption, in most cases.

Optimal Site Requirements

The ideal soil is a moist loam soil with less than 10% dwb of stones or cobbles. The water table should be far below the depth of treatment. RF heating technique can be used above or below the water table; however, costs would be substantially higher in applications below the water table due to the increased depths and difficulty in obtaining the required heating. No large pockets of gravel should exist within the treatment zone. No explosives should be in the treatment zone.

Standard 60-Hz electrical power must be available for instrumentation and general power needs. A minimum of approximately a 200-A, 3-phase service, or the 1-phase equivalent, assuming off-gas treatment uses natural or propane as fuel, is required. Additional power and larger transformers may be required for multiple trailer unit use.

To perform this technology, the work area should be a minimum of 30 × 30 m (100 × 100 ft). The area also must be secure, free of utility lines within a minimum of 3 m (10 ft) both above and below the ground, and have a level treatment zone with vehicle access. Deeper RF heating wells may require more clearance.

6.2.2 Contaminants and Concentration Ranges

DROs or other organic compounds that are not treatable by conventional SVE at 10°C, yet have vapor pressures ≥1 mm Hg at 100 to 150°C, are primary candidates for the RF-SVE technology.

6.3 ECONOMIC FACTORS

The effectiveness of the RF-SVE system is highly dependent on the specific soil and chemical properties of the contaminated media. Furthermore, cost is related to the efficiency of the RF transmission lines and antennae to deliver heat as a function of the AC electrical power supplied to the RF units. Also, locations where electricity costs are high may find this technology cost limiting.

6.4 REGULATORY REQUIREMENTS

The specific cleanup goals mandated or negotiated with the regulatory agencies significantly affect treatment cost. The remediation time allotted by regulatory agencies for completion of remediation may be the most important factor in proposing the RF-SVE technology. Certain air-monitoring requirements or more stringent off-gas treatment standards could be cost limiting for some RF-SVE applications.

6.5 RF HEATING IN CONJUNCTION WITH OTHER TECHNOLOGIES

6.5.1 RF Heating with Bioventing

RF heating can be coupled with bioventing for increased *in situ* bioremediation. This application of RF heating is a low temperature application aimed at enhancing natural biological degradation of contaminants. Many contaminant-utilizing microbes are most effective at temperatures ranging from 15 to 40°C. If the ambient temperature is elevated to an optimum temperature for bioventing, the remedial efficiency can be increased. Heating the soil with RF energy could be especially effective in climates that have soil temperatures at 5°C or less.

6.5.2 RF Heating with SVE-Air Sparging

RF heating applied at the capillary fringe coupled with soil vapor extraction and air sparging can be used to improve the removal of dissolved phase contaminants. The remediation system can be further enhanced or adapted to include *in situ* groundwater sparging to improve the removal of dissolved phase contaminants.

6.5.3 RF to Heat NAPL or DNAPL

RF heating can be applied to heavier-end, nonaqueous phase liquid (NAPL) or dense NAPL (DNAPL) volumes to enhance free-phase product recovery systems. NAPL and DNAPL typically become less viscous or increase in solubility when heated. The RF heating can be specifically directed to form a focused pattern to enable greater NAPL flow toward installed recovery points.

Appendix A

Demonstration Mass Balance Calculations: Contaminant and Water Mass Calculations

Appendix A Listing and Description of Example Calculations, Figures, and Tables

Item	Title	Description
Calculations	Contaminant and water mass calculations	Example calculations for mass balances using the **Triangular Grouping Method**
Figure A1	Pre RF-SVE soil borings	Location of pre-demonstration boreholes for soil sample analyses
Figure A2	Post RF-SVE soil borings	Location of post-demonstration boreholes for soil sample analyses
Table A1	TPH Analyses for the *Sampled* Volume Pre-treatment TPH analyses Post-treatment TPH analyses,	TPH analyses of soil samples for the *sampled* volume at 0.61-meter depth intervals for each pre- and post-demonstration borehole Pre- and post-demonstration TPH borehole averages: 1. Average TPH — average analyses per borehole 2. Bottom 3.05 m — average TPH analyses for bottom 3.05 m of each borehole
Table A2	DRO Analyses for the *Sampled* Volume Pre-treatment DRO analyses Post-treatment DRO analyses	DRO analyses of soil samples for the *sampled* volume at 0.61-meter depth intervals for each pre- and post-demonstration borehole Pre- and post-demonstration TPH borehole averages: 1. Average DRO – average analyses per borehole 2. Bottom 3.05 m – average DRO analyses for bottom 3.05 m of each borehole
Table A3	DRO Analyses for the *Extended Design* Volume Pre-treatment DRO analyses, Post-treatment extended design DRO analyses	DRO analyses of soil samples for the *extended design* volume at 0.61-meter depth intervals for each pre- and post-demonstration borehole Pre- and post-demonstration DRO borehole averages: 1. Average DRO – average analyses per borehole
Table A4	Water Content Analyses for the *Sampled* Volume Pre-treatment water content analyses Post-treatment water content analyses	Water analyses of soil samples for the *sampled* volume at 0.61-meter depth intervals for each pre- and post-demonstration borehole Pre- and post-demonstration water content borehole averages: 1. Average H_2O content – average analyses per borehole
Table A5	Water Content Analysis for the *Extended Design* Volume Pre-treatment water content analyses Post-treatment water content analyses	Water analyses of soil samples for the *extended design* volume at 0.61 m depth intervals for each pre- and post-demonstration borehole Pre- and post-demonstration water content summaries: 1. Average H_2O content – average analyses per borehole
Table A6	Triangular grouping method (TPH mass balance)	TPH mass balance calculations using the triangular grouping method within the *sampled* volume (Parks, 1949)
Table A7	Triangular grouping method (DRO mass balance)	DRO mass balance calculations using the triangular grouping method within the *sampled* volume (Parks, 1949)
Table A8	Triangular grouping method (DRO mass balance)	DRO mass balance calculations using the triangular grouping method within the *extended design* volume (Parks, 1949)
Table A9	Triangular grouping method (water mass balance)	Water mass balance calculations using the triangular grouping method within the *sampled* volume (Parks, 1949)
Table A10	Triangular grouping method (water mass balance)	Water mass balance calculations using the triangular grouping method within the *extended design* volume (Parks, 1949)

CONTAMINANT AND WATER MASS CALCULATIONS
TRIANGULAR GROUPING METHOD

1. The layouts, the average TPH, DRO, and H_2O values, and the depths for the pre- and post-demonstration RF-SVE soil borings are shown in Figures A1 and A2.
2. The drill-hole patterns are different for the pre- and post-demonstration sampling. However, in the calculations, an attempt was made to obtain the same total values and orientations for the pre- and post-demonstration surface areas (and volumes). Therefore, the pre-demonstration samples from BH 01 and BH 12 and the post-demonstration samples from BH 32 were not used in the mass calculations.
3. The Triangular Grouping Method (Parks, 1949) for sampling calculations was used to estimate soil volumes and TPH, DRO, and H_2O masses. The calculation method is described in the following sections:

   A. Initially, the average values for the TPH, DRO, and H_2O concentrations are calculated using the following formula:

   AVERAGE = SUM[(Sample Concentration, mg/kg)(Sample Interval, m)]/[Well Depth, m]

   B. An example for a TPH sample from Well 8 is as follows:

   Sample TPH = (2,620 mg/kg)(0.610 m − 0.000 m)/(9.144 m) = 174.7 mg/kg.

   Repeat the calculation for each sample and sum the results.
   For Well 8, the average TPH value was 6343 mg/kg.

   C. Then triangles are drawn by connecting the well points that are defined in the "Area Points" column in Tables A6 through A10. The "Side Lengths" for each triangle (a, b, and c) are measured in centimeters (cm). The measurements are converted to their meter equivalents (Figures A1 and A2, Scale: 4 cm = 1 m).

   D. The area, volume, and TPH mass within each triangle are calculated as shown by the following example for Area 1 in Table A6. Area 1 is formed by connecting lines to BH 08, BH 09, and BH 15 in Figure A1. The formulas used in these calculations are as follows:

   $$S = (a + b + c)/2$$

   $$\text{Area, m}^2 = [S(S - a)(S - b)(S - c)]^{0.5}$$

   $$\text{Volume, m}^3 = (\text{Area, m}^2)(\text{Average Depth, m})$$

   E. Calculate the area for Area 1:

   $$a = 1.55 \text{ m, } b = 2.31 \text{ m, } c = 3.04 \text{ m}$$

   $$S = (1.55 + 2.31 + 3.04)/2 = 3.45 \text{ m}$$

   $$\text{Area} = [3.45(3.45 - 1.55)(3.45 - 2.31)(3.45 - 3.04)]^{0.5} = 1.75 \text{ m}^2$$

   F. Calculate average borehole depth and volume of influence:

BH	Depth (m)	TPH_{av} (mg/kg)
08	9.144	6343
09	6.100	3617
15	9.144	3275

   Average depth = (9.144 + 6.100 + 9.144)/3 = 8.128 m

Volume within area 1 = $(1.75 \text{ m}^2)(8.13 \text{ m}) = 14.26 \text{ m}^3$

G. Calculate the TPH_{av} (mg/kg) for the volume:

$TPH_{av} = [(6,343)(9.144) + (3,617)(6.100) + (3,275)(9.144)]/(9.144 + 6.100 + 9.144)$

$TPH_{av} = 4,511$ mg/kg

H. Calculate the TPH mass below Area 1:

TPH, kg = [(Soil Density, kg/m^3)(Volume, m^3)(TPH_{av}, mg/kg]/(10^6 mg/kg)

TPH, kg = $(1,780)(14.26)(4,511)/(10^6) = 114.5$ kg

I. The procedure is repeated for all areas in Figures A1 and A2.
J. The total TPH mass for the pre- and post-demonstration is the sums of the individual mass for each area.

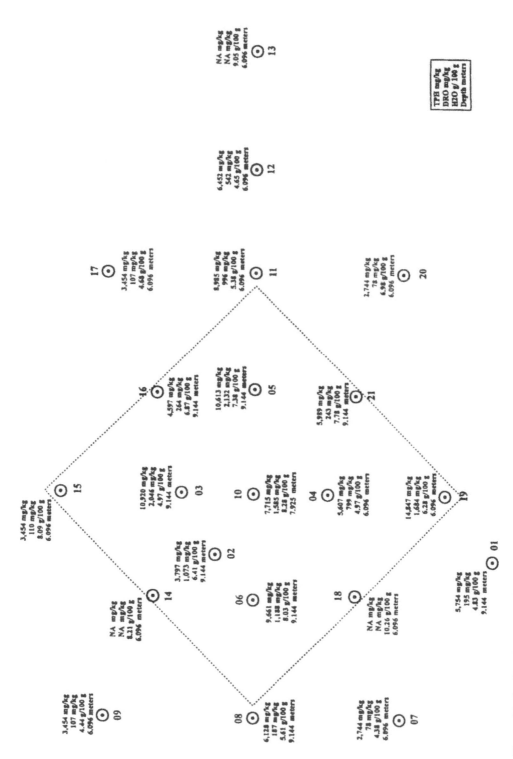

Figure A1 Pre-Demonstration RF-SVE soil borings. Scale: 4 cm = 1 meter.

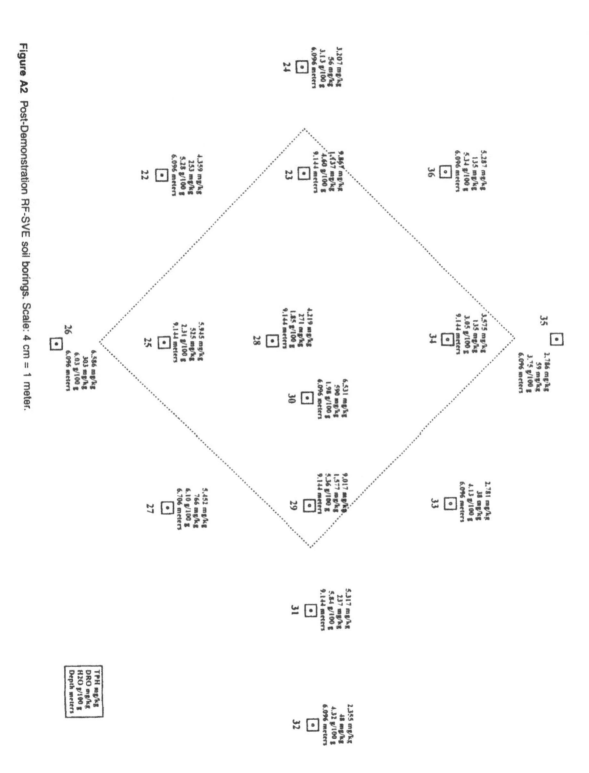

Figure A2 Post-Demonstration RF-SVE soil borings. Scale: 4 cm = 1 meter.

Table A1 TPH Analyses for the *Sampled* Volume

Pre-treatment TPH Analyses

TPH (mg/kg-dry weight)

Depth m	m	1	2	3	4	5	6	7	8	9	10	11	12	15	16	19	21	Total
0.000	0.610	8,170	1,320	5,920	10,200	12,400	24,900	1,690	2,620	2,400	2,020	4,690	1,220	2,780	2,580	5,170	5,880	
0.610	1.219	2,620	1,130	3,560	3,100	11,700	3,550	2,050	1,930	3,710	2,030	5,780	7,790	3,460	3,290	4,100	7,010	
1.219	1.829	2,830	4,870	4,540	13,300	13,700	3,410	3,030	6,610	3,600	2,810	4,400	2,800	8,230	2,770	9,540	7,930	
1.829	2.438	2,900	2,850	10,100	8,690	16,100	17,900	3,330	7,240	1,330	9,790	5,430	2,660	4,360	6,270	16,000	8,830	
2.438	3.048	15,300	4,330	15,900	11,900	9,680	13,800	3,820	9,480	2,550	15,900	17,900	5,930	4,030	5,060	20,500	10,800	
3.048	3.658	19,100	2,820	11,800	9,630	14,900	16,600	3,000	9,480	2,750	13,500	13,100	3,250	2,430	4,600	25,700	7,160	
3.658	4.267	21,400	6,560	11,900	7,640	16,200	16,000	2,480	11,600	5,100	6,630	18,800	9,340	2,920	1,650	29,900	5,870	
4.267	4.877	3,300	1,130	17,400	6,090	17,600	14,200	3,080	9,570	7,550	17,200	6,620	19,320	4,080	4,880	21,100	7,120	
4.877	5.486	4,800	1,920	21,000	8,170	9,590	13,700	3,000	7,170	3,590	4,630	6,730	10,250	2,750	12,200	14,500	5,150	
5.486	6.096	1,960	8,290	19,800	8,270	11,400	13,700	3,460	5,190	3,590	7,700	13,800	7,800	3,630	12,500	9,500	4,290	
6.096	6.706	668	16,300	14,900	660	8,570	3,020		7,880		6,400	15,300		971	3,070		5,040	
6.706	7.315	688	6,310	16,400	640	12,000	2,590		3,870		4,120	9,290		2,680	3,590		3,730	
7.315	7.925	530	2,980	18,300	540	8,710	3,980		5,310		13,300	11,000		3,610	7,190		4,170	
7.925	8.534	881	1,190	3,190	1,170	2,860	1,090		4,240			9,720		1,640	4,530		4,940	
8.534	9.144	1,160	1,280	6,930	420	3,230	8,220		2,960			4,050		1,560	5,310		4,250	
Average TPH		5,754	4,219	12,109	6,028	11,243	10,444	2,894	6,343	3,617	8,156	9,774	7,036	3,275	5,299	15,601	6,145	N.A.
Check values		5,754	4,219	12,109	6,028	11,243	10,444	2,894	6,343	3,617	8,156	9,774	7,036	3,275	5,299	15,601	6,145	N.A.
Bottom 3.05 m		785	5,612	11,944	686	7,074	3,780	0	4,852	0	7,940	9,872	0	2,092	4,738	0	4,426	5,317

Post-treatment TPH Analyses

Borehole Number

m	m	22	23	24	25	26	27	28	29	30	31	32	33	34	35	36	Total
0.000	0.610	2,020	1,100	3,520	1,340	4,740	2,940	1,300	2,250	1,490	1,490	1,220	4,470	2,130	1,790	4,380	
0.610	1.219	1,230	4,390	3,390	3,280	4,340	2,480	5,800	3,070	3,120	3,190	1,170	2,240	2,660	1,590	3,130	
1.219	1.829	3,440	3,190	2,890	4,160	3,700	2,020	4,630	3,180	1,920	2,400	1,090	2,930	3,130	2,530	3,180	
1.829	2.438	4,130	5,480	3,300	6,220	6,590	2,150	6,590	3,300	3,520	3,500	1,770	2,710	3,460	3,300	5,160	
2.438	3.048	2,620	6,780	2,470	2,760	10,100	2,560	5,090	6,330	11,500	4,950	1,860	2,520	3,070	4,110	2,910	
3.048	3.658	3,160	8,580	6,120	2,660	10,000	4,560	5,090	7,060	11,000	5,200	2,550	3,060	2,930	3,710	4,480	
3.658	4.267	12,690	13,800	3,540	3,410	14,700	10,700	4,250	11,400	11,000	8,680	1,590	1,790	*3,400*	2,540	4,070	
4.267	4.877	6,900	17,000	1,670	4,010	5,000	12,500	*5,470*	11,600	*10,265*	7,250	4,500	2,610	3,870	3,370	12,900	
4.877	5.486	5,440	15,200	3,210	4,310	4,730	5,910	6,690	15,200	9,530	10,820	5,840	3,520	10,300	2,960	10,700	
5.486	6.096	2,970	10,700	2,620	5,010	1,260	13,100	2,420	15,600	9,120	9,850	4,110	6,350	5,830	2,300	3,600	
6.096	6.706		11,300		15,400		11,900	3,960	13,200		7,350			4,810			
6.706	7.315		7,320		11,400			5,070	14,100		3,830			2,590			
7.315	7.925		14,900		8,550			4,050	12,300		7,260			2,800			
7.925	8.534		14,600		13,800			1,570	14,200		6,670			3,980			
8.534	9.144		22,400		5,920			1,760	16,100		5,200			2,540			
Average TPH		4,460	10,449	3,273	6,149	6,516	6,438	4,249	9,926	7,247	5,843	2,570	3,220	3,833	2,820	5,451	N.A.
Check values		4,460	10,449	3,273	6,149	6,516	6,438	4,249	9,926	7,247	5,843	2,570	3,220	3,833	2,820	5,451	N.A.
Bottom 3.05 m		0	14,104	0	11,014	0	11,900	3,282	13,980	0	6,062	0	0	3,344	0	0	9,098

Table A2 DRO Analyses for the Sampled Volume

Pre-treatment DRO Analyses

DRO (mg/kg-dry weight) Borehole Number

Depth (m)	1	2	3	4	5	6	7	8	9	10	11	12	15	16	19	21	Total
0.000	131	20	114	301	20	20	96	115	41	20	97	47	74	20	20	20	
0.610	20	813	20	150	1390	20	20	222	20	68	55	803	20	20	100	20	
1.219	20	940	124	1480	1860	20	92	20	20	192	20	175	527	20	557	77	
1.829	108	808	1530	1020	2990	2340	20	53	20	1480	341	119	59	46	1360	146	
2.438	2090	1340	5060	1970	2070	1990	20	205	76	2970	1780	249	121	51	2840	749	
3.048	150	1140	1910	1520	3340	2780	79	429	65	942	1440	148	86	20	3760	88	
3.658	153	1220	2220	1840	3690	3200	153	497	143	1030	1720	556	72	20	3960	401	
4.267	20	3950	4040	1600	4250	3220	96	267	423	3670	361	139	139	42	2380	1210	
4.877	20	2180	3620	1890	4210	3000	89	161	151	1830	365	2250	63	2310	1750	643	
5.486	110	2060	4660	3050	3270	2930	85	197	205	1610	2590	963	82	3150	811	74	
6.096	20	3180	4160	20	3250	182		20		1880	2720	543	61	126		20	
6.706	20	334	4950	20	2670	279		20		3350	2110		52	63		20	
7.315	20	20	2650	20	2090	625		143		3060	2450		164	772		94	
7.925	20	20	165	20	20	20		207			1010		76	197		20	
8.534	20	20	20	20	20	20		332			358		20	136		20	
Average DRO	195	1203	2350	995	2343	1376	75	193	116	1700	1161	585	108	466	1754	240	N.A.
Check values	195	1203	2350	995	2343	1376	75	193	116	1700	1161	585	108	466	1754	240	N.A.
Bottom 3.05 m	20	715	2389	20	1610	225	0	144	0	2763	1730	0	75	259	0	35	832

Post-treatment DRO Analyses

Borehole Number

Depth (m)	22	23	24	25	26	27	28	29	30	31	32	33	34	35	36	Total
0.000	20	20	38	20	20	20	20	20	20	20	20	20	20	20	48	
0.610	32	20	20	43	20	20	201	56	42	86	20	20	20	20	20	
1.219	54	20	43	76	49	62	166	74	20	55	20	20	20	20	39	
1.829	62	134	61	187	108	20	507	130	248	51	20	20	60	52	87	
2.438	51	507	20	79	398	86	551	1530	1460	96	20	20	36	81	40	
3.048	143	560	203	223	662	399	433	1340	1250	108	20	44	51	65	59	
3.658	1710	1640	20	154	1520	1650	196	1730	1150	327	20	20	442	55	69	
4.267	266	2630	20	279	85	3130	149	1550	916	603	20	45	832	114	584	
4.877	84	2080	20	267	59	632	101	2240	682	1090	208	65	42	48	291	
5.486	89	1240	105	492	20	2650	110	2660	716	519	55	207	139	65	93	
6.096		1280		1520		2300	121	2110		383			109			
6.706		830		1600			956	2600		92			43			
7.315		2370		945			520	3040		187			69			
7.925		1870		1730			20	3390		174			128			
8.534		2990		636			20	3740		179			41			
Average DRO	251	1213	55	550	294	997	271	1747	650	265	42	48	137	54	133	N.A.
Check values	251	1213	55	550	294	997	243	1747	650	265	42	48	137	54	133	N.A.
Bottom 3.05 m	0	1868	0	1286	0	2300	327	2976	0	203	0	0	78	0	0	1291

Table A3 DRO Analyses for the Extended Design Volume

Pre-treatment Extended Design-Volume DRO Analyses

Depth									DRO (mg/kg-dry weight) Borehole Number										
m	m	1	2	3	4	5	6	7	8	9	10	11	12	15	16	19	21	Total
3.048	3.658	150	1140	1910	1520	3340	2780	79	429	65	942	1440	148	86	20	3760	88	
3.658	4.267	153	1220	2220	1840	3690	3200	153	497	143	1030	1720	556	72	20	3960	401	
4.267	4.877	20	3950	4040	1600	4250	3220	96	267	423	3670	361	2250	139	42	2380	1210	
4.877	5.486	20	2180	3620	1890	4210	3000	89	161	151	1830	365	963	63	2310	1750	643	
5.486	6.096	110	2060	4660	3050	3270	2930	85	197	205	1610	2590	543	82	3150	811	74	
Average DRO		91	2110	3290	1980	3752	3026	100	310	197	1816	1295	892	88	1108	2532	483	N.A.
Check values		91	2110	3290	1980	3752	3026	100	310	197	1816	1295	892	88	1108	2532	483	N.A.

Post-treatment Extended Design-Volume DRO Analyses

Depth									DRO (mg/kg-dry weight) Borehole Number								
m	m	22	23	24	25	26	27	28	29	30	31	32	33	34	35	36	Total
3.048	3.658	143	560	203	223	662	399	433	1340	1250	108	20	44	51	65	59	
3.658	4.267	1710	1640	20	154	1520	1650	196	1730	1150	327	20	20	442	55	69	
4.267	4.877	266	2630	20	279	85	3130	149	1550	916	603	20	45	832	114	584	
4.877	5.486	84	2080	20	267	59	632	101	2240	682	1090	208	65	42	48	291	
5.486	6.096	89	1240	105	492	20	2650	110	2660	716	519	55	207	139	65	93	
Average DRO		458	1630	74	283	469	1692	198	1904	943	529	65	76	301	69	219	N.A.
Check values		458	1630	74	283	469	1692	198	1904	943	529	65	76	301	69	219	N.A.

Table A4 Water Content Analyses for the *Sampled* Volume

Pre-treatment Water Content Analyses

H_2O (g/100 g-dry weight) Borehole Number

Depth m	m	1	2	3	4	5	6	7	8	9	10	11	12	15	16	19	21	Total
0.000	0.610	7.0	8.2	1.6	6.6	6.9	6.2	2.3	8.2	3.4	7.2	8.5	4.1	5.2	3.7	6.0	4.1	
0.610	1.219	1.9	7.8	2.5	6.5	5.4	7.1	3.3	7.8	6.2	8.7	3.3	5.9	3.9	4.6	6.6	8.3	
1.219	1.829	3.1	6.8	5.2	4.8	6.2	9.0	7.9	3.5	5.4	12.3	4.8	4.4	14.5	11.3	8.5	6.2	
1.829	2.438	6.7	7.0	5.9	4.1	7.4	13.1	4.3	4.1	3.7	11.8	4.4	4.2	8.4	13.1	5.1	12.4	
2.438	3.048	8.9	7.6	7.6	6.6	6.4	6.3	6.2	4.0	2.4	8.4	8.4	3.8	6.2	6.6	7.4	8.6	
3.048	3.658	5.8	6.1	4.4	5.8	6.7	5.6	5.5	5.0	6.2	7.1	7.1	7.2	12.2	6.4	8.4	8.7	
3.658	4.267	6.4	6.8	5.6	8.8	9.6	6.8	5.7	4.5	4.9	7.4	6.2	4.1	7.9	6.4	6.8	8.3	
4.267	4.877	3.6	7.3	7.4	8.6	7.3	7.8	2.5	7.0	4.7	10.2	4.1	6.3	14.5	6.2	6.6	8.6	
4.877	5.486	7.2	5.8	5.5	6.8	13.7	7.3	3.3	6.4	4.7	9.0	4.2	3.7	6.9	8.9	4.6	8.6	
5.486	6.096	2.8	5.8	4.6	6.6	4.6	6.2	3.3	6.8	4.5	6.9	8.0	4.8	10.4	7.4	6.7	9.4	
6.096	6.706	2.7	7.3	6.1	5.9	6.3	9.3		6.5		7.7	4.7		15.8	6.1		10.9	
6.706	7.315	5.4	7.7	6.8	8.6	6.7	8.1		8.5		8.2	7.0		10.7	7.3		8.9	
7.315	7.925	3.9	3.8	5.6	6.9	9.0	18.6		6.3		6.8	6.1		5.6	7.7		5.0	
7.925	8.534	3.6	6.2	4.6	6.4	7.5	7.7		3.6			5.5		3.5	6.6		8.7	
8.534	9.144	3.5	4.9	3.0	10.5	8.8	4.7		5.9			3.6		3.2	5.4		6.6	
Average %H₂O		4.83	6.61	5.09	6.90	7.50	8.25	4.43	5.87	4.61	8.59	5.73	4.85	8.59	7.18	6.67	8.22	N.A.
Check values		4.83	6.61	5.09	6.90	7.50	8.25	4.43	5.87	4.61	8.59	5.73	4.85	8.59	7.18	6.67	8.22	N.A.

m	m	1-12	13	14	17	18	20	Total
0.000	0.610		13.0	5.2	4.0	3.8	5.3	
0.610	1.219		9.0	4.7	6.3	11.4	4.0	
1.219	1.829		7.3	6.9	7.5	19.0	6.2	
1.829	2.438		14.8	9.1	3.5	8.2	5.0	
2.438	3.048		9.0	8.8	4.0	11.3	8.3	
3.048	3.658		3.0	9.1	2.4	9.2	9.1	
3.658	4.267		7.5	14.0	2.4	15.4	5.3	
4.267	4.877		9.6	8.1	2.9	10.0	8.6	
4.877	5.486		5.1	10.2	4.4	11.0	11.0	
5.486	6.096		11.8	11.1	3.6	9.9	7.0	
Average %H₂O		N.A.	9.01	8.72	4.10	10.92	6.98	N.A.
Check values		N.A.	9.01	8.72	4.10	10.92	6.98	N.A.

Table A4 Water Content Analyses for the Sampled Volume (continued)

Post-treatment Water Content Analyses

Depth		H₂O (g/100 g-dry weight) Borehole Number															
m	m	22	23	24	25	26	27	28	29	30	31	32	33	34	35	36	Total
0.000	0.610	4.3	5.5	3.1	4.8	7.2	5.4	2.3	6.1	3.5	4.2	5.3	5.8	4.6	4.2	6.4	
0.610	1.219	5.1	5.6	3.3	2.9	7.9	4.5	0.9	6.1	1.4	7.0	5.9	4.5	2.2	2.9	5.7	
1.219	1.829	6.7	7.4	4.1	2.9	4.8	8.3	2.1	7.0	3.3	5.0	4.6	4.6	3.0	5.8	1.9	
1.829	2.438	6.1	11.0	3.0	1.7	7.6	7.2	0.9	8.2	2.0	4.9	3.3	5.1	2.9	6.8	8.4	
2.438	3.048	3.1	7.2	2.5	3.2	4.8	7.9	1.0	8.3	1.7	5.1	4.4	3.5	2.3	4.5	3.8	
3.048	3.658	5.9	3.3	2.9	0.5	7.4	6.7	0.8	9.5	1.9	6.9	3.5	7.6	2.2	3.7	5.9	
3.658	4.267	5.8	2.7	2.9	0.9	6.8	6.2	0.5	5.8	1.6	5.8	4.0	1.3	1.4	2.8	5.5	
4.267	4.877	5.3	2.5	4.1	0.6	5.4	5.4	0.4	1.4	1.1	5.9	5.2	1.4	0.5	2.5	6.8	
4.877	5.486	7.7	1.5	2.6	0.6	5.6	8.8	0.3	2.3	0.5	6.4	4.2	4.7	0.6	1.5	6.2	
5.486	6.096	4.2	2.0	3.0	0.8	5.1	3.2	0.4	2.2	0.6	5.1	3.7	4.4	0.5	3.8	3.6	
6.096	6.706		1.1		0.4		3.9	0.4	1.4		6.3			1.2			
6.706	7.315		1.2		1.5			1.2	3.5		6.9			5.8			
7.315	7.925		5.4		2.0			4.4	6.2		7.3			6.3			
7.925	8.534		5.9		5.7			4.1	6.0		7.1			6.7			
8.534	9.144		5.9		4.2			5.5	5.8		6.0			3.3			
Average %H₂O		5.42	4.55	3.15	2.18	6.26	6.14	1.68	5.32	1.76	5.99	4.41	4.29	2.90	3.85	5.42	N.A.
Check values		5.42	4.55	3.15	2.18	6.26	6.14	1.68	5.32	1.76	5.99	4.41	4.29	2.90	3.85	5.42	N.A.

Table A5 Water Content Analyses for the *Extended-Design* Volume

Pre-treatment Extended Design-Volume Water Analyses

H_2O (g/100 g-dry weight)
Borehole Number

Depth																		
m	m	1	2	3	4	5	6	7	8	9	10	11	12	15	16	19	21	Total
3.048	3.658	5.8	6.1	4.4	5.8	6.7	5.6	5.5	5.0	6.2	7.1	7.1	7.2	12.2	6.4	8.4	8.7	
3.658	4.267	6.4	6.8	5.6	8.8	9.6	6.8	5.7	4.5	4.9	7.4	6.2	4.1	7.9	6.4	6.8	8.3	
4.267	4.877	3.6	7.3	7.4	8.6	7.3	7.8	2.5	7.0	4.7	10.2	4.1	6.3	14.5	6.2	6.6	8.6	
4.877	5.486	7.2	5.8	5.5	6.8	13.7	7.3	3.3	6.4	4.7	9.0	4.2	3.7	6.9	8.9	4.6	8.6	
5.486	6.096	2.8	5.8	4.6	6.6	4.6	6.2	3.3	6.8	4.5	6.9	8.0	4.8	10.4	7.4	6.7	9.4	
Average %H_2O		5.16	6.36	5.50	7.32	8.38	6.74	4.06	5.94	5.00	8.12	5.92	5.22	10.38	7.06	6.62	8.72	N.A.
Check values		5.16	6.36	5.50	7.32	8.38	6.74	4.06	5.94	5.00	8.12	5.92	5.22	10.38	7.06	6.62	8.72	N.A.

m	m	1-21	13	14	17	18	20	Total
3.048	3.658		3.0	9.1	2.4	9.2	9.1	
3.658	4.267		7.5	14.0	2.4	15.4	5.3	
4.267	4.877		9.6	8.1	2.9	10.0	8.6	
4.877	5.486		5.1	10.2	4.4	11.0	11.0	
5.486	6.096		11.8	11.1	3.6	9.9	7.0	
Average %H_2O	N.A.		7.40	10.50	3.14	11.10	8.20	N.A.
Check values	N.A.		7.40	10.50	3.14	11.10	8.20	N.A.

Post-treatment Extended Design-Volume Water Analyses

Borehole Number

m	m	22	23	24	25	26	27	28	29	30	31	32	33	34	35	36	Total
3.048	3.658	5.9	3.3	2.9	0.5	7.4	6.7	0.8	9.5	1.9	6.9	3.5	7.6	2.2	3.7	5.9	
3.658	4.267	5.8	2.7	2.9	0.9	6.8	6.2	0.5	5.8	1.6	5.8	4.0	1.3	1.4	2.8	5.5	
4.267	4.877	5.3	2.5	4.1	0.6	5.4	5.4	0.5	1.4	1.1	5.9	5.2	1.4	0.5	2.5	6.8	
4.877	5.486	7.7	1.5	2.6	0.6	5.6	8.8	0.4	2.3	0.5	6.4	4.2	4.7	0.6	1.5	6.2	
5.486	6.096	4.2	2.0	3.0	0.8	5.1	3.2	0.3	2.2	0.6	5.1	3.7	4.4	0.5	3.8	3.6	
Average %H_2O		5.78	2.40	3.10	0.68	6.06	6.06	0.49	4.24	1.13	6.02	4.12	3.88	1.03	2.86	5.60	N.A.
Check values		5.78	2.40	3.10	0.68	6.06	6.06	0.49	4.24	1.13	6.02	4.12	3.88	1.03	2.86	5.60	

Table A6   Triangular Grouping Method (TPH Mass Balance)

Area No	Area Points	Side Lengths (cm)			Conversion Lengths (m)			S (m)	Calculated Area (m²)	Drill-Hole Depth (m)	Vol (m³)	TPH$_w$ Analysis mg/kg	TPH kg
		a	b	c	a	b	c						
					Pre-RF-SVE TPH Balance								
1	08-09-15	6.200	9.250	12.150	1.550	2.313	3.038	3.450	1.754	8.128	14.255	4,511	114.5
2	08-15-02	12.150	6.850	6.800	3.038	1.713	1.700	3.225	1.181	9.144	10.799	4,612	88.7
3	08-02-06	6.800	2.400	4.850	1.700	0.600	1.213	1.756	0.249	9.144	2.279	7,002	28.4
4	08-06-04	4.850	5.250	9.600	1.213	1.313	2.400	2.463	0.470	9.144	4.301	7,605	58.2
5	08-04-19	9.600	4.850	12.100	2.400	1.213	3.025	3.319	1.374	8.128	11.164	8,540	169.7
6	08-19-07	12.100	9.300	6.100	3.025	2.325	1.525	3.438	1.737	7.112	12.353	8,003	176.0
7	02-15-03	6.850	4.950	2.850	1.713	1.238	0.713	1.831	0.380	9.144	3.475	6,534	40.4
8	02-03-10	2.850	3.000	2.950	0.713	0.750	0.738	1.100	0.233	8.738	2.032	8,162	29.5
9	06-02-10	2.400	2.950	4.400	0.600	0.738	1.100	1.219	0.208	8.738	1.814	10,333	33.4
10	06-10-04	4.400	3.100	5.250	1.100	0.775	1.313	1.594	0.426	8.738	3.719	8,212	54.4
11	03-15-16	4.950	5.700	4.400	1.238	1.425	1.100	1.881	0.657	9.144	6.008	6,895	73.7
12	03-16-05	4.400	4.100	5.250	1.100	1.025	1.313	1.719	0.547	9.144	5.006	9,550	85.1
13	10-03-05	3.000	5.250	4.350	0.750	1.313	1.088	1.575	0.408	8.738	3.563	6,690	42.4
14	04-10-05	3.100	4.350	5.300	0.775	1.088	1.325	1.594	0.421	8.738	3.68	8,490	55.6
15	04-05-21	5.300	4.150	4.250	1.325	1.038	1.063	1.713	0.540	9.144	4.93	7,805	68.5
16	19-04-21	4.850	4.250	5.550	1.213	1.063	1.388	1.831	0.622	8.128	5.05	8,465	76.1
17	16-15-17	5.750	9.300	5.350	1.438	2.325	1.338	2.550	0.880	8.128	7.15	4,079	51.9
18	16-17-11	5.350	6.100	6.450	1.338	1.525	1.613	2.238	0.947	8.128	7.70	6,516	89.3
19	05-16-11	4.100	6.450	4.850	1.025	1.613	1.213	1.925	0.621	9.144	5.68	8,772	88.7
20	21-05-11	4.150	4.850	6.550	1.038	1.213	1.638	1.944	0.628	9.144	5.74	9,054	92.6
21	21-11-20	6.550	6.150	5.450	1.638	1.538	1.363	2.269	0.974	8.128	7.92	6,656	93.8
22	19-21-20	5.600	5.450	9.350	1.400	1.363	2.338	2.550	0.860	7.112	6.12	7,875	85.8
Totals/Av									16.116	8.361	134.74	7074	1697

Pre - Post TPH (kg)   300.8

Post-RF-SVE TPH Balance

#	ID												
1	24-36-23	7.350	5.950	4.350	1.838	1.488	1.088	2.206	0.809	7.112	5.752	6971	71.4
2	24-23-22	4.350	5.900	7.350	1.088	1.475	1.838	2.200	0.802	7.112	5.704	6688	67.9
3	36-35-34	8.150	4.450	6.850	2.038	1.113	1.713	2.431	0.953	7.112	6.775	4006	48.3
4	23-36-34	5.950	6.850	9.100	1.488	1.713	2.275	2.738	1.274	8.128	10.352	6719	123.8
5	23-34-28	9.100	7.350	7.000	2.275	1.838	1.750	2.931	1.576	9.144	14.415	6177	158.5
6	22-23-28	5.900	7.000	8.200	1.475	1.750	2.050	2.638	1.264	8.128	10.277	6627	121.2
7	22-28-25	8.200	4.500	6.850	2.050	1.125	1.713	2.444	0.963	8.128	7.830	5014	69.9
8	22-25-26	6.850	4.450	8.200	1.713	1.113	2.050	2.438	0.953	7.112	6.774	5771	69.6
9	34-35-33	4.450	8.150	6.850	1.113	2.038	1.713	2.431	0.953	7.112	6.775	3369	40.6
10	34-33-30	6.850	7.450	6.400	1.713	1.863	1.600	2.588	1.273	7.112	9.055	4633	74.7
11	28-34-30	7.350	6.400	2.300	1.838	1.600	0.575	2.006	0.444	8.128	3.606	4843	31.1
12	28-30-25	2.300	6.400	4.500	0.575	1.600	1.125	1.650	0.216	8.128	1.754	5711	17.8
13	25-30-27	6.400	7.450	6.850	1.600	1.863	1.713	2.588	1.273	7.315	9.313	6542	108.5
14	26-25-27	4.450	6.850	8.150	1.113	1.713	2.038	2.431	0.953	7.315	6.968	6339	78.6
15	30-33-29	7.450	5.900	4.500	1.863	1.475	1.125	2.231	0.830	7.112	5.901	7244	76.1
16	30-29-27	4.500	5.950	7.450	1.125	1.488	1.863	2.238	0.837	7.315	6.121	8116	88.4
17	29-33-31	5.900	7.350	4.300	1.475	1.838	1.075	2.194	0.793	8.128	6.443	6718	77.1
18	27-29-31	5.950	4.300	9.140	1.488	1.075	2.285	2.424	0.652	8.331	5.429	7496	72.4
	Totals/Av								16.815	7.686	129.24	6068	1396

Table A7   Triangular Grouping Method (DRO Mass Balance)

Area No	Area Points	Side Lengths (cm) a	b	c	Conversion Lengths (m) a	b	c	S (m)	Calculated Area (m²)	Drill-Hole Depth (m)	Vol (m³)	DRO_av Analysis (mg/kg)	DRO (kg)
						Pre-RF-SVE DRO Balance							
1	08-09-15	6.200	9.250	12.150	1.550	2.313	3.038	3.450	1.754	8.128	14.255	142	3.6
2	08-15-02	12.150	6.850	6.800	3.038	1.713	1.700	3.225	1.181	9.144	10.799	501	9.6
3	08-02-06	6.800	2.400	4.850	1.700	0.600	1.213	1.756	0.249	9.144	2.279	924	3.7
4	08-06-04	4.850	5.250	9.600	1.213	1.313	2.400	2.463	0.470	9.144	4.301	855	6.5
5	08-04-19	9.600	4.850	12.100	2.400	1.213	3.025	3.319	1.374	8.128	11.164	884	17.6
6	08-19-07	12.100	9.300	6.100	3.025	2.325	1.525	3.438	1.737	7.112	12.353	605	13.3
7	02-15-03	6.850	4.950	2.850	1.713	1.238	0.713	1.831	0.380	9.144	3.475	1220	7.5
8	02-03-10	2.850	3.000	2.950	0.713	0.750	0.738	1.100	0.233	8.738	2.032	1753	6.3
9	06-02-10	2.400	2.950	4.400	0.600	0.738	1.100	1.219	0.208	8.738	1.814	1814	5.9
10	06-10-04	4.400	3.100	5.250	1.100	0.775	1.313	1.594	0.426	8.738	3.719	1341	8.9
11	03-15-16	4.950	5.700	4.400	1.238	1.425	1.100	1.881	0.657	9.144	6.008	974	10.4
12	03-16-05	4.400	4.100	5.250	1.100	1.025	1.313	1.719	0.547	9.144	5.006	1719	15.3
13	10-03-05	3.000	5.250	4.350	0.750	1.313	1.088	1.575	0.408	8.738	3.563	2151	13.6
14	04-10-05	3.100	4.350	5.300	0.775	1.088	1.325	1.594	0.421	8.738	3.68	1678	11.0
15	04-05-21	5.300	4.150	4.250	1.325	1.038	1.063	1.713	0.540	9.144	4.93	1193	10.5
16	19-04-21	4.850	4.250	5.550	1.213	1.063	1.388	1.831	0.622	8.128	5.05	902	8.1
17	16-15-17	5.750	9.300	5.350	1.438	2.325	1.338	2.550	0.880	8.128	7.15	1079	13.7
18	16-17-11	5.350	6.100	6.450	1.338	1.525	1.613	2.238	0.947	8.128	7.70	1474	20.2
19	05-16-11	4.100	6.450	4.850	1.025	1.613	1.213	1.925	0.621	9.144	5.68	1323	13.4
20	21-05-11	4.150	4.850	6.550	1.038	1.213	1.638	1.944	0.628	9.144	5.74	1248	12.8
21	21-11-20	6.550	6.150	5.450	1.638	1.538	1.363	2.269	0.974	8.128	7.92	1211	17.1
22	19-21-20	5.600	5.450	9.350	1.400	1.363	2.338	2.550	0.860	7.112	6.12	1388	15.1
Totals/Av									16.116	8.361	134.74	1018	244.2

Pre - Post DRO (kg)   115.9

Post-RF-SVE DRO Balance

1	24-36-23	7.350	5.950	4.350	1.838	1.488	1.088	2.206	0.809	7.112	5.752	573	5.9
2	24-23-22	4.350	5.900	7.350	1.088	1.475	1.838	2.200	0.802	7.112	5.704	607	6.2
3	36-35-34	8.150	4.450	6.850	2.038	1.113	1.713	2.431	0.953	7.112	6.775	112	1.4
4	23-36-34	5.950	6.850	9.100	1.488	1.713	2.275	2.738	1.274	8.128	10.352	539	9.9
5	23-34-28	9.100	7.350	7.000	2.275	1.838	1.750	2.931	1.576	9.144	14.415	540	13.9
6	22-23-28	5.900	7.000	8.200	1.475	1.750	2.050	2.638	1.264	8.128	10.277	619	11.3
7	22-28-25	8.200	4.500	6.850	2.050	1.125	1.713	2.444	0.963	8.128	7.830	371	5.2
8	22-25-26	6.850	4.450	8.200	1.713	1.113	2.050	2.438	0.953	7.112	6.774	392	4.7
9	34-35-33	4.450	8.150	6.850	1.113	2.038	1.713	2.431	0.953	7.112	6.775	88	1.1
10	34-33-30	6.850	7.450	6.400	1.713	1.863	1.600	2.588	1.273	7.112	9.055	258	4.2
11	28-34-30	7.350	6.400	2.300	1.838	1.600	0.575	2.006	0.444	8.128	3.606	316	2.0
12	28-30-25	2.300	6.400	4.500	0.575	1.600	1.125	1.650	0.216	8.128	1.754	471	1.5
13	25-30-27	6.400	7.450	6.850	1.600	1.863	1.713	2.588	1.273	7.315	9.313	715	11.8
14	26-25-27	4.450	6.850	8.150	1.113	1.713	2.038	2.431	0.953	7.315	6.968	616	7.6
15	30-33-29	7.450	5.900	4.500	1.863	1.475	1.125	2.231	0.830	7.112	5.901	948	10.0
16	30-29-27	4.500	5.950	7.450	1.125	1.488	1.863	2.238	0.837	7.315	6.121	1213	13.2
17	29-33-31	5.900	7.350	4.300	1.475	1.838	1.075	2.194	0.793	8.128	6.443	767	8.8
18	27-29-31	5.950	4.300	9.140	1.488	1.075	2.285	2.424	0.652	8.331	5.429	1004	9.7
Totals/Av									16.815	7.686	129.24	558	128.3

Table A8 Triangular Grouping Method (DRO Mass Balance)

Area No	Area Points	Side Lengths (cm) a	b	c	Conversion Lengths (m) a	b	c	S (m)	Calculated Area (m²)	Drill-Hole Depth (m)	Vol (m³)	DRO_av Analysis (mg/kg)	DRO (kg)
					Pre-RF-SVE Extended Design-Volume DRO Balance								
1	08-09-15	6.200	9.250	12.150	1.550	2.313	3.038	3.450	1.754	3.048	5.346	199	1.9
2	08-15-02	12.150	6.850	6.800	3.038	1.713	1.700	3.225	1.181	3.048	3.600	836	5.4
3	08-02-06	6.800	2.400	4.850	1.700	0.600	1.213	1.756	0.249	3.048	0.760	1815	2.5
4	08-06-04	4.850	5.250	9.600	1.213	1.313	2.400	2.463	0.470	3.048	1.434	1772	4.5
5	08-04-19	9.600	4.850	12.100	2.400	1.213	3.025	3.319	1.374	3.048	4.186	1607	12.0
6	08-19-07	12.100	9.300	6.100	3.025	2.325	1.525	3.438	1.737	3.048	5.294	981	9.2
7	02-15-03	6.850	4.950	2.850	1.713	1.238	0.713	1.831	0.380	3.048	1.158	1829	3.8
8	02-03-10	2.850	3.000	2.950	0.713	0.750	0.738	1.100	0.233	3.048	0.709	2405	3.0
9	06-02-10	2.400	2.950	4.400	0.600	0.738	1.100	1.219	0.208	3.048	0.633	2711	3.1
10	06-10-04	4.400	3.100	5.250	1.100	0.775	1.313	1.594	0.426	3.048	1.297	2274	5.3
11	03-15-16	4.950	5.700	4.400	1.238	1.425	1.100	1.881	0.657	3.048	2.003	1496	5.3
12	03-16-05	4.400	4.100	5.250	1.100	1.025	1.313	1.719	0.547	3.048	1.669	2717	8.1
13	10-03-05	3.000	5.250	4.350	0.750	1.313	1.088	1.575	0.408	3.048	1.243	2953	6.5
14	04-10-05	3.100	4.350	5.300	0.775	1.088	1.325	1.594	0.421	3.048	1.28	2516	5.8
15	04-05-21	5.300	4.150	4.250	1.325	1.038	1.063	1.713	0.540	3.048	1.64	2072	6.1
16	19-04-21	4.850	4.250	5.550	1.213	1.063	1.388	1.831	0.622	3.048	1.89	1665	5.6
17	16-15-17	5.750	9.300	5.350	1.438	2.325	1.338	2.550	0.880	3.048	2.68	1550	7.4
18	16-17-11	5.350	6.100	6.450	1.338	1.525	1.613	2.238	0.947	3.048	2.89	1953	10.0
19	05-16-11	4.100	6.450	4.850	1.025	1.613	1.213	1.925	0.621	3.048	1.89	2052	6.9
20	21-05-11	4.150	4.850	6.550	1.038	1.213	1.638	1.944	0.628	3.048	1.91	1843	6.3
21	21-11-20	6.550	6.150	5.450	1.638	1.538	1.363	2.269	0.974	3.048	2.97	1507	8.0
22	19-21-20	5.600	5.450	9.350	1.400	1.363	2.338	2.550	0.860	3.048	2.62	1920	9.0
Totals/Av									16.116	3.048	49.12	1549	135.5

Pre - Post DRO (kg) 72.4

Post-RF-SVE Extended Design-Volume DRO Balance

1	24-36-23	7.350	5.950	4.350	1.838	1.488	1.088	2.206	0.809	3.048	2.465	641	2.8
2	24-23-22	4.350	5.900	7.350	1.088	1.475	1.838	2.200	0.802	3.048	2.445	721	3.1
3	36-35-34	8.150	4.450	6.850	2.038	1.113	1.713	2.431	0.953	3.048	2.903	197	1.0
4	23-36-34	5.950	6.850	9.100	1.488	1.713	2.275	2.738	1.274	3.048	3.882	717	5.0
5	23-34-28	9.100	7.350	7.000	2.275	1.838	1.750	2.931	1.576	3.048	4.805	710	6.1
6	22-23-28	5.900	7.000	8.200	1.475	1.750	2.050	2.638	1.264	3.048	3.854	762	5.2
7	22-28-25	8.200	4.500	6.850	2.050	1.125	1.713	2.444	0.963	3.048	2.936	313	1.6
8	22-25-26	6.850	4.450	8.200	1.713	1.113	2.050	2.438	0.953	3.048	2.903	404	2.1
9	34-35-33	4.450	8.150	6.850	1.113	2.038	1.713	2.431	0.953	3.048	2.903	149	0.8
10	34-33-30	6.850	7.450	6.400	1.713	1.863	1.600	2.588	1.273	3.048	3.881	440	3.0
11	28-34-30	7.350	6.400	2.300	1.838	1.600	0.575	2.006	0.444	3.048	1.352	481	1.2
12	28-30-25	2.300	6.400	4.500	0.575	1.600	1.125	1.650	0.216	3.048	0.658	475	0.6
13	25-30-27	6.400	7.450	6.850	1.600	1.863	1.713	2.588	1.273	3.048	3.881	973	6.7
14	26-25-27	4.450	6.850	8.150	1.113	1.713	2.038	2.431	0.953	3.048	2.903	815	4.2
15	30-33-29	7.450	5.900	4.500	1.863	1.475	1.125	2.231	0.830	3.048	2.529	974	4.4
16	30-29-27	4.500	5.950	7.450	1.125	1.488	1.863	2.238	0.837	3.048	2.550	1513	6.9
17	29-33-31	5.900	7.350	4.300	1.475	1.838	1.075	2.194	0.793	3.048	2.416	837	3.6
18	27-29-31	5.950	4.300	9.140	1.488	1.075	2.285	2.424	0.652	3.048	1.986	1375	4.9
	Totals/Av								16.815	3.048	51.25	692	63.1

Table A9   Triangular Grouping Method (Water Mass Balance)

Area No	Area Points	Side Lengths (cm)			Conversion Lengths (m)			S (m)	Calculated Area (m²)	Drill-Hole Depth (m)	Vol (m³)	DRO_sw Analysis (mg/kg)	DRO (kg)
		a	b	c	a	b	c						
					Pre-RF-SVE Water Balance								
1	08-09-15	6.200	9.250	12.150	1.550	2.313	3.038	3.450	1.754	8.128	14.255	6.58	1669
2	08-15-02	12.150	6.850	6.800	3.038	1.713	1.700	3.225	1.181	9.144	10.799	7.02	1350
3	08-02-06	6.800	2.400	4.850	1.700	0.600	1.213	1.756	0.249	9.144	2.279	6.91	280
4	08-06-04	4.850	5.250	9.600	1.213	1.313	2.400	2.463	0.470	9.144	4.301	7.01	537
5	08-04-19	9.600	4.850	12.100	2.400	1.213	3.025	3.319	1.374	8.128	11.164	6.46	1283
6	08-19-07	12.100	9.300	6.100	3.025	2.325	1.525	3.438	1.737	7.112	12.353	5.69	1251
7	02-15-03	6.850	4.950	2.850	1.713	1.238	0.713	1.831	0.380	9.144	3.475	6.76	418
8	02-03-10	2.850	3.000	2.950	0.713	0.750	0.738	1.100	0.233	8.738	2.032	6.68	242
9	06-02-10	2.400	2.950	4.400	0.600	0.738	1.100	1.219	0.208	8.738	1.814	7.25	234
10	06-10-04	4.400	3.100	5.250	1.100	0.775	1.313	1.594	0.426	8.738	3.719	7.88	522
11	03-15-16	4.950	5.700	4.400	1.238	1.425	1.100	1.881	0.657	9.144	6.008	6.95	744
12	03-16-05	4.400	4.100	5.250	1.100	1.025	1.313	1.719	0.547	9.144	5.006	6.59	587
13	10-03-05	3.000	5.250	4.350	0.750	1.313	1.088	1.575	0.408	8.738	3.563	6.99	443
14	04-10-05	3.100	4.350	5.300	0.775	1.088	1.325	1.594	0.421	8.738	3.68	7.62	499
15	04-05-21	5.300	4.150	4.250	1.325	1.038	1.063	1.713	0.540	9.144	4.93	7.54	662
16	19-04-21	4.850	4.250	5.550	1.213	1.063	1.388	1.831	0.622	8.128	5.05	7.34	660
17	16-15-17	5.750	9.300	5.350	1.438	2.325	1.338	2.550	0.880	8.128	7.15	7.08	902
18	16-17-11	5.350	6.100	6.450	1.338	1.525	1.613	2.238	0.947	8.128	7.70	6.01	823
19	05-16-11	4.100	6.450	4.850	1.025	1.613	1.213	1.925	0.621	9.144	5.68	6.80	688
20	21-05-11	4.150	4.850	6.550	1.038	1.213	1.638	1.944	0.628	9.144	5.74	7.15	731
21	21-11-20	6.550	6.150	5.450	1.638	1.538	1.363	2.269	0.974	8.128	7.92	6.98	983
22	19-21-20	5.600	5.450	9.350	1.400	1.363	2.338	2.550	0.860	7.112	6.12	7.42	808
Totals/Av									16.116	8.361	134.74	6.80	16,317
Pre-Post Water (kg)	7,528												

Post-RF-SVE Water Balance

1	24-36-23	7.350	5.950	4.350	1.838	1.488	1.088	2.206	0.809	7.112	5.752	4.40	450
2	24-23-22	4.350	5.900	7.350	1.088	1.475	1.838	2.200	0.802	7.112	5.704	4.40	446
3	36-35-34	8.150	4.450	6.850	2.038	1.113	1.713	2.431	0.953	7.112	6.775	3.89	469
4	23-36-34	5.950	6.850	9.100	1.488	1.713	2.275	2.738	1.274	8.128	10.352	4.15	764
5	23-34-28	9.100	7.350	7.000	2.275	1.838	1.750	2.931	1.576	9.144	14.415	3.04	781
6	22-23-28	5.900	7.000	8.200	1.475	1.750	2.050	2.638	1.264	8.128	10.277	3.69	675
7	22-28-25	8.200	4.500	6.850	2.050	1.125	1.713	2.444	0.963	8.128	7.830	2.80	391
8	22-25-26	6.850	4.450	8.200	1.713	1.113	2.050	2.438	0.953	7.112	6.774	4.27	515
9	34-35-33	4.450	8.150	6.850	1.113	2.038	1.713	2.431	0.953	7.112	6.775	3.57	430
10	34-33-30	6.850	7.450	6.400	1.713	1.863	1.600	2.588	1.273	7.112	9.055	2.97	478
11	28-34-30	7.350	6.400	2.300	1.838	1.600	0.575	2.006	0.444	8.128	3.606	2.16	138
12	28-30-25	2.300	6.400	4.500	0.575	1.600	1.125	1.650	0.216	8.128	1.754	1.89	59
13	25-30-27	6.400	7.450	6.850	1.600	1.863	1.713	2.588	1.273	7.315	9.313	3.27	542
14	26-25-27	4.450	6.850	8.150	1.113	1.713	2.038	2.431	0.953	7.315	6.968	4.52	561
15	30-33-29	7.450	5.900	4.500	1.863	1.475	1.125	2.231	0.830	7.112	5.901	4.01	421
16	30-29-27	4.500	5.950	7.450	1.125	1.488	1.863	2.238	0.837	7.315	6.121	4.58	499
17	29-33-31	5.900	7.350	4.300	1.475	1.838	1.075	2.194	0.793	8.128	6.443	5.32	610
18	27-29-31	5.950	4.300	9.140	1.488	1.075	2.285	2.424	0.652	8.331	5.429	5.79	559
Totals/Av									16.815	7.686	129.24	3.82	8789

Table A10 Triangular Grouping Method (Water Mass Balance)

Area No	Area Points	Side Lengths (cm) a	b	c	Conversion Lengths (m) a	b	c	S (m)	Calculated Area (m²)	Drill-Hole Depth (m)	Vol (m³)	DRO_sv Analysis (mg/kg)	DRO (kg)
					Pre-RF-SVE Extended Design-Volume Water Balance								
1	08-09-15	6.200	9.250	12.150	1.550	2.313	3.038	3.450	1.754	3.048	5.346	7.11	676
2	08-15-02	12.150	6.850	6.800	3.038	1.713	1.700	3.225	1.181	3.048	3.600	7.56	484
3	08-02-06	6.800	2.400	4.850	1.700	0.600	1.213	1.756	0.249	3.048	0.760	6.35	86
4	08-06-04	4.850	5.250	9.600	1.213	1.313	2.400	2.463	0.470	3.048	1.434	6.67	170
5	08-04-19	9.600	4.850	12.100	2.400	1.213	3.025	3.319	1.374	3.048	4.186	6.63	494
6	08-19-07	12.100	9.300	6.100	3.025	2.325	1.525	3.438	1.737	3.048	5.294	5.54	522
7	02-15-03	6.850	4.950	2.850	1.713	1.238	0.713	1.831	0.380	3.048	1.158	7.41	153
8	02-03-10	2.850	3.000	2.950	0.713	0.750	0.738	1.100	0.233	3.048	0.709	6.66	84
9	06-02-10	2.400	2.950	4.400	0.600	0.738	1.100	1.219	0.208	3.048	0.633	7.07	80
10	06-10-04	4.400	3.100	5.250	1.100	0.775	1.313	1.594	0.426	3.048	1.297	7.39	171
11	03-15-16	4.950	5.700	4.400	1.238	1.425	1.100	1.881	0.657	3.048	2.003	7.65	273
12	03-16-05	4.400	4.100	5.250	1.100	1.025	1.313	1.719	0.547	3.048	1.669	6.98	207
13	10-03-05	3.000	5.250	4.350	0.750	1.313	1.088	1.575	0.408	3.048	1.243	7.33	162
14	04-10-05	3.100	4.350	5.300	0.775	1.088	1.325	1.594	0.421	3.048	1.28	7.94	182
15	04-05-21	5.300	4.150	4.250	1.325	1.038	1.063	1.713	0.540	3.048	1.64	8.14	238
16	19-04-21	4.850	4.250	5.550	1.213	1.063	1.388	1.831	0.622	3.048	1.89	7.55	255
17	16-15-17	5.750	9.300	5.350	1.438	2.325	1.338	2.550	0.880	3.048	2.68	6.86	327
18	16-17-11	5.350	6.100	6.450	1.338	1.525	1.613	2.238	0.947	3.048	2.89	5.37	276
19	05-16-11	4.100	6.450	4.850	1.025	1.613	1.213	1.925	0.621	3.048	1.89	7.12	240
20	21-05-11	4.150	4.850	6.550	1.038	1.213	1.638	1.944	0.628	3.048	1.91	7.67	261
21	21-11-20	6.550	6.150	5.450	1.638	1.538	1.363	2.269	0.974	3.048	2.97	7.61	402
22	19-21-20	5.600	5.450	9.350	1.400	1.363	2.338	2.550	0.860	3.048	2.62	7.85	366
Totals/Av									16.116	3.048	49.12	6.99	6110

Pre-Post Water (kg)   3354

Post-RF-SVE Extended Design-Volume Water Balance

#														
1	24-36-23	7.350	5.950	4.350	1.838	1.488	1.088	2.206	0.809	3.048	2.465	3.70	162
2	24-23-22	4.350	5.900	7.350	1.088	1.475	1.838	2.200	0.802	3.048	2.445	3.76	164
3	36-35-34	8.150	4.450	6.850	2.038	1.113	1.713	2.431	0.953	3.048	2.903	3.16	163
4	23-36-34	5.950	6.850	9.100	1.488	1.713	2.275	2.738	1.274	3.048	3.882	3.01	208
5	23-34-28	9.100	7.350	7.000	2.275	1.838	1.750	2.931	1.576	3.048	4.805	1.31	112
6	22-23-28	5.900	7.000	8.200	1.475	1.750	2.050	2.638	1.264	3.048	3.854	2.89	198
7	22-28-25	8.200	4.500	6.850	2.050	1.125	1.713	2.444	0.963	3.048	2.936	2.32	121
8	22-25-26	6.850	4.450	8.200	1.713	1.113	2.050	2.438	0.953	3.048	2.903	4.17	216
9	34-35-33	4.450	8.150	6.850	1.113	2.038	1.713	2.431	0.953	3.048	2.903	2.59	134
10	34-33-30	6.850	7.450	6.400	1.713	1.863	1.600	2.588	1.273	3.048	3.881	2.01	139
11	28-34-30	7.350	6.400	2.300	1.838	1.600	0.575	2.006	0.444	3.048	1.352	0.88	21
12	28-30-25	2.300	6.400	4.500	0.575	1.600	1.125	1.650	0.216	3.048	0.658	0.77	9
13	25-30-27	6.400	7.450	6.850	1.600	1.863	1.713	2.588	1.273	3.048	3.881	2.62	181
14	26-25-27	4.450	6.850	8.150	1.113	1.713	2.038	2.431	0.953	3.048	2.903	4.27	221
15	30-33-29	7.450	5.900	4.500	1.863	1.475	1.125	2.231	0.830	3.048	2.529	3.08	139
16	30-29-27	4.500	5.950	7.450	1.125	1.488	1.863	2.238	0.837	3.048	2.550	3.81	173
17	29-33-31	5.900	7.350	4.300	1.475	1.838	1.075	2.194	0.793	3.048	2.416	4.71	203
18	27-29-31	5.950	4.300	9.140	1.488	1.075	2.285	2.424	0.652	3.048	1.986	5.44	192
Totals/Av									16.815	3.048	51.25	3.02	2756

Appendix B

Summary of Demonstration Recorded Data

Appendix B Listing and Description of Tables

Item	Title	Description
Table B1	P/T, SVE, and PITT well temperatures	Temperature data from field measurements in P/T, SVE, and PITT wells (Figure 3.16). Example data point (T-014-5'): T-014 = well number, 5' = depth temperature measurement. True temperature reading = field measurement, °C + 6°C (temperature probe reading error)
Table B2	RF-SVE field data summary	Summary of field measured data: RF energy (kWh), ambient atmospheric conditions at site, airflow rates from each SVE well (slpm), data measurement time (h), P/T well temperature at a depth of 4.27 m (14 ft) and 5.18 m (17 ft), SVE well temperature at a depth of 3.35 m (11 ft) or 4.57 m (15 ft), and applicator (antenna) well temperatures at a depth of 4.57 m (15 ft)
Table B3	RF-SVE water and free DRO product recovered	Field measured data: start and end dates for each 208-L (55-gal) collection drum, total days of operation, barrel (drum) number, and water and DRO liquid collected each day or after each drum was full

Table B1   P/T, SVE, and PITT Well Temperatures[a,b]

Date & Time	T-012 5'	T-012 8'	T-012 11'	T-012 14'	T-012 15'	T-012 17'	T-012 20'	T-013 5'	T-013 8'	T-013 11'	T-013 14'	T-013 15'	T-013 17'	T-013 20'	T-014 5'	T-014 8'	T-014 11'	T-014 14'	T-014 15'	T-014 17'	T-014 20'
12/02/96 10:01	17	19	22	24		25	24	14	17	19	20		21	21	20	23	26	27		28	27
12/02/96 15:31	18	20	22	24		25	25	15	18	20	21		22	22	21	25	27	28		28	27
12/03/96 11:00										Begin Heating											
12/03/96 15:45	18	22	23	24		25	25	18	19	21	22		23	23	19	22	25	27		27	26
12/04/96 10:53	17	19	22	23		24	24	14	16	19	20		22	22	18	22	24	26		26	25
12/04/96 16:45	16	17	19	21		22	22	12	15	17	19		20	21	17	21	24	26		26	25
12/05/96 9:48	17	19	22	24		25	24	13	17	20	21		22	22	17	20	25	26		26	26
12/06/96 7:43	17	20	21	23		25	24	16	18	20	21		22	22	18	22	26	28		28	27
12/07/96 8:29	16	19	25	26		28	28	17	20	22	23		23	23	18	23	26	29		29	28
12/09/96 13:43	21	22	24	25		26	26	19	20	21	23		23	23	20	23	29	32		32	31
12/10/96 12:16	19	21	23	24		25	25	17	18	20	22		23	23	18	22	28	33		33	32
12/11/96 12:15	17	20	23	24		25	25	16	18	20	22		23	23	19	24	29	34		34	31
12/12/96 15:00	20	22	24	25		26	26	18	19	21	23		23	22	21	25	30	36		35	32
12/13/96 14:14	20	22	24	26		26	25	17	19	21	23		23	23	20	26	32	35		35	32
12/14/96 14:35	18	21	24	25		26	25	17	18	21	23		23	23	22	28	34	36		27	27
12/16/96 16:13	20	21	23	24		25	26	17	18	19	21		22	22	20	27	38	43		41	36
12/17/96 14:24	17	21	24	25		26	25	17	18	20	22		23	23	19	27	37	44		41	37
12/18/96 15:17	19	20	23	24		25	26	15	17	19	21		22	22	22	30	42	45		44	38
12/20/96 13:00	19	21	23	25		26	26	14	17	19	21		22	22	22	29	39	47		47	42
12/21/96 12:40	21	22	25	26		27	27	15	17	19	21		22	22	21	30	41	48		48	47
12/23/96 15:30	22	22	24	26		27	27	16	17	19	22		23	24	20	35	46	54		52	46
12/25/96 8:38	19	20	22	25		26	27	17	18	21	21		22	23	22	31	47	54		54	48
12/27/96 15:22	20	21	23	25		27	27	18	19	20	21		22	22	22	36	47	56		56	49
12/30/96 14:57	19	20	22	25		27	26	17	19	20	21		22	21	21	41	52	61		59	52
1/02/97 14:00	20	23	27	30		31	30	17	18	20	23		24	24	20	44	58	64		63	56
1/08/97 15:49	21	23	26	30		32	31	16	17	19	21		22	21	25	46	56	61		59	54
1/09/97 12:00				32		33					21		22		29			58		58	
1/10/97 10:41				33		34					22		23		32			58		58	
1/13/97 12:40				33		34					22		23		35			72		71	
1/15/97 15:40				33		34					21		22					70		72	
1/20/97 13:00				32		33					21		21					73		75	
1/22/97 13:12				31		35					24		26					68		68	
1/24/97 10:50				36		37					26		26					72		76	
1/27/97 12:07				36		38					25		26					75		78	

The following table lists recorded temperature data (columns are unlabeled measurement points in the source).

Date/Time																					
1/28/97 12:15	26	29	38	41		41	40	22	22	27	28		29	30	52	68	77	83		83	71
1/29/97 12:00	23	29	37	41		43	41	19	22	26	29		31	30	60	69	77	85		86	71
1/30/97 12:12	27	31	37	40		42	41	23	26	28	30		31	31	64	70	76	84		85	71
1/31/97 13:12				44		44					33		33					85		86	
2/01/97 10:12				42		42					31		32					81		81	
2/03/97 10:10				42		42					30		31					86		87	
2/04/97 12:50				42		45					32		33					89		88	
2/05/97 13:00				45		42					31		31					87		87	
2/06/97 13:00				42		44					31		32					87		88	
2/07/97 11:33				44		43					38	37	36					86		86	
2/10/97 12:00				42		49					33		34					93		93	
2/11/97 14:00				48	49	49					36		37					92	95	93	
2/12/97 15:00	32	35	41	48		45	43	25	28	31	34		34	32	60	72	81	88		89	82
2/13/97 11:43				44		46					31		33					88		88	
2/14/97 8:30				47		45					35		34					90		88	
2/15/97 8:30				44		43					35		36					95		89	
2/17/97 10:35	32	38	44	43		49	48	26	29	33	36		37	37	71	81	90	100		95	90
2/18/97 12:35				46		50					36		38					102		100	
2/19/97 13:23				49		49					35		36					101		99	
2/20/97 11:45	31	36	45	49		48	46	26	27	33	36		36	35	65	75	84	100		96	85
2/21/97 12:30				46		50					35		36					100		99	
2/22/97 10:50				49		50					35		36					101		99	
2/24/97 12:50				49		49					34		35					100		98	
2/25/97 12:30				49		51					33		35					96		94	
2/26/97 10:50				49		48					36		35					97		97	
2/27/97 12:20				47	50	51					36	37 Stop Heating	37					102	101	99	
2/28/97 14:04	33	39	46	50		50	47	23	27	32			37	36	65	75	86	102		99	86
2/28/97 11:50				49																	
3/01/97 12:45				50		51					36		37					100		97	
3/03/97 9:45				49		49					35		36					94		91	
3/04/97 14:24				51		52					36		37					89		90	
3/05/97 11:25				49		50					36		36					90		88	
3/07/97 11:30	32	41	47	50		51	48	21	28	32	36		38	37	58	69	78	88		84	71
3/12/97 9:55				44		44					32		32					73		72	
3/17/97 12:15	31	34	40	43		43	41	20	23	29	31		32	31	44	53	64	67		65	56

[a] Temperatures recorded in °C.
[b] Add 6°C to each recorded temperature to obtain actual temperature measurement.

Table B1 P/T, SVE, and PITT Well Temperatures[a,b] (continued)

Date & Time	T-015 5'	T-015 8'	T-015 11'	T-015 14'	T-015 15'	T-015 17'	T-015 20'	T-016 5'	T-016 8'	T-016 11'	T-016 14'	T-016 15'	T-016 17'	T-016 20'	T-018 5'	T-018 8'	T-018 11'	T-018 14'	T-018 15'	T-018 17'	T-018 20'
12/02/96 10:01	19	21	23	24		26	25	18	23	25	26		27	26	20	25	28	28		27	27
12/02/96 15:31	19	22	24	26		27	26	21	25	27	28		29	28	22	26	28	28		28	27
12/30/96										Begin Heating											
12/03/96 15:45	19	22	24	26		26	26	22	25	28	28		29	28	24	26	27	28		27	28
12/04/96 10:53	19	22	24	26		26	26	21	23	27	28		29	28	21	25	28	28		28	27
12/04/96 16:45	17	20	22	25		26	26	20	22	24	27		28	28	21	23	26	28		28	27
12/05/96 9:48	19	21	24	26		27	27	19	22	25	27		29	29	22	24	27	28		28	27
12/06/96 7:43	19	22	25	27		29	28	20	23	28	30		31	30	22	25	27	29		29	28
12/07/96 8:29	18	21	25	28		29	29	19	24	28	31		33	32	22	25	29	30		30	29
12/09/96 13:43	22	24	29	34		35	33	24	27	32	36		38	35	25	27	30	33		32	31
12/10/96 12:16	21	23	29	34		35	33	22	26	32	37		38	36	25	28	31	33		33	31
12/11/96 12:15	19	24	29	34		34	31	23	26	33	36		38	35	23	27	31	34		33	31
12/12/96 15:00	21	24	29	36		36	33	22	28	35	39		40	36	25	29	33	36		35	33
12/13/96 14:14	22	24	29	34		36	34	24	29	36	42		41	37	27	29	33	37		37	34
12/14/96 14:35	21	26	34	42		39	33	23	30	38	47		43	36	25	29	35	40		39	34
12/16/96 16:13	23	28	38	48		46	40	24	31	43	57		42	42	27	30	42	45		44	40
12/17/96 14:24	25	31	40	48		47	40	21	32	43	58		54	44	22	30	42	47		45	42
12/18/96 15:17	22	28	42	52		50	42	24	32	46	61		58	48	25	33	43	47		49	43
12/20/96 13:00	21	29	41	51		52	46	23	30	40	61		56	54	27	33	41	46		49	48
12/21/96 12:40	24	30	41	51		54	50	28	34	45	61		65	61	33	38	45	52		54	51
12/23/96 15:30	23	31	45	58		59	51	27	37	52	68		67	57	37	43	52	59		59	53
12/25/96 8:38	26	34	46	60		61	52	26	38	55	73		70	61	33	44	54	61		64	57
12/27/96 15:22	28	35	49	63		63	53	30	40	59	71		72	61	35	48	60	65		67	60
12/30/96 14:57	31	39	53	64		66	56	33	45	62	72		74	64	44	56	62	67		69	63
1/02/97 14:00	31	43	57	68		69	60	35	49	64	77		78	68	46	59	66	71		73	65
1/08/97 15:49	31	41	53	60		60	55	36	48	61	68		71	61	51	63	70	72		71	64
1/09/97 12:00				60		61					71		70					74		80	
1/10/97 10:41				60		60					70		69					76		81	
1/13/97 12:40				82		78					80		81					76		78	
1/15/97 15:40				76		79					79		83					77		79	
1/20/97 13:00				72		81					74		83					80		78	
1/22/97 13:12				75		78					77		75					77		78	
1/24/97 10:50				74		79					73		72					65		75	
1/27/97 12:07				72		79					81		85					72		69	
1/28/97 12:15	58	67	70	75		86	73	54	71	76	85		88	75	57	67	74	79		80	70

Table B1 P/T, SVE, and PITT Well Temperatures[a,b] (continued)

1/29/97 12:00	53	67	75	83		86	71	63	72	77	88		93	78	59	67	74	78		83	71
1/30/97 12:12	58	69	75	83		86	77	61	73	82	92		95	77	61	70	78	84		85	74
1/31/97 13:12				85		89					93		97					84		84	
2/01/97 10:12				81		85					92		94					81		83	
2/03/97 10:10				86		89					95		99					83		83	
2/04/97 12:50				92		90					97		100					85		84	
2/05/97 13:00				90		90					97		101					83		84	
2/06/97 13:00				92		92					98		102					83		84	
2/07/97 11:33				91		91					98		102					86		84	
2/10/97 12:00				97		101					101		108					88		88	
2/11/97 14:00				96	101	99					101	109	108					91	93	91	
2/12/97 15:00	65	71	78	97		98	83	61	71	86	103		108	91	60	68	80	89		90	78
2/13/97 11:43				94		96					103		106					88		87	
2/14/97 8:30				94		98					103		106					88		87	
2/15/97 8:30				96		99					107		110					92		90	
2/17/97 10:35				102		105					108		113					94		93	
2/18/97 12:35	70	82	87	106		108	91	69	83	97	109		116	99	69	80	88	101		100	86
2/19/97 13:23				105		106					110		115					95		92	
2/20/97 11:45				102		103					108		113					96		93	
2//21/97 12:30	65	74	86	103		105	86	65	78	96	110		114	96	65	73	81	95		93	76
2/22/97 10:50				103		105					109		114					95		93	
2/24/97 12:50				103		104					109		114					93		93	
2/25/97 12:30				100		100					108		110					95		92	
2/26/97 10:50				101		102					111		114					92		92	
2/27/97 12:20				103		106					113		116					97		94	
2/28/97 14:04	67	74	86	102	105	104	90	69	81	89	110	112	116	101	66	75	86	96	95	95	80
2/28/97 11:50											**Stop Heating**										
3/01/97 12:45				102		102					108		113					94		92	
3/03/97 9:45				94		94					106		106					89		87	
3/04/97 14:24				93		95					105		103					89		86	
3/05/97 11:25				90		90					101		99					88		83	
3/07/97 11:30	55	68	79	93		89	78	58	70	88	97		96	83	56	67	78	85		81	71
3/12/97 9:55				74		72					84		79					71		68	
3/17/97 12:15	41	51	60	69		67	59	43	54	70	76		73	56	43	53	63	66		64	55

[a] Temperatures recorded in °C.
[b] Add 6°C to each recorded temperature to obtain actual temperature measurement.

Table B1 P/T, SVE, and PITT Well Temperatures[a,b] (continued)

Date & Time	T-019 5'	T-019 8'	T-019 11'	T-019 14'	T-019 15'	T-019 17'	T-019 20'	T-021 5'	T-021 8'	T-021 11'	T-021 14'	T-021 15'	T-021 17'	T-021 20'	SVEB 18'	SVEC 15'	SVEC 17'	SVEC 21'	SVEA 11'
12/02/96 10:01	21	27	28	28		27	27	18	22	25	27		28	27					
12/02/96 15:31	22	25	28	28		28	27	22	25	27	28		29	28					
12/03/96										Begin Heating									
12/03/96 15:45	22	25	27	27		27	28	20	24	26	27		28	28					
12/04/96 10:53	21	23	25	26		26	26	20	23	26	27		28	27					
12/04/96 16:45	20	23	26	27		28	27	18	22	25	27		28	28					
12/05/96 9:48	22	25	27	28		28	28	22	25	27	28		29	28					
12/06/96 7:43	22	25	27	28		29	28	19	23	28	29		30	29					
12/06/96 8:29	23	25	28	30		30	29	20	25	29	30		31	30					
12/09/96 13:43	24	26	28	30		31	31	26	27	31	34		35	33					
12/10/96 12:16	24	28	30	32		33	31	24	27	31	34		35	33					
12/11/96 12:15	23	27	30	33		34	31	22	26	31	34		35	33					
12/12/96 15:00	24	29	34	39		38	34	25	31	34	39		41	37					
12/13/96 14:14	27	29	34	38		39	36	23	28	34	40		42	39					
12/14/96 14:35	25	29	36	42		43	32	20	28	35	40		43	39					
12/16/96 16:13	25	29	38	45		47	42	23	29	40	47		49	45					
12/17/96 14:24	23	32	41	50		49	44	23	31	41	52		53	46					
12/18/96 15:17	27	33	43	51		52	43	25	34	44	55		56	50					
12/20/96 13:00	34	36	41	50		53	49	20	31	42	49		56	55					
12/21/96 12:40	29	34	44	54		60	52	23	32	42	52		59	60					
12/23/96 15:30	33	41	53	64		67	55	31	42	54	65		71	65					
12/25/96 8:38	35	44	57	66		73	62	33	45	56	69		75	69					
12/27/96 15:22	33	48	61	69		75	63	38	50	62	72		77	71					
12/30/96 14:57	41	54	65	71		78	64	44	54	64	72		79	72					
1/02/97 14:00	44	57	68	74		78	67	48	60	69	75		81	74					
1/08/97 15:49	49	60	69	76		77	70	51	63	70	73		81	74	80	76			
1/09/97 12:00				81		84					82		84		88	74			80
1/10/97 10:41				78		84					82		90						
1/13/97 12:40				74		78					76		81		88	77			84
1/15/97 15:40				76		80					76		85						
1/20/97 13:00				79		81					78		85		98				
1/22/97 13:12				70		71					72		69		124				
1/24/97 10:50				76		78					76		74		127	75	74		99
1/27/97 12:07				73		80					74		78		133	77		69	109

Table B1 P/T, SVE, and PITT Well Temperatures[a,b] (continued)

Date/Time	54	63	70		79	85	68	54	64	76	82		88	76	119	81	90	98
1/28/97 12:15	54	63	70		79	85	68	54	64	76	82		88	76	119	81	90	98
1/29/97 12:00	51	63	73		82	85	66	55	68	74	83		89	76	122	93		103
1/30/97 12:12	53	64	74		83	85	71	61	70	75	82		88	82	124	99		105
1/31/97 13:12					85	88					85		93		127	103		105
2/01/97 10:12					83	85					81		90					
2/03/97 10:10					80	86					83		93		125	102		102
2/04/97 12:50					87	88					86		96		126	107		105
2/05/97 13:00					84	87					85		95		125	105		106
2/06/97 13:00					83	87					86		95		122	106		100
2/07/97 11:33					85	89					84		96		122	107		92
2/10/97 12:00					85	89					88		98		131	113		98
2/11/97 14:00				96	92	94					94	99	102		133	113		109
2/12/97 15:00	56	70	77		87	89	75	60	69	79	92		101		125	113		110
2/13/97 11:43					88	89					90		98		125	110		108
2/14/97 8:30					87	87					91		97		125	110		100
2/15/97 8:30					91	92					95		101					
2/17/97 10:35					94	92					96		105		132	117		103
2/18/97 12:35	56	68	80		96	95	77	66	78	81	98		106	94	131	118	89	106
2/19/97 13:23					92	92					100		105		132	117	88	104
2/20/97 11:45					90	91					96		105		131	114		115
2/21/97 12:30	56	67	78		94	90	77	66	78	88	96		104	92	130	116		106
2/22/97 10:50					93	93					96		103					
2/24/97 12:50					93	94					94		102					
2/25/97 12:30					90	92					92		98		125	105		104
2/26/97 10:50					91	92					97		102		124	113		111
2/27/97 12:20					94	95					98		104		127	116		115
2/28/97 14:04	57	70	80	96	95	95	77	61	72	82	96	101	104	94	128	117	112	110
2/28/97 11:50										Stop Heating								
3/01/97 12:45					91	92					97		102		125	114		113
3/03/97 9:45					85	85					86		93		127	113	87	113
3/04/97 14:24					84	85					91		92		119	107		109
3/05/97 11:25					82	82					88		92		116	105		104
3/07/97 11:30	53	62	74		82	81	78	55	66	77	87		88	78	114	103	98	106
3/12/97 9:55					71	68					66		65		95	86		89
3/17/97 12:15	40	48	57		61	61	53	41	50	59	67		67	61	83	73	77	78

a Temperatures recorded in °C.
b Add 6°C to each recorded temperature to obtain actual temperature measurement.

Table B1   P/T, SVE, and PITT Well Temperatures[a,b] (continued)

Date & Time	TT1B 15'	TT2C 15'	TT3C 15'	TT5B 15'	TT6B 15'	Av (all)[c]	Av (12 &13)
12/02/96 10:01						24.83	20.25
12/02/96 15:31						25.94	21.00
12/03/96 11:00		Begin Heating				25.39	20.63
12/03/96 15:45						25.50	
12/04/96 10:53						24.86	
12/04/96 16:45						24.33	
12/05/96 9:48						25.25	
12/06/96 7:43						26.06	
12/07/96 8:29						26.81	
12/09/96 13:43						29.72	
12/10/96 12:16						29.86	
12/11/96 12:15						29.75	
12/12/96 15:00						32.06	
12/13/96 14:14						32.61	
12/14/96 14:35						33.33	
12/16/96 16:13						37.61	
12/17/96 14:24						38.92	
12/18/96 15:17						41.08	
12/20/96 13:00						41.39	
12/21/96 12:40						44.25	
12/23/96 15:30						49.28	
12/25/96 8:38						51.47	
12/27/96 15:22						54.08	
12/30/96 14:57						57.56	
1/02/97 14:00						60.86	
1/08/97 15:49						60.00	
1/09/97 12:00						71.92	
1/10/97 10:41						72.17	
1/13/97 12:40						77.25	
1/15/97 15:40						77.67	
1/20/97 13:00						78.25	
1/22/97 13:12						72.92	
1/24/97 10:50						73.75	

Table B1 P/T, SVE, and PITT Well Temperatures[a,b] (continued)

Date/Time						
1/27/97 12:07						76.33
1/28/97 12:15						72.17
1/29/97 12:00						73.64
1/30/97 12:12						75.50
1/31/97 13:12						87.83
2/01/97 10:12						84.75
2/03/97 10:10						87.50
2/04/97 12:50						90.17
2/05/97 13:00						89.17
2/06/97 13:00						89.75
2/07/97 11:33						89.83
2/10/97 12:00						94.08
2/11/97 14:00						
2/12/97 15:00						
2/13/97 11:43						
2/14/97 8:30						
2/15/97 8:30						
2/17/97 10:35						
2/18/97 12:35						
2/19/97 13:23						
2/20/97 11:45	57	68	60	67	55	
2/21/97 12:30						
2/22/97 10:50						
2/24/97 12:50						
2/25/97 12:30						
2/26/97 10:50						
2/27/97 12:20	61	71	66	68	61	
2/28/97 14:04		Stop Heating				
2/28/97 11:50						
3/01/97 12:45						
3/03/97 9:45						
3/04/97 14:24						
3/05/97 11:25	57	66	60	66	57	
3/07/97 11:30						
3/12/97 9:55						
3/17/97 12:15	49	53	50	52	48	

[a] Temperatures recorded in °C.
[b] Add 6°C to each recorded temperature to obtain actual temperature measurement.
[c] Average of all temperatures except for P/T 12 and P/T 13.

Table B2   RF-SVE Field Data Summary

Date	RF Energy (kWh)	Ambient Values °C	R.H. %	Airflows (slpm) SVEB	SVEA	SVEC	Sample Time (h)	Water Vol (L)	P/T Well Temperatures (°C) (Average of 14 ft & 17 ft depth) PT12	PT14	PT15	PT16	PT18	PT19	PT21	Air (or Well) Temperatures (°C) SVEB (15 ft)	SVEA (11 ft)	SVEC (15 ft)	Before GAC	Antennae Wells °C A1	A2
									Recovery of Air from Wells A, B, and C (Airflow Measured Near Wellhead)												
12/03/96	29.00	5.6	24	147.2	184.1	99.1	0.00	0.0	25	27	26	29	28	27	28	26.7	21.7	26.7	7.8		
12/06/96	115.46	6.1	56	147.2	198.2	124.6			24	27	26	29	28	27	28	30.0	22.8	26.7	4.4		
12/09/96	140.24	12.2	45	124.6	184.1	25.5			26	32	35	37	33	31	35	37.8	30.0	24.4	14.4	69.0	61.6
12/12/96	208.39	12.2	38	62.3	110.4	87.8			26	36	36	40	36	39	40	46.1	35.6	33.3	13.3	72.9	74.5
12/13/96	195.62	11.7	38	62.3	110.4	87.8			26	35	35	42	37	39	41	40.6	23.9	26.7	8.3	113.1	44.7
12/14/96	159.63	10.0	46	11.3	0.0	11.3			32	38	47	51	46	49	48	49.3	38.8	33.2	10.0	98.3	112.8
12/17/96	310.73	-3.3	22	25.5	62.3	36.8			32	49	54	62	52	56	54	60.4	47.1	38.2	-3.3	211.6	111.7
12/18/96	157.43	-3.3	24	11.3	25.5	36.8			31	51	57	66	54	58	62	63.2	49.3	27.1	-3.3	114.7	102.9
12/20/96	338.41	10.0	20	48.1	36.8	62.3	336.00	195.7								74.3	63.2	49.3	15.6	127.9	121.9
12/23/96	552.48	8.9	38	48.1	25.5	87.8			33	59	65	74	65	72	74	82.7	68.8	52.1	8.9	137.9	118.0
12/25/96	412.98	-5.6	55	25.5	0.0	0.0			32	60	67	67	69	76	78	82.7			-5.6	141.1	146.0
12/26/96	155.82	6.7	33	87.8	0.0	0.0										85.4			3.9	146.3	141.9
12/27/96	177.18	7.8	25	48.1	36.8	36.8			32	62	69	78	72	78	81	88.2	67.1	38.2	8.9	149.7	132.0
12/28/96	221.40	4.4	54	48.1	25.5	36.8										88.2	67.1	38.2	3.3	143.1	127.3
12/30/96	332.38	12.2	40	48.1	11.3	25.5	240.00	188.9	32	66	71	79	74	81	82	88.2	73.8	46.0	15.6	148.2	149.3
1/01/97	368.71	14.4	49	25.5	2.8	25.5										88.2	70.4	54.3	20.0	141.3	122.6
1/02/97	157.35	8.9	62	25.5	0.0	25.5			37	70	75	84	78	82	84	89.3	70.4	57.1	8.9	142.8	135.9
1/03/97	342.08						96.00	186.6												149.1	147.6
1/07/97	645.44	-2.2	60	70.8	0.0	0.0			37	66	66	76	78	83	83	93.8			12.8	157.1	96.1
1/08/97	227.62	-3.3	60	19.8	0.0	0.0	120.00	182.5	39	64	67	77	83	89	89	88.2			-2.2	147.8	108.3
1/09/97	187.90	0.0	30	51.0	0.0	0.0	N.A.	7.9								93.8			37.8	162.0	86.1
									Recovery of Air from Well B (Airflow Measured Before GAC Drums)												
1/10/97	190.85	-0.6	65	22.7	0.0	0.0	N.A.	4.9	40	64	66	76	87	87	92	91.0			12.2	170.2	85.9
1/11/97	244.18	2.2	62	37.7	0.0	0.0	2.67	6.4								96.6			10.0	148.6	121.4
1/14/97	551.84	-2.2	75	56.6	0.0	0.0	4.17	18.9	40	78	86	87	83	82	87	96.6			7.8	148.8	108.2
1/15/97	260.88	-3.9	63	63.7	0.0	0.0	24.83	60.2	40	77	84	87	84	84	87	94.9			-1.1	141.2	146.6
1/16/97	269.65	-2.8	65	113.3	0.0	0.0	23.50	63.6								98.2			20.0	150.5	143.3
1/17/97	76.19	-3.3	67	254.8	0.0	0.0	23.92	152.2								99.3			48.9	152.9	144.2
Totals	7030							1067.8													
									Recovery of Air from Wells A and C (Airflow Measured Near Wellhead)												
1/18/97	122.94	-1.1	62	0.0	370.9	124.6	21.23	123.8									79.9	75.4	13.3	133.2	117.7
1/20/97	490.39	4.4	61	0.0	320.0	99.1	47.65	165.0	39	80	83	85	86	86	88	104.0	82.1	77.7	21.1	152.2	147.1

Recovery of Air from Wells 14, 16, 18, & 21 (Airflow Measured Before GAC Drums)

Date																				
1/21/97	271.35	6.7	368.1	42	0.0	22.33	171.1	40	74	82	82	83	77	77	130.0			33.3	156.6	147.4
1/22/97	196.02	1.1	424.7	62	0.0	21.17	171.1											46.7	157.1	156.2
1/23/97	259.08	2.2	396.4	69	0.0	22.50	168.1											25.0	153.6	150.0
1/24/97	272.48	1.1	424.7	74	0.0	23.07	182.8	43	80	79	74	83	83	81	133.0	105.0	81.0	28.1	156.9	174.1
1/25/97	244.20	-0.6	453.1	80	0.0	24.33	186.6											38.3	150.8	164.5
1/27/97	583.93	7.8	453.1	57	0.0	51.67	418.7	43	83	89	77	82	83	82	139.0	115.0	83.0	30.6	169.5	164.7

Recovery of Air from Well B (Airflow Measured Before GAC Drums)

Date																				
1/28/97	252.62	7.8	481.4	57	0.0	18.17	174.9	47	89	87	93	86	88	91	125.0	104.0	87.0	45.0	174.7	162.4
1/29/97	238.95	4.4	538.0	48	0.0	22.92	158.6	48	92	91	97	87	90	92	128.0	109.0	99.0	31.7	169.7	168.2
1/30/97	240.66	6.7	481.4	51	0.0	25.75	205.5	47	91	91	100	91	90	91	130.0	111.0	105.0	48.9	168.7	157.0
1/31/97	142.06	6.7	481.4	53	0.0	22.58	202.9	50	92	93	101	90	93	95	133.0	111.0	109.0	46.7	174.1	164.0
2/01/97	273.69	7.8	226.5	43	0.0	22.63	133.2	48	87	89	99	88	90	92		108.0		20.6	166.8	150.7
2/03/97	604.10	6.7	226.5	54	0.0	24.50	221.8	48	93	94	103	89	89	94	131.0	108.0	108.0	15.0	170.4	164.5
2/04/97	299.00	6.1	368.1	44	0.0	23.87	202.9	51	95	97	105	91	94	97	132.0	111.0	113.0	44.4	174.7	172.2
2/05/97	208.39	3.9	283.2	45	0.0	23.17	145.7	48	93	96	105	90	92	96	131.0	112.0	111.0	33.9	173.7	170.8
2/06/97	267.29	3.9	283.2	68	0.0	23.17	133.2	50	94	98	106	90	91	97	128.0	106.0	112.0	36.1	173.7	162.1
2/07/97	215.71	-1.1	283.2	74	0.0	23.25	133.2	49	92	97	106	91	93	96	128.0	98.0	113.0	24.7	167.9	169.6
2/08/97	296.66	-4.4	311.5	80	0.0	22.33	114.3												170.4	168.5
2/10/97	502.97	4.4	283.2	47	0.0	45.75	289.2	55	99	105	111	94	93	99	137.0	104.0	119.0	37.2	174.1	168.6
2/11/97	276.61	10.6	311.5	28	0.0	25.50	144.6	55	99	105	111	97	99	104	139.0	115.0	119.0	44.4	174.9	164.3
2/12/97	235.99	6.7	311.5	48	0.0	23.58	144.6	51	95	104	112	96	94	103	131.0	116.0	119.0	40.0	170.0	170.8
2/13/97	181.62	2.2	254.8	78	0.0	23.67	125.7	53	94	101	111	94	95	100	131.0	114.0	116.0	6.1	163.3	161.0
2/14/97	279.01	-1.7	254.8	66	0.0	19.25	106.4	51	95	102	111	94	93	100	131.0	106.0	116.0	18.9	161.9	172.8
Totals	13,986						5,291.8													

Recovery of Air from Well B (Airflow Measured Before GAC Drums)

Date																				
2/15/97	240.46	1.1	283.2	56	0.0	23.75	118.1	49	98	104	115	97	98	104	138.0	109.0	123.0	33.9	173.0	174.5
2/17/97	554.72	11.1	254.8	29	0.0	49.50	262.7	54	104	110	117	100	99	107	137.0	112.0	124.0	25.0	167.0	165.3
2/18/97	210.12	11.1	311.5	38	0.0	23.25	125.7	56	107	113	119	107	102	108	138.0	110.0	123.0	44.4	175.2	175.0
2/19/97	215.22	7.2	283.2	60	0.0	24.95	156.0	55	106	112	119	100	98	109	137.0	121.0	120.0	22.2	162.3	166.3
2/20/97	262.95	6.7	254.8	62	0.0	22.25	102.6	53	104	109	117	101	97	107	136.0	112.0	122.0	31.1	165.7	167.3
2/21/97	271.34	4.4	254.8	30	0.0	27.00	125.7	56	106	110	118	100	98	106				25.6		
2/22/97	269.87	1.1	254.8	56	0.0	20.00	104.9	56	106	110	118	100	99	106				29.7	169.9	174.9
2/24/97	131.03	-1.1	254.8	70	0.0	48.83	224.5	55	105	110	118	99	100	104				3.3	163.4	167.7
2/25/97	284.60	-1.1	254.8	67	0.0	23.83	110.5	56	101	106	115	100	97	101	131.0	110.0	111.0	12.8	150.0	152.2
2/26/97	291.40	6.1	254.8	28	0.0	30.00	114.3	54	103	108	119	98	98	106	130.0	117.0	119.0	16.7	165.8	173.4
2/27/97	259.86	7.2	254.8	32	0.0	19.25	125.7	57	107	111	121	102	101	107	133.0	121.0	122.0	11.7	161.7	163.0

Table B2   RF-SVE Field Data Summary (continued)

Date	RF Energy (kWh)	Ambient Values °C	Ambient Values R.H., %	Airflows (slpm) SVEB	Airflows (slpm) SVEA	Airflows (slpm) SVEC	Sample Time (h)	Water Vol (L)	P/T Well Temperatures (°C) (Average of 14 ft & 17 ft depth) PT12	PT14	PT15	PT16	PT18	PT19	PT21	Air (or Well) Temperatures (°C) SVEB (15 ft)	SVEA (11 ft)	SVEC (15 ft)	Before GAC	Antennae Wells °C A1	A2
2/28/97	OFF	3.3	82	0.0	254.8	0.0	19.58	98.8	56	107	109	119	102	101	106	134.0	116.0	123.0	3.9	173.5	173.6
3/01/97	OFF	2.2	44	0.0	254.8	0.0	26.58	114.3	57	105	108	117	99	98	106	131.0	119.0	120.0	11.1	135.0	140.0
3/03/97	OFF	10.0	25	0.0	254.8	0.0	47.33	129.5	54	99	100	112	94	91	96	133.0	119.0	119.0	21.7	122.0	124.0
3/04/97	OFF	12.8	20	0.0	269.0	0.0	24.92	87.4	58	96	100	110	94	91	98	125.0	115.0	113.0	20.6	118.0	126.0
3/05/97	OFF	0.0	42	0.0	283.2	0.0	21.50	57.2	56	95	96	106	92	88	96	122.0	110.0	111.0	10.8	115.0	125.0
3/07/97	OFF	6.7	45	0.0	297.3	0.0	46.50	118.1	57	92	97	103	89	88	94	120.0	112.0	109.0	19.4	108.0	114.0
3/12/97	OFF	20.0	22	0.0	283.2	0.0	126.58	174.9	50	79	79	88	76	76	72	101.0	95.0	92.0	26.7	86.0	92.0
3/17/97	OFF	14.4	30	0.0	283.2	0.0	114.83	178.7	49	72	74	81	71	67	73	89.0	84.0	79.0	16.7	77.0	82.0
Totals	16,977							7,821.2													

Note: %RH = percent relative humidity; SVEA = soil vapor extraction (SVE) well A; SVEB = soil vapor extraction (SVE) well B; SVEC = soil vapor extraction (SVE) well C; PT# = pressure-temperature (P/T) wells; GAC = granulated activated carbon cells.

Table B3   RF-SVE Water and Free DRO Product Recovered

Start Date	End Date	Total Days	Drum	Water (drum)[a] (L) per Drum	Total[d]	Water (daily)[b] (L) per Day	Total[d]	DRO Product[c] (L) per Drum	Total[d]
12/03/96	12/03/96	0	0	0.0	0.0	0.0	0.0	0.00	0.00
12/03/96	12/06/96	3	0	0.0	0.0	0.0	0.0	0.00	0.00
12/06/96	12/20/96	17	1	195.8	195.8	195.7	195.7	0.79	0.79
12/20/96	12/30/96	27	2	188.7	384.5	188.9	384.6	1.59	2.38
12/30/96	1/03/97	31	3	186.7	571.3	186.6	571.2	0.40	2.77
1/03/97	1/08/97	36	4	182.4	753.6	182.5	753.7	4.76	7.53
1/08/97	1/16/97	44	5	186.3	939.9	161.9	915.6	3.96	11.50
1/16/97	1/18/97	46	6	205.8	1145.7	276.0	1191.6	0.40	11.89
1/18/97	1/20/97	48	7	186.3	1332.0	165.0	1356.6	3.96	15.86
1/20/97	1/22/97	50	8	186.3	1518.3	342.2	1698.8	0.79	16.65
1/22/97	1/23/97	51	9	186.3	1704.7	168.1	1866.9	0.79	17.44
1/23/97	1/24/97	52	10	187.9	1892.6	182.8	2049.7	2.38	19.82
1/24/97	1/25/97	53	11	186.3	2078.9	186.6	2236.3	0.79	20.61
1/25/97	1/26/97	54	12	188.7	2267.6		2236.3	1.59	22.20
1/26/97	1/27/97	55	13	188.7	2456.3	418.7	2655.0	1.59	23.79
1/27/97	1/28/97	56	14	186.3	2642.6	174.9	2829.9	3.96	27.75
1/28/97	1/30/97	58	15	185.1	2827.8	364.1	3194.0	1.98	29.73
1/30/97	1/31/97	59	16	183.9	3011.7	202.9	3396.9	3.17	32.90
1/31/97	2/01/97	60	17	186.3	3198.0	133.2	3530.1	3.96	36.87
2/01/97	2/03/97	62	18	188.7	3386.8	221.8	3751.9	1.59	38.45
2/03/97	2/04/97	63	19	183.9	3570.7	202.9	3954.8	6.34	44.80
2/04/97	2/05/97	64	20	177.6	3748.3	145.7	4100.5	6.34	51.14
2/05/97	2/07/97	66	21	180.8	3929.1	266.4	4366.9	3.17	54.31
2/07/97	2/09/97	68	22	179.2	4108.3		4366.9	4.76	59.07
2/09/97	2/10/97	69	23	188.7	4297.0	403.5	4770.4	1.59	60.65
2/10/97	2/11/97	70	24	186.3	4483.3	144.6	4915.0	4.00	64.65
2/11/97	2/12/97	71	25	180.1	4663.4	144.6	5059.6	3.81	68.45
2/12/97	2/14/97	73	26	187.1	4850.5	232.1	5291.7	3.17	71.63
2/14/97	2/16/97	75	27	197.4	5047.9		5291.7	2.38	74.00
2/16/97	2/17/97	76	28	196.6	5244.6	380.8	5672.5	6.34	80.35
2/17/97	2/19/97	78	29	194.3	5438.8	281.7	5954.2	2.38	82.73
2/19/97	2/21/97	80	30	196.6	5635.5	228.3	6182.5	3.17	85.90
2/21/97	2/23/97	82	31	207.7	5843.2		6182.5	4.76	90.65
2/23/97	2/24/97	83	32	209.3	6052.5	329.4	6511.9	6.34	97.00
2/24/97	2/26/97	85	33	222.0	6274.5	224.8	6736.7	3.17	100.17
2/26/97	2/28/97	87	34	224.5	6499.1	224.5	6961.2	3.81	103.97
2/28/97	3/01/97	88				114.3	7075.5		
3/01/97	3/03/97	90				129.5	7205.0		
3/03/97	3/04/97	91				87.4	7292.4		
3/04/97	3/05/97	92				57.2	7349.6		
3/05/97	3/07/97	94				118.1	7467.7		
3/07/97	3/12/97	99				174.9	7642.6		
3/12/97	3/17/97	104				178.7	7821.3		
2/28/97	3/03/97	90	35	206.1	6705.2			3.17	107.15
3/03/97	3/06/97	93	36	205.3	6910.5			4.00	111.14
3/06/97	3/10/97	97	37	199.8	7110.3			3.17	114.31
3/10/97	3/17/97	104	38	201.4	7311.7			1.59	115.90
Total				7311.7		7821.3		115.9	

Note: The volumes of DRO liquid and water per drum (or volume of water collected per day) were measured using the height of organic liquid and water phases in a modified Caliwasa sampler.

[a] Volume of water after each 208-L (55-gal) drum was nearly full.
[b] Volume of water that was collected each day by measuring the increase in total volume in each drum.
[c] Volume of liquid DRO after each 208-L (55-gal) drum was nearly full.
[d] Cumulative volume.

Appendix C

Demonstration Heat Balance and Calculations

Appendix C Listing and Description of Example Calculations and Tables

Item	Title	Description
Calculations	Heat balance example calculations	Calculations for Table C1: Soil temperature for sampled volume #4 Calculations for Table C2: RF-SVE heat utilization Calculations for Table C3: SVE calculated dry and wet airflow rates (slpm) and water vapor content of dry air (humidity, kg water vapor per kg dry air) Calculations for Table C4: RF-SVE heat requirement calculations for air and water vapor at each measurement time in Table B2, Appendix B and the total heat requirement for air and water vapor during each period specified in Table 3.8 Macro used to calculate the volume for each temperature interval and the average temperature for each volume in Table C1 Macro for calculating the heat loss to the surroundings (Boundary Loss) in Table C1, using NO-e-SYS and HDF data set
Table C1	Soil temperatures for sampled volume #4	Average soil vol (m³), mass (kg), and temperature (°C) for the temperature intervals below 40°C, 40–60°C, 60–80°C, 80–94°C, 94–110°C, 110–130°C, above 130°C Heat capacity values for soil and water (c_p), soil density (d), and soil heat conductivity (k) that were used to calculate the data in Tables C1 and C2
Table C2	RF-SVE heat utilization	Summary of heat balance data calculated in Table C1
Table C3	SVE calculated airflows	Calculation of wet and dry airflow rates for the SVE well(s). The measured and calculated (bold, italic letters) airflow rates are in standard liters per minute (slpm). 1. Temperature SVE well flow, °C and K: temperature (T) of air-water vapor mixture in the SVE well at 3.4–4.6 m depth 2. Humidity of dry air, kg/kg: water in dry air at standard temperature (20°C) and pressure (101.31 kPa) 3. Density of dry air, kg/m³: density of air at SVE well temperature (T) 4. Wet air SVE well, slpm: wet airflow rate measured at the SVE wellhead and specified temperature (T) 5. Dry air C-drums, slpm: dry airflow rate measured by a rotameter at the knockout drum at ambient conditions 6. Time, minutes: time difference between current and previous field data measurements 7. Water removed: water condensed from hot, wet air on a daily basis (Appendix B, Table B3) 8. Water (gas) at T&P, liters: volume of water vapor at the specified temperature (T) 9. Dry air at T&P (L): volume of dry air at specified well temperature 10. Dry air at T&P (L) (kg): mass of dry air at the specified well temperature 11. Dry (or wet) air at T&P, slpm: dry (or wet) airflow rate calculated (or measured) at the specified well temperature 12. Standard atmospheric pressure

Appendix C Listing and Description of Example Calculations and Tables (continued)

Item	Title	Description
Table C4	RF-SVE heat balance	Calculation of heat requirements for SVE air and water vapor at the wellhead temperature (T) 1. Temperature SVE well flow, °C and K: temperatures (T) of air-water vapor mixture in SVE B, C, and/or A well(s) at 3.4 to 4.6 m depth and date of field data measurements (example: 12/03/96 data) 2. Ambient air temperature, °C and K: current atmospheric temperature at the demonstration site 3. Water enthalpy, kJ/kg-°C: enthalpy values at the specified temperature (T) for liquid water (H_l) and for the vaporization of water (H_{lg}) from Table C5 4. Heat capacity of air, kJ/kg-°C: heat capacity values for air at the specified temperature (T) from Table C6 5. Component mass, kg: mass of water recovered and of air used from Table C3 6. Heat content, kJ/day: heat content of liquid water, water vapor, and air at the specified temperature (T). For a given time period, the total values are underlined and in bold type
Table C5	Thermodynamic properties of water	1. Boiling point vs. elevation: boiling point of water (°C) and atmospheric pressure (kPa) at increasing elevations 2. Heat content of water: heat capacity (kJ/kg-K) and heat content (kJ/kg) of water at increasing temperatures
Table C6	Thermodynamic properties of air	1. Properties of dry air: standard pressure, kPa; specific volume, m³/kg; density, kg/m³; and heat capacity, kJ/kg-°C at increasing temperatures 2. Humidity-saturated air temperature: humidity of water-saturated air in kg water per kg dry air at increasing temperatures
Table C7	Heat capacity (Cp) of minerals	Heat capacity values for minerals found in the soil at the Kirtland AFB demonstration site from Kubaschewski (1993) and Saxena (1993) 1. Mineral name and composition, formula weight of mineral, density of mineral, average mineral content at Kirtland AFB site 2. Heat capacity, Cp (j/[gm-mol]-K (°C): heat capacity of individual minerals and inorganic oxides 3. Heat capacity, Cp (kJ/[kg]-K (°C): heat capacity of individual minerals used in calculations 4. Heat capacity, Cp (kJ/[kg]-K (°C): average heat capacity of soil at increasing temperatures
Table C8	Mass water vapor in exhaust air	Table calculating the mass of water and heat content of the wet air entering (ambient air) and leaving (blower) the SVE system

HEAT BALANCE EXAMPLE CALCULATIONS

Note: Standard Temperature and Pressure (STP) — Standard atmospheric pressure (P) for airflow measurements at the rotameter was 101.31 kPa (14.70 lb/in²). The standard temperature (T) for the airflow measurements at the rotameter was 20.0°C (68°F). The elevation at Kirtland AFB site is 1686 m. The atmospheric pressure adjusted for elevation equals 82.60 kPa or 12.54 lb/in.² [Table C5]. The air-and-water vapor volumes should be corrected for the lower atmospheric pressure. The calculated data in Appendix C tables were obtained using macros and the computer programs, NO-e-SYS and MS Excel. In most cases, the values within the example calculations have been rounded to an appropriate significant digit.

CALCULATIONS FOR TABLE C1:
SOIL TEMPERATURES FOR SAMPLED VOLUME
AND TABLE C2: RF-SVE HEAT UTILIZATION

All Calculations for "2-14-97, temp 48 K" Column
Period covered: January 27 to February 14, 1997
Number of days in time period = 18

Table C1 data
    Heat conductivity of soil, k-soil = 1.70 W/m-°C or 1.70 (J/s)/m-°C
    Total *heated* volume of soil = (8.778 m)(8.778 m)(4.206 m) = 324.1 m^3
    Soil density (dry basis) = 1778 kg of dry soil per m^3
    Total soil mass = (324.1 m^3)(1778 kg/m^3) = 576,300 kg of dry soil

I. Data Calculated by NO-e-SYS Program in Table C1
    A. NO-e-SYS uses the macro (BASIC) programs to calculate the average temperature and the boundary-heat-loss factor, A(dT/dx).
    B. The program also calculates the shape and size of the voxel volume (soil volume) for each temperature range, for example, soil temperature between 60 and 80°C:
        1. Voxel volume defined in the Average Temperature macro
           a. Total voxel volume = (47)(49)(49) = 112,847 voxels
           b. Soil volume per voxel = 324.1 m^3/112,847 voxels = 0.002872 m^3/voxel
        2. Average Temperature = 69.1°C
        3. Voxel vol = 22,147
        4. Vol = (22,147 voxels)(0.002872 m^3/voxel) = 63.61 m^3
        5. Mass = (1778 kg/m^3)(63.61 m^3) = 113,105 kg of dry soil
    C. Total *heated* vol:
        1. Average temperature of 324.1-m^3 vol = 57.5°C
    D. Boundary loss:
        1. Soil A(dT/dx) = 1.4512 kilo-m^2/m-°C
II. Data calculated in Table C1 and Table C2:
    A. Residual water in soil
        1. The water removed by vaporization, water (daily), can be found in Table B3, RF-SVE water and free DRO product recovered.
        2. Initial moisture content of soil = 7.0% dwb or 7.0 kg water per 100 kg of dry soil
        3. Total soil mass = 576,300 kg of dry soil
        4. Initial moisture in soil = (576,300 kg)(7.0 kg/100 kg) = 40,341 kg
        5. Water collected (end date: 2-14 -97) = 5293 kg
        6. Residual water = 40,341 kg – 5293 kg = 35,049 kg
    B. RF heat produced
        1. Table D1, summary of KAI RF energy data
        2. RF operating time period: 1-27-97 to 2-14-97
        3. RF energy delivered = 900,852 kJ/day
        4. Total RF energy delivered = (900,850 kJ/day)(18 days) = 16,215,300 kJ
    C. Heat to soil
        1. Heat to soil = (c_p, specific heat of soil)(average temperature difference)(soil mass)
        2. Average temperatures for total soil mass from Table C1
        3. Average temperature previous period (temp 33 K) = 47.7°C
        4. Average temperature current period (temp 48 K) = 57.5°C
        5. Average temperature difference = 57.5°C – 47.7°C = 9.8°C

6. Specific heat values for soil (c_p) from Table C7: c_p (50°C) = 0.873 kJ/kg-K and c_p (75°C) = 0.911 kJ/kg-K(°C)

7. c_p (57.5°C) = 0.873 kJ/kg-°C + (57.5°C – 50.0°C)/(75.0°C – 50.0°C)(0.911 kJ/kg-°C – 0.873 kJ/kg-°C) = 0.884 kJ/kg-°C

8. Heat to soil = (0.884 kJ/kg-°C)(9.8°C)(576,300 kg) = 4,984,456 kJ

D. Heat to residual water
1. Specific heat of water at 60°C (c_p) from Table C5: c_p = 4.195 kJ/kg-K(°C)
2. Residual water = 38,085 kg
3. Heat to residual water = (4.195 kJ/kg-°C)(9.8°C)(35,049 kg) = 1,436,613 kJ

E. Heat to water vapor
1. From Table C4: Heat content, totals for water (L) plus water (g)
Totals = 805,568 kJ + 6,103,879 kJ = 6,909,447 kJ

F. Heat to SVE air
1. From Table C4: heat content, totals for air = 1,255,750 kJ

G. Heat to water in exhaust air
1. From Table C8: heat content, total in exhaust air = 727,847 kJ

H. Boundary loss, Table C1
1. From Table C1: k = 1.70 (J/s)/m-°C
2. Date & time: 1-27-97 to 2-14 -97
3. dt = (18 days)(24 h/day)(60 min/h)(60 s/min) = 1,555,200 s
4. A(dT/dx) = 1.4512 kilo-m²-°C/m
5. Boundary loss = (k)(A(dT/dx))(dt)
Boundary loss = [1.70 (J/s)/m-°C](1.4512 kilo-m²-°C/m)(1,555,200 s)
Boundary loss = 3,836,700 kJ

I. Total heat utilized, Table C1 = 4,984,456 kJ + 1,436,613 kJ + 6,911,691 kJ + 1,255,750 kJ +727,847 kJ + 3,836,700 kJ = 19,153,061 kJ

CALCULATIONS FOR (1/28/97) DATA POINT IN TABLE C3:
SVE CALCULATED AIRFLOWS

Use (1/28/97) data from Table B2, Appendix B:

Air was extracted from Well-B.
Air temperature measured in Well-B at 3.81 m and 5.18 m (15 ft and 17 ft).
Average air temperature recorded: $t_C = (t_{3.8} + t_{5.2})/2$.
Well-B air temperature: t_C = 125.0°C $(t_K$ = 398.1 K)
Airflow measured by rotameter attached to the outlet of the knock-out drum.
Number of days between each measurement of the water and DRO volumes = 1.0 day.
Operating Time (OT) = 18.17 h/day = (18.17 h/day)(60 min/h) = 1090 min/day
Measured airflow at STP = 481 slpm or 17 scfm
Water collected = 175 kg over 1.0 day

TABLE C3: SVE CALCULATED AIRFLOW
(WATER-VAPOR CONTENTS IN DRY AIR)

I. Dry air at actual temperature and standard pressure
A. Temperature, $t_K = (t_C,$ °C + 273.15), K
B. Dry air vol = (1090 min/day)(481 slpm)(125.0°C + 273.15)/(20°C + 273.15) = 712,767 L/day

C. Density dry air = 0.8864 kg/m³ (Table C6, interpolated at 125°C)

D. Dry air weight = (712,767 L/day)(0.8864 kg/m³)/(1000 L/m³) = 632 kg/day

E. Actual humidity of dry air = (175 kg water per day)/(632 kg air per day) = 0.2768 kg/kg

II. Water vapor at actual T&P

A. Water vapor (gas) vol = [(175 kg water)(22,420 L/kg-mol)/(18.015 kg/kg-mol)][(125.0°C + 273.15)/273.15)]

B. Water vapor (gas) volume = 317,279 L/day

III. Total wet air volume at actual T&P

A. Wet air volume = (712,767 L/day + 317,279 L/day) = 1,030,046 L/day

B. Wet airflow = (1,030,046 L/day)/(1090 min/day) = 944.8 Lpm

IV. Saturated humidity at T&P

A. The calculated humidity results for the saturated dry air for temperatures between 273.15 K and 37 K are in Table C6.

B. At 96.7°C, water-saturated air will contain 5.186 kg of water per kg of dry air [Table C6].

C. Above 100°C, the air can contain an unlimited amount of water vapor. The amount of water vapor in the air is limited by the heat input into the treatment volume.

D. Below 100°C, the amount of water in the air is limited. The partial pressure of water at a given temperature determines the amount of water vapor in the air as follows:

1. Standard atmospheric pressure = 101.3 kPa

2. Data from Table C5:

a. Air and water-vapor temperature = 86.85°C (360 K)

b. Water vapor pressure at 360 K = 62.15 kPa

3. Water in saturated dry air at 360 K = ((62.15 kPa)/(101.3 kPa – 62.15 kPa))(18.02 kg water/kg-mol)/(28.96 kg air/kg-mol) = 0.9872 kg/kg

4. Example, refer to (1/08/97) data in Table B2, and in Table C3:

5. (1/02/97) temperature = 89.3°C at 0 days
(1/07/97) temperature = 93.8°C at 4 days
(1/08/97) temperature = 88.2°C at 1 day
Total = 5 days

6. Weighted well-B air temperature for five-day period:
t_C = (((89.3°C + 93.8°C)/2)(4 days) + (88.2°C)(1 day))/(5 days) = 90.9°C
t_K = 90.9°C + 273.15 = 364.0 K

E. Humidity calculation at a temperature below 100°C

1. Saturated humidity (dry air) at 360.0 K, Table C6 = 0.9872 kg/kg

2. Saturated humidity (dry air) at 365.0 K, Table C6 = 1.7920 kg/kg

3. Saturated humidity (dry air) at 364.0 K = 0.9872 + ((364 – 360)/(365 – 360))(1.7920 – 0.9872) = 1.6343 kg/kg

CALCULATIONS FOR (1/28/97) DATA POINT IN TABLE C4: RF-SVE HEAT BALANCES

Use (01/28/97) data from Appendix B and Appendix C:
Data from Appendix B, Table B2:

Air temperature measured in well B at 3.81 m and 5.18 m (15 ft and 17 ft).
Well-B air temperature = 125.0°C (398.1 K)
Ambient temperature = 7.8°C (280.9 K)
Dry airflow = 481 std, L/min (slpm) or 17 std, ft³/min (scfm)
Water (mass) removed = 175 kg/day

Data from Appendix C, Table C3

Mass dry airflow, ID = 632 kg/day

I. Enthalpy of Water
   A. Interpolated results using the enthalpy data in Table C5
   B. Final temperature of water vapor in air = 125.0°C (398.1 K)
   C. Initial temperature of liquid water in soil = 27.4°C (300.5 K)
   D. Enthalpy of liquid water (H_l) at 100°C = 419.5 kJ/kg
   E. Enthalpy of liquid water (H_l) at 27.4°C = 111.7 kJ/kg + ((300.5 K) – 300 K)/(305 K – 300 K)(132.8 kJ/kg – 111.7 kJ/kg) = 114.0 kJ/kg
   F. Heat of vaporization at 100°C (H_{lg}) = 2256.1 kJ/kg
   G. Enthalpy of water vapor (K_g) = (33.46 + (0.688*10–2)(Δt°C) + (0.7604*10–5)(Δt°C)^2 – (3.593*10–9)(Δt°C)^3 *kJ/kg-mol-°C*)(Δt°C)/(18.016 kg water/kg-mol)
   H. Enthalpy of water vapor (K_g) = (33.46 + (0.688*10–2)(125°C – 100°C) + (0.7604*10–5)(125°C –100°C)^2 – (3.593*10–9)(125°C – 100°C)^3)(125°C – 100°C)/(18.016)
      Enthalpy of water vapor (K_g) = 46.7 kJ/kg
   I. Total enthalpy of water vapor
      1. Liquid at 100°C = (419.5 kJ/kg – 114.0 kJ/kg) = 305.5 kJ/kg
      2. Vapor from 100°C–125°C = (2256.1 kJ/kg + 46.7 kJ/kg) = 2302.8 kJ/kg
II. Total heat content of water vapor
   A. Mass of water vapor in air = 175 kg/day
   B. Heat content of liquid = (305.5 kJ/kg)(175 kg/day) = 53,436 kJ/day
   C. Heat content of vapor = (2302.8 kJ/kg)(175 kg/day) = 402,756 kJ/day
   D. Total heat content of water vapor = 53,436 kJ/day + 402,756 kJ/day = 456,191 kJ/day
III. Heat capacity of air
   A. Interpolated results from heat capacity data in Table C6
   B. Temperatures between 395.0 K and 400.0 K
   C. Heat capacity of air = 1.019 kJ/kg-°C
IV. Total heat content of air
   A. Mass of dry air = 632 kg/day
   B. Total heat content of air = (1.019 kJ/kg-°C)(632 kg/day)(125°C – 7.8°C) = 75,470 kJ/day
V. Total heat added to air and water
   A. Total heat added = 456,191 kJ/day + 75,470 kJ/day = 531,661 kJ/day

CALCULATIONS FOR TABLE C1: MACRO FOR AVERAGE TEMPERATURE USING NO-e-SYS PROGRAM

```
! This routine calculates the volume and temperature for various temperature ranges within
! a soil volume centered at the RF antennae:
!              x(depth: bottom to top): –9.4 to +4.4
!              y(south to north): –14.4 to +14.4
!              z(west to east): –14.4 to +14.4
! Data interpolated within a 47 × 49 × 49 volume.
!
! Program: Version 1.0 (jwj)
integer*1 i,j,k
integer*4 cnt40, cnt60, cnt80, cnt94, cnt110, cnt130, cnt130plus, cnttotal, cnt94plus
real*4 avg40, avg60, avg80, avg94, avg110, avg130, avg130plus, avgtotal, avg94plus
```

```
do i=3,49
     do j = 2,50
     do k = 2, 50
          if (DATA(i,j,k) <= 40.0) then
               cnt40 = cnt40 + 1
               avg40 = avg40 + DATA(i,j,k)
          else if (DATA(i,j,k)> 40.0 .AND. DATA(i,j,k) <= 60.0) then
               cnt60 = cnt60 +1
               avg60 = avg60 + DATA(i,j,k)
          else if (DATA(i,j,k)> 60.0 .AND. DATA(i,j,k) <= 80.0) then
               cnt80 = cnt80 +1
               avg80 = avg80 + DATA(i,j,k)
          else if (DATA(i,j,k)> 80.0 .AND. DATA(i,j,k) <= 94.0) then
               cnt94 = cnt94 +1
               avg94 = avg94 + DATA(i,j,k)
          else if (DATA(i,j,k)>94.) then
               cnt94plus = cnt94plus+1
               avg94plus = cnt94plus + DATA(i,j,k)
               if (DATA (i,j,k)> 94.0 .AND. DATA(i,j,k) <= 110.0) then
                    cnt110 = cnt110 +1
                    avg110 = avg110 + DATA(i,j,k)
               else if (DATA(i,j,k)> 110.0 .AND. DATA(i,j,k) <= 130.0) then
                    cnt130 = cnt130 +1
                    avg130 = avg130 + DATA(i,j,k)
               else if (DATA(i,j,k)> 130.0 ) then
                    cnt130plus = cnt130plus +1
                    avg130plus = avg130plus + DATA(i,j,k)
               endif
          endif
          cnttotal = cnttotal + 1
          avgtotal = avgtotal + DATA(i,j,k)
     enddo
  enddo
  enddo
  if (cnt40>0) then
       avg40 = avg40/cnt40
  endif
  if (cnt60>0) then
       avg60 = avg60/cnt60
  endif
  if (cnt80>0) then
       avg80 = avg80/cnt80
  endif
  if(cnt94>0) then
       avg94 = avg94/cnt94
  endif
  if(cnt110>0)then
       avg110 = avg110/cnt110
  endif
  if(cnt130>0) then
       avg130 = avg130/cnt130
```

```
endif
if (cnt130plus>0) then
      avg130plus = avg130plus/cnt130plus
endif
avgtotal = avgtotal/cnttotal
if (cnt94plus>0) then
 avg94plus = avg94plus/cnt94plus
endif
print ""
print "Temperature Volume Averages for File TEMP26k.HDF"
print ""
print "avg40 ", avg40
print "cnt40 = ", cnt40
print ""
print "avg60 ", avg60
print "cnt60 = ", cnt60
print ""
print "avg80 ", avg80
print "cnt80 = ", cnt80
print ""
print "avg94 = ",avg94
print "cnt94 = ", cnt94
print ""
print "avg110 = ", avg110
print "cnt110 = ", cnt110
print ""
print "avg130 = ", avg130
print "cnt130 = ", cnt130
print ""
print "avg130plus = ", avg130plus
print "cnt130plus = ", cnt130plus
print ""
print "avg94plus = ", avg94plus
print " cnt94plus = ", cnt94plus
print ""
print "avgtotal = ", avgtotal
print "cnttotal = ", cnttotal
print ""
```

Sample Output
Temperature Volume Averages for File TEMP26k.HDF

```
avg40 =        2.759489e+01
cnt40 =        71511
avg60 =        4.889531e+01
cnt60 =        23634
avg80 =        6.910997e+01
cnt80 =        13981
avg94 =        8.416115e+01
cnt94 =        3721
avg110 =       0.000000e+00
```

```
cnt110 =        0
avg130 =        0.000000e+00
cnt130 =        0
avg130plus =    0.000000e+00
cnt130plus =    0
avg94plus =     0.000000e+00
cnt94plus =     0
avgtotal =      3.906395e+01
cnttotal =      112847
```

CALCULATIONS FOR TABLE C3: MACRO FOR HEAT LOSS TO SURROUNDINGS USING NO-e-SYS HDF DATASET

```
! Calculation of energy lost at the surfaces of the temperature dataset.
! Temperature of volume (13.8 ft × 28.8 ft × 28.8 ft or 4.206 m × 8.778 m × 8.778 m) = 324 m³
!
! Energy loss at a boundary is calculated using the following equation
!               Qs = k*dT/dx*A*dt
! where
!               Qs = Heat loss at a surface (kJ)
!               k  = Thermal conductivity of the soil (kW/m-°C)
!               dT = Rate of change of temperature at the surface, °C,
!                    estimated by T(+0.6 feet from surface) – T(-0.6 feet from surface)
!               A  = Area of surface (m²)
!               dt = Period of heat (seconds) = days * 24 hours/day * 60 min/h * 60 sec/min
!
! A * dT/dx is calculated for the six rectangular surfaces and summed.
! The sum is multiplied by k * dt in the energy balance spreadsheet to get total heat loss at the
! surface:
!
!               Top surface at x = 4.4 ft (49)
!               Bottom surface at x = –9.4 ft (3)
!               North surface at y = 14.4 ft (50)
!               South surface at y = –14.4 ft (2)
!               East surface at z = 14.4 ft (50)
!               West surface at z = –14.4 ft (2)
!
! Data interpolated within a 51 x 51 x 51 volume:
!               delta_x = 0.3 feet/index, dx = 1.2 feet = 1.2/3.28902 = 0.3658 m
!               delta_y, delta_z = 0.6 feet/index, dy, dz = 1.2 feet/3.28902 = 0.3658 (3.66?) m
!
! Program: Version 1.0 (jwj)
integer*1 m,n
real*4 dT, Tmax, Tmin, At, As, vol
real*4 dx, dy, dz, Qt_k, Qb_k, Qn_k, Qs_k, Qe_k, Qw_K, Qtotal_k
At = 28.8*28.8 /(3.2808 * 3.2808) ! m²
As = 13.8*28.8/(3.2808 * 3.2808) ! m²
dx = 0.3658 !m
```

```
dy = 0.3658  !m
dz = 0.3658  !m
vol = 13.8 * 28.8 * 28.8 /(3.2808 * 3.2808 * 3.2808) ! m³

! Calculation of top energy loss (Qt)
Tmax=0
Tmin=0
do m=2, 50
          do n=2, 50
                Tmax = DATA(47,m,n) + Tmax
                Tmin = DATA(51,m,n) + Tmin
          enddo
enddo
dT = (Tmax − Tmin)/(49*49)
Qt_k = ((dT/dx)*At)/1000

! Calculation of bottom energy loss (Qb)
Tmax=0
Tmin=0
do m=2, 50
          do n=2, 50
                Tmax = DATA(5,m,n) + Tmax
                Tmin = DATA(1,m,n) + Tmin
          enddo
enddo
dT = (Tmax − Tmin)/(49*49)
Qb_k = ((dT/dx)*At)/1000

! Calculation of North side energy loss(Qn)
Tmax=0
Tmin=0
do m=3, 49
          do n=2, 50
                Tmax = DATA(m,49,n) + Tmax
                Tmin = DATA(m,51,n) + Tmin
          enddo
enddo
dT = (Tmax − Tmin)/(47*49)
Qn_k = ((dT/dy)*As)/1000

! Calculation of South side energy loss (Qs)
Tmax=0
Tmin=0
do m=3, 49
          do n=2, 50
                Tmax = DATA(m,3,n) + Tmax
                Tmin = DATA(m,1,n) + Tmin
          enddo
enddo
dT = (Tmax − Tmin)/(47*49)
Qs_k = ((dT/dy)*As)/1000
```

```
! Calculation of East side energy loss (Qe)
Tmax=0
Tmin=0
do m=3, 49
            do n=2, 50
                  Tmax = DATA(m,n,49) + Tmax
                  Tmin = DATA(m,n,51) + Tmin
            enddo
enddo
dT = (Tmax - Tmin)/(47*49)
Qe_K = ((dT/dz)*As)/1000

! Calculation of West side energy loss (Qw)
Tmax=0
Tmin=0
do m=3, 49
            do n=2, 50
                  Tmax = DATA(m,n,3) + Tmax
                  Tmin = DATA(m,n,1) + Tmin
            enddo
enddo
dT = (Tmax - Tmin)/(47*49)
Qw_k = ((dT/dz)*As)/1000
Qtotal_k = Qt_k+Qb_k+Qn_k+Qs_k+Qe_k+Qw_k
print ""
print "At = ", At
print "As = ", As
print "dx = ", dx
print "Qt_k = ", Qt_k
print "Qb_k = ", Qb_k
print "Qn_k = ", Qn_k
print "Qs_k = ", Qs_k
print "Qe_k = ", Qe_k
print "Qw_k = ", Qw_k
Print "Sum of (dT/dx)*A over six surfaces for Temp26k = ", Qtotal_k
```

Sample Output
```
At = 7.705937e+01
As = 3.692428e+01
dx = 3.658000e-01
Qt_k = 1.912484e-01
Qb_k = 3.336670e-01
Qn_k = 1.430703e-01
Qs_k = 1.646180e-01
Qe_k = 1.529155e-01
Qw_k = 1.553209e-01
Sum of (dT/dx)*A over six surfaces for Temp26k = 1.140840e+00
```

Table C1 Soil Temperatures for Sampled Volume

Date & Data File	12/3/96 Temp. 2 K	12/12/96 Temp. 12 K	12/20/96 Temp. 18 K	1/3/97 Temp. 25 K	1/9/97 Temp. 26 K	1/17/96 Temp. 30 K	1/27/97 Temp. 33 K	2/14/97 Temp. 48 K	2/28/97 Temp. 60 K	3/12/97 Temp. 66 K	3/18/97 Temp. 68 K
below 40°C											
Av temp (°C)	28	29	28	28	28	30	34	36	37	34	34
Vol (voxels)	112,847	107,311	90,174	72,106	71,511	61,969	50,388	23,151	12,978	26,153	41,593
Vol (m³)	0	0	0	0	0	0	0	0	0	0	0
Mass (kg)	0	0	0	0	0	0	0	0	0	0	0
40–60°C											
Av temp (°C)		42	48	49	49	49	48	48	49	49	48
Vol (voxels)	0	5,536	19,445	22,852	23,634	29,979	38,514	49,237	53,220	48,579	51,917
Vol (m³)	0	0	0	0	0	0	0	0	0	0	0
Mass (kg)	0	0	0	0	0	0	0	0	0	0	0
60–80°C											
Av temp (°C)			63	69	69	69	69	69	69	69	68
Vol (voxels)	0	0	3,228	12,995	13,981	14,179	16,410	22,147	23,744	23,195	17,302
Vol (m³)	0	0	0	0	0	0	0	0	0	0	0
Mass (kg)	0	0	0	0	0	0	0	0	0	0	0
80–94°C											
Av temp (°C)				85	84	87	86	86	87	87	83
Vol (voxels)	0	0	0	4,894	3,721	5,939	5,126	9,392	10,309	9,080	2,035
Vol (m³)	0	0	0	0	0	0	0	0	0	0	0
Mass (kg)	0	0	0	0	0	0	0	0	0	0	0
94–110°C											
Av temp (°C)						95	100	101	101	101	
Vol (voxels)	0	0	0	0	0	781	2,051	6,240	7,384	5,057	0
Vol (m³)	0	0	0	0	0	0	0	0	0	0	0
Mass (kg)	0	0	0	0	0	0	0	0	0	0	0
110–130°C											
Av temp (°C)							115	116	118	113	
Vol (voxels)	0	0	0	0	0	0	356	2,678	5,000	783	0
Vol (m³)	0	0	0	0	0	0	0	0	0	0	0
Mass (kg)	0	0	0	0	0	0	0	0	0	0	0
above 130°C											
Av temp (°C)	0	0	0	0	0	0	131	130	131	0	0
Vol (voxels)	0	0	0	0	0	0	2	2	212	0	0

Vol (m³)	0	0	0	0	0.00	0.00	0.00	0	0
Mass (kg)	0	0	0	0	0	0	0	0	0
Totals									
Vol (m³)	324	324	324	324	324	324	324	324	324
Mass (kg)	576,300	576,300	576,300	576,300	576,300	576,300	576,300	576,300	576,300
Av temp (°C)	32	40	39	44	48	57	62	55	47

≥ 80°C vol

Vol, m³	30	14	11	19	22	53	66	43	6
Ave. T,°C		85	84	88	91	96	99	93	83

Constants

Total vol(m³)	324
Total voxels	112,847
Vol/voxel	0.002872
Density-soil (kg/m³)	1,778
Total mass (kg)	576,311

c_p-water (kJ/kg-°C)

c_p (30°C)	4.163
c_p (40°C)	4.175
c_p (50°C)	4.185
c_p (60°C)	4.195
c_p (70°C)	4.205
c_p (80°C)	4.215
c_p (90°C)	4.226

(1.0 W = 1.0 J/sec)

c_p-soil (kJ/kg-°C)

c_p (0°C)	0.777
c_p (25°C)	0.830
c_p (50°C)	0.874
c_p (75°C)	0.911
c_p (100°C)	0.944
c_p (125°C)	0.974
c_p (150°C)	1.000
k-soil (W/m-°C)*	1.70
ft/m	3.2808

Table C1 Soil Temperatures for Sampled Volume (continued)

Date & Data File	12/3/96 Temp. 2 K	12/12/96 Temp. 12 K	12/20/96 Temp. 18 K	1/3/97 Temp. 25 K	1/9/97 Temp. 26 K	1/17/96 Temp. 30 K	1/27/97 Temp. 33 K	2/14/97 Temp. 48 K	2/28/97 Temp. 60 K	3/12/97 Temp. 66 K	3/18/97 Temp. 68 K
Residual water (kg)	40,238	40,146	39,769	39,574	39,273	37,686	35,049	33,379	33,379	0	0
Water collected (kg)	104	92	377	195	301	1,587	2,637	1,670	6,963	681	179
Total collected (kg)	104	196	573	768	1,069	2,656	5,293	6,963	6,963	7,644	7,141
RF heat produced[b]	1,485,400	3,917,900	9,116,200	4,374,500	6,139,100	8,095,100	16,215,300	11,774,100	61,117,600	0	0
Heat to soil[d]	922,676	1,345,879	3,552,705	–265,977	2,231,932	2,080,372	4,984,456	2,224,898	17,044,941	–3,258,642	–4,293,309
Heat to res. water	319,926	462,915	1,185,381	–89,211	736,133	654,574	1,436,613	606,369	5,312,700	–877,223	–1,167,776
Heat to water vapor[b]	253,852	231,621	953,845	497,216	769,814	4,015,198	6,911,691	4,387,370	2,922,752	1,771,263	454,611
Heat to SVE air[b]	87,401	47,723	64,087	48,859	79,383	500,753	1,255,750	838,796	2,922,752	611,643	177,552
Heat to water in exhaust	67,098	13,140	10,601	10,010	16,078	365,410	727,847	229,315	1,439,499	204,816	114,729
Boundary loss											
Soil A(dT/dx)	0.4668	0.6843	1.0892	1.1408	1.0734	1.0010	1.4512	1.5342	8.4409	2.0382	1.5204
Days	9	8	14	6	8	10	18	14	87	12	5
Boundary loss[c]	0	0	0	0	0	0	0	0	0	0	0
Total	2,268,087	2,905,347	7,974,364	1,206,296	5,094,673	9,086,522	19,153,061	11,441,464	59,129,815	2,044,388	–3,597,623
Difference	–782,687	1,012,553	1,141,836	3,168,204	1,044,427	–991,422	–2,937,761	332,636	1,987,785	–2,044,388	3,597,623
% of RF energy in	152.7%	74.2%	87.5%	27.6%	83.0%	112.2%	118.1%	97.2%	96.7%		

Note: Temperature averages and energy balance for total temperature vol (4.21 m [13.8 ft] × 8.78 m [28.8 ft] × 8.78 m [28.8 ft]). Cartesian coordinate system centered at the middle of the design volume. Total vol = 324.09 m³. X(depth): –3.05 m to +1.52 m; Y(North), and Z(East): –4.57 m to +4.57 m. Calculations in SI units.

a The average temperature was calculated by summing temperatures at each point within the design volume and dividing by the total number of point.

b Values taken from Table C4 for the specific time period.

c Total energy loss at boundary = sum of energy lost at each surface of the design volume. Energy loss at a surface = kA(dT/dx)dt (all units converted to SI prior to making any calculations). k = thermal conductivity for the soil, W/m-°C or (J/s)/m-°C; A = area of the surface, m²; dT = average temperature (°C) of surface 0.6 ft inside the design surface, average temperature of surface 0.6 feet outside design surface; dx = 0.366 m (1.2 ft); dt = (number of days in time period × 24 × 60 × 60) s; A(dT/dx), kilo-m²/m-°C.

d Heat to soil = (specific heat for soil)*(average temperature for period/average temperature for last period)*(soil mass of design volume).

e k-soil (W/m-°C), Table A6 (Incropera et al., 1990).

	12/3/96 Temp. 2 K	12/12/96 Temp. 12 K	12/20/96 Temp. 18 K	1/3/97 Temp. 25 K	1/9/97 Temp. 26 K	1/17/97 Temp. 30 K
RF Heat Input, kJ						
Total	0	1,485,000	3,918,000	9,116,000	4,375,000	6,139,000
Operating days	0	9	8	14	6	8
Heat Utilization, kJ						
Dry soil	0	923,000	1,346,000	3,521,000	−266,000	2,232,000
Residual soil water	0	320,000	463,000	1,185,000	−89,000	736,000
Condensed water vapor	0	254,000	232,000	954,000	497,000	770,000
SVE airflow	0	87,000	48,000	64,000	49,000	79,000
Water vapor in exhaust air	0	67,000	13,000	11,000	10,000	16,000
Loss to surroundings	0	617,000	804,000	2,240,000	1,005,000	1,261,000
Total Utilized, kJ	0	2,268,000	2,906,000	7,975,000	1,206,000	5,094,000
% of Input	0.0	153	74	87	28	83

	1/27/97 Temp. 33 K	2/14/97 Temp. 48 K	2/28/97 Temp. 60 K	2/28/97 Total	3/12/97 Temp. 66 K	3/18/97 Temp. 68 K
Heat Input, kJ						
RF unit	8,095,000	16,215,000	11,774,000	61,118,000	0	0
Operating days	10	18	14	87	12	6
Heat Utilization, kJ						
Dry soil	2,080,000	4,984,000	2,225,000	17,045,000	−3,259,000	−4,293,000
Residual soil water	655,000	1,437,000	606,000	5,313,000	−877,000	−1,168,000
Condensed water vapor	4,015,000	6,912,000	4,387,000	18,021,000	1,771,000	455,000
SVE airflow	501,000	1,256,000	839,000	2,923,000	612,000	178,000
Water vapor in exhaust air	365,000	728,000	229,000	1,439,000	205,000	115,000
Loss to surroundings	1,470,000	3,837,000	3,155,000	14,389,000	3,593,000	1,117,000
Total Utilized, kJ	9,086,000	19,154,000	11,441,000	59,130,000	2,045,000	−3,596,000
% of Input	112	118	97	97	0.0	0.0

	12/3/96 Temp. 2 K	12/12/96 Temp. 12 K	12/20/96 Temp. 18 K	1/3/97 Temp. 25 K	1/9/97 Temp. 26 K	1/17/97 Temp. 30 K
RF Heat Input, kJ						
Total	0	1,485,000	5,403,000	14,519,000	18,894,000	25,033,000
Total operating days	0	9	17	31	37	45
Heat Utilization, kJ						
Dry soil	0	923,000	2,269,000	5,790,000	5,524,000	7,756,000

Table C2 RF-SVE Heat Utilization (continued)

	12/3/96 Temp. 2 K	12/12/96 Temp. 12 K	12/20/96 Temp. 18 K	1/3/97 Temp. 25 K	1/9/97 Temp. 26 K	1/17/97 Temp. 30 K
Residual soil water	0	320,000	783,000	1,968,000	1,879,000	2,615,000
Condensed water vapor	0	254,000	486,000	1,440,000	1,937,000	2,707,000
SVE airflow	0	87,000	135,000	199,000	248,000	327,000
Water vapor in exhaust air	0	67,000	80,000	91,000	101,000	117,000
Loss to surroundings	0	617,000	1,421,000	3,661,000	4,666,000	5,927,000
Total Utilized, kJ	0	2,268,000	5,174,000	13,149,000	14,355,000	19,449,000
% of Input	0.0	153	96	91	76	78

	1/27/97 Temp. 33 K	2/14/97 Temp. 48 K	2/28/97 Temp. 60 K	2/28/97 Total	3/12/97 Temp. 66 K	3/18/97 Temp. 68 K
Heat Input, kJ						
RF unit	33,128,000	49,343,000	61,117,000	61,118,000	0	0
Operating days	55	73	87	87	99	105
Heat Utilization, kJ						
Dry soil	9,836,000	14,820,000	17,045,000	17,045,000	13,786,000	12,752,000
Residual soil water	3,270,000	4,707,000	5,313,000	5,313,000	4,436,000	4,145,000
Condensed water vapor	6,722,000	13,634,000	18,021,000	18,021,000	19,792,000	18,476,000
SVE airflow	828,000	2,084,000	2,923,000	2,923,000	3,535,000	3,101,000
Water vapor in exhaust air	482,000	1,210,000	1,439,000	1,439,000	1,644,000	1,554,000
Loss to surroundings	7,397,000	11,234,000	14,389,000	14,389,000	17,982,000	15,506,000
Total Utilized, kJ	28,535,000	47,689,000	59,130,000	59,130,000	61,175,000	55,534,000
% of input	86	97	97	97	0.0	0.0

	12/03/96 1/17/97			12/03/96 1/03/96		
		% of Heat Input	kJ/kJ_{input}		% of Heat Input	kJ/kJ_{input}
RF Heat Input, kJ						
Total	25,033,000			14,519,000		
Operating days	45			31		
Heat Utilization, kJ						
Dry soil	7,756,000	31.0	0.310	5,790,000	39.9	0.399
Residual soil water	2,615,000	10.4	0.104	1,968,000	13.6	0.136
Condensed water vapor	2,707,000	10.8	0.108	1,440,000	9.9	0.099
SVE airflow	327,000	1.3	0.013	199,000	1.4	0.014
Water vapor in exhaust air	117,000	0.5	0.005	91,000	0.6	0.006
Loss to surroundings (calc)	5,927,000	23.7	0.237	3,661,000	25.2	0.252
Loss to surroundings (diff)	11,511,000	46.0	0.460	5,031,000	34.7	0.347
Total Utilized, kJ	19,449,000	77.7		13,149,000	90.6	
of input						

	1/17/97 2/28/97			12/03/96 2/28/97		
		% of Heat Input	kJ/kJ_{input}		% of Heat Input	kJ/kJ_{input}
Heat Input, kJ						
RF unit	36,084,000			61,118,000		
Operating days	42			87		
Heat Utilization, kJ				19,460,000		
Dry soil	9,289,000	25.7	0.257	17,045,000	27.9	0.279
Residual soil water	2,698,000	7.5	0.075	5,313,000	8.7	0.087
Condensed water vapor	15,314,000	42.4	0.424	18,021,000	29.5	0.295
SVE airflow	2,596,000	7.2	0.072	2,923,000	4.8	0.048
Water vapor in exhaust air	1,322,000	3.7	0.037	1,439,000	2.4	0.024
Loss to surroundings (calc)	8,462,000	23.5	0.235	14,389,000	23.5	0.235
Loss to surroundings (diff)	4,865,000	13.5	0.135	16,377,000	26.8	0.268
Total Utilized, kJ	39,681,000	110.0		59,130,000	96.7	
% of input						

Table C3   SVE Calculated Airflows

Temperature SVE Well Flow[a] °C		K	Dry Air Humidity[c] (kg/kg)	Dry Air Density[a] (kg/m³)	Wet Air SVE Well[a] (L/min)	Time[a] (min)	Water (L) Removed[a] (kg)	Water (g) at T&P[b] (L)	Dry Air at T&P[b] (L)	Dry Air at T&P[b] (kg)	Dry Air at T&P[b] (L/min)
					Recovery of Air from SVE A, B, and C Wells (Airflow Measured Near SVE Wellheads)						
25.03 (12/03/96 data)	T	298.17	0.02031 *0.02031*	1.18360	430.4	0	0.0	0	0	0	430.4
28.35	B	301.49	0.02485	1.17057	147.2	4,320	34.5	47,392	588,512	689	136.2
26.70	C	299.84	0.02231	1.17695	124.6	4,320	0.0	0	538,272	634	124.6
22.25 (12/06/96 data)	A	295.39	0.01698 *0.01471*	1.19470	198.2 470.0	4,320	0.0	0	856,224	1,023	198.2 459.0
33.90	B	307.04	0.03462	1.14942	124.6	4,320	34.6	48,405	489,867	563	113.4
25.55	C	298.69	0.02093	1.18154	25.5	4,320	0.0	0	110,160	130	25.5
32.80	A	305.94	0.03234	1.15352	184.1	4,320	0.0	0	795,312	917	184.1
(12/09/96 data)			*0.02148*		334.2						323.0
41.95	B	315.09	0.05470	1.11999	62.3	4,320	34.5	49,530	219,606	246	50.8
28.85	C	301.99	0.02564	1.17632	87.7	4,320	0.0	0	378,864	446	87.7
32.80	A	305.94	0.03234	1.15352	110.4	4,320	0.0	0	476,928	550	110.4
(12/12/96 data)			*0.02778*		260.4						248.9
50.89	B	324.03	0.09130	1.08912	25.8	7,200	57.6	85,040	100,432	109	13.9
33.40	C	306.54	0.03358	1.15128	34.3	7,200	0.0	0	246,600	284	34.3
38.31	A	311.45	0.04471	1.13313	40.8	7,200	0.0	0	293,544	333	40.8
(12/17/96 data)			*0.07935*		100.8						89.0
					Recovery of Air from SVE A, B, and C Wells (Airflow Measured Near SVE Wellheads)						
70.60	B	343.74	0.29015	1.02667	35.8	4,320	34.5	54,034	100,766	103	23.3
41.90	C	315.04	0.05452	1.12016	53.8	4,320	0.0	0	232,416	260	53.8
58.57	A	331.71	0.15236	1.06393	33.0	4,320	0.0	0	142,704	152	33.0
(12/20/96 data)			*0.06691*		122.7						110.2

	Well										
85.17	B	358.31	0.86706	0.98493	47.6	188.7	308,068	14,400	376,652	371	26.2
44.01	C	317.15	0.06206	1.11277	38.8	0.0	0	14,400	558,720	622	38.8
68.95	A	342.09	0.26277	1.03164	16.1	0.0	0	14,400	232,416	240	16.1
(12/30/96 data)			0.15311		102.5						81.1
89.03	B	362.17	1.33667	0.97443	25.5	186.7	308,088	5,760	-161,208	-157	-28.0
55.70	C	328.84	0.12002	1.07319	25.5	0.0	0	5,760	146,880	158	25.5
70.40	A	343.54	0.28683	1.02728	1.4	0.0	0	5,760	8,064	8	1.4
(1/03/97 data)			21.14646		52.4						-1.1
90.88	B	364.02	1.63426	0.96947	60.6	182.4	302,528	7,200	436,320	423	60.6
(1/08/97 data)			0.43121								
92.40	B	365.54	2.15856	0.96542	36.9	12.8	21,319	2,880	106,128	102	36.9
(1/10/97 data)			0.12493								
Recovery of Air from SVE B Well (Airflow Measured Before GAC Drums)											
96.60		369.74	5.00950	0.95445	37.7	6.4	10,782	158	7,524	7	115.6
(1/11/97 data)			0.89118								
96.60		369.74	5.00950	0.95445	56.6	18.9	31,840	250	17,872	17	198.7
(1/14/97 data)			1.10800								
94.90		368.04	3.85555	0.95889	63.7	60.2	100,950	1,490	119,169	114	147.8
(1/15/97 data)			0.52682								
98.20		371.34	5.18598	0.95034	113.3	63.2	106,931	1,410	202,305	192	219.3
(1/16/97 data)			0.32872								
99.30		372.44	5.18598	0.94753	254.8	152.2	258,278	1,435	464,695	440	503.7
(1/17/97 data)			0.34566								
Recovery of Air from SVE A and C Wells (Airflow Measured Near SVE A and C Wellheads)											
77.65		350.79	0.46565	1.00603	284.3	123.8	197,872	1,274	433,296	436	495.5
(1/18/97 data)			0.28401								
79.90		353.04	0.55446	0.99964	270.9	165.0	265,414	2,859	932,793	932	419.1
(1/20/97 data)			0.17695								

Table C3 SVE Calculated Airflows (continued)

Temperature SVE Well Flow[a] °C	K	Dry Air Humidity[c] (kg/kg)	Dry Air Density[a] (kg/m³)	Wet Air SVE Well[a] (L/min)	Time[a] (min)	Water (L) Removed[e] (kg)	Water (g) at T&P[b] (L)	Dry Air at T&P[b] (L)	Dry Air at T&P[b] (kg)	Dry Air at T&P[b] (L/min)
				Recovery of Air from P/T Wells 14, 16, 18, & 21 (Airflow Measured Before GAC Drums)						
81.63 (1/21/97 data)	354.77	0.62255 *0.28818*	0.99474	368.1	1,340	171.1	276,571	596,872	594	651.9
78.75 (1/22/97 data)	351.89	0.50907 *0.26343*	1.00290	424.7	1,270	171.1	274,330	647,630	650	725.8
78.63 (1/23/97 data)	351.77	0.50414 *0.26091*	1.00326	396.4	1,350	168.1	269,424	642,201	644	675.3
78.50 (1/24/97 data)	351.64	0.49920 *0.25826*	1.00361	424.7	1,384	182.8	292,881	705,253	708	721.1
80.63 (1/25/97 data)	353.77	0.58308 *0.23436*	0.99758	453.1	1,460	186.6	330,776	798,151	796	752.8
82.75 (1/27/97 data)	355.89	0.69508 *0.24762*	0.99161	453.1	3,100	418.7	678,945	1,705,227	1,691	769.0
				Recovery of Air from SVE B Well (Airflow Measured Before GAC Drums)						
125.00 (1/28/97 data)	398.14	5.18598 *0.27683*	0.88639	481.4	1,090	174.9	317,279	712,767	632	944.8
Totals (1/27–2/14)					24,845		4,861,028	11,183,355		645.8
128.00 (1/29/97 data)	401.14	5.18598 *0.17808*	0.87968	538.0	1,375	158.6	289,878	1,012,447	891	947.0
130.00 (1/30/97 data)	403.14	5.18598 *0.22957*	0.87521	481.4	1,545	205.5	337,471	1,022,799	895	906.3

Recovery of Air from SVE B Well (Airflow Measured Before GAC Drums)

Date										
133.00 (1/31/97 data)	406.14	5.18598 / 0.2585	0.86869	481.4	1,355	202.9	375,469	903,559	785	944.1
132.00 (2/01/97 data)	405.14	5.18598 / 0.35985	0.87075	226.5	1,358	133.2	245,881	425,097	370	494.2
131.00 (2/03/97 data)	404.14	5.18598 / 0.27957	0.87297	226.5	2,910	221.8	408,422	908,807	793	452.7
132.00 (2/04/97 data)	405.14	5.18598 / 0.31980	0.87075	368.1	1,432	202.9	374,544	728,633	634	770.3
131.00 (2/05/97 data)	404.14	5.18598 / 0.30754	0.87297	283.2	1,390	145.7	268,292	542,708	474	583.4
128.00 (2/06/97 data)	401.14	5.18598 / 0.28109	0.87968	283.2	1,390	133.2	243,454	538,679	474	562.6
128.00 (2/07/97 data)	401.14	5.18598 / 0.28012	0.87968	283.2	1,395	133.2	243,454	540,539	476	562.0
132.50 (2/08/97 data)	405.64	5.18598 / 0.22755	0.86972	311.5	1,340	114.3	211,253	577,549	502	588.7
137.00 (2/10/97 data)	410.14	5.18598 / 0.30907	0.86043	283.2	2,745	289.2	540,439	1,087,505	936	593.1
139.00 (2/11/97 data)	412.14	5.18598 / 0.25204	0.85628	311.5	1,530	144.6	271,537	670,017	574	615.4
131.00 (2/12/97 data)	404.14	5.18598 / 0.27264	0.87297	311.5	1,415	144.6	266,266	607,542	530	617.6
131.00 (2/13/97 data)	404.14	5.18598 / 0.28861	0.87297	254.8	1,420	125.7	231,464	498,907	436	514.3

Table C3   SVE Calculated Airflows (continued)

Recovery of Air from SVE B Well (Airflow Measured Before GAC Drums)

Temperature SVE Well Flow[a] °C	K	Dry Air Humidity[a] (kg/kg)	Density[a] (kg/m³)	Wet Air SVE Well[a] (L/min)	Time[a] (min)	Water (L) Removed[a] (kg)	Water (g) at T&P[b] (L)	Dry Air at T&P[b] (L)	Dry Air at T&P[b] (kg)	Dry Air at T&P[b] (L/min)
131.00 (2/14/97 data)	404.14	5.18598 / 0.30035 / 0.27028	0.87297 = Average Humidity	254.8	1,155	106.4 / 2,637	195,925	405,801	354 / 9,756	521.0
134.50 (2/15/97 data)	407.64	5.18598 / 0.24316	0.86559 = Average Humidity	283.2	1,425	118.1	219,353	561,111	486	547.7
138.00 (2/17/97 data)	411.14	5.18598 / 0.28830	0.85835	254.8	2,970	262.7	492,114	1,061,562	911	523.1
137.00 (2/18/97 data)	410.14	5.18598 / 0.24031	0.86043	311.5	1,395	125.7	234,900	607,933	523	604.2
138.00 (2/19/97 data)	411.14	5.18598 / 0.31005	0.85835	283.2	1,476	156.0	292,234	586,183	503	595.1
137.00 (2/20/97 data)	410.14	5.18598 / 0.25051	0.86043	254.8	1,335	102.6	191,732	476,006	410	500.2
136.00 (2/21/97 data)	409.14	5.18598 / 0.25293	0.86249	254.8	1,620	125.7	234,327	576,217	497	500.3
136.00 (2/22/97 data)	409.14	5.18598 / 0.28495	0.86249	254.8	1,200	104.9	195,553	426,828	368	518.7
136.00 (2/24/97 data)	409.14	5.18598 / 0.24978	0.86249	254.8	2,930	224.5	418,509	1,042,100	899	498.5
131.00 (2/25/97 data)	404.14	5.18598 / 0.25194	0.87297	254.8	1,430	110.5	203,475	502,413	439	493.6
130.00 (2/26/97 data)	403.14	5.18598 / 0.20702	0.87521	254.8	1,800	114.3	209,951	630,852	552	467.1

133.00 (2/27/97 data)	406.14	5.18598 *0.35482*	0.86869	254.8	1,155	125.7	*232,609*	*407,809*	*354*	*554.5*
134.00 (2/28/97 data)	407.14	5.18598 *0.27417*	0.86662	254.8	1,175	98.8	*183,281*	*415,822*	*360*	*510.0*
131.00 (3/01/97 data)	404.14	5.18598 *0.23367*	0.87297	254.8	1,595	114.3	*210,472*	*560,322*	*489*	*483.3*
133.00 (3/03/97 data)	406.14	5.18598 *0.14867*	0.86869	254.8	2,840	129.5	*261,019*	*1,022,751*	*871*	*445.0*
125.00 (3/04/97 data)	398.14	5.18598 *0.18050*	0.88639	269.0	1,495	87.4	*158,549*	*546,208*	*484*	*471.4*
122.00 (3/05/97 data)	395.14	5.18598 *0.13008*	0.89309	283.2	1,290	57.2	*102,982*	*492,377*	*440*	*461.5*
120.00 (3/07/97 data)	393.14	5.18598 *0.11827*	0.89767	297.3	2,790	118.1	*211,550*	*1,112,426*	*999*	*474.5*
101.00 (3/12/97 data)	374.14	5.18598 *0.06756*	0.94322	283.2	7,595	174.9	*298,154*	*2,744,670*	*2,589*	*400.7*
89.00 (3/17/97 data)	362.14	1.33164 *0.07609*	0.97452	283.2	6,890	178.7	*294,861*	*2,410,022*	*2,349*	*392.6*

a Measured or Table C6 data — standard text.

b Calculated data — bold, italic text.

c Humidity: maximum humidity from Table C6 — standard text; actual calculated humidity — bold, italic text.

Table C4 Actual Airflow Data

Temperature SVE Well Flow °C		K	Ambient Air Temperature °C	K	Water Enthalpy (kJ/kg) H_i	H_{ig}	Heat Capacity of Air (kJ/kg·°C)	Component Mass (kg/day) Water	Air	Heat Content (kJ/day) Water (L)	Water (g)	Air	Total
\multicolumn{14}{l}{Recovery of Air from SVE A, B, and C Wells (Airflow Measured Near SVE Wellheads)}													
25.03	T	298.17	5.60	278.74	−10.0	2,442.7	1.003	0.0	0.0	0	0	0	0
(12/03/96 data)													
											0	0	0
28.35	B	301.49	5.85	278.99	4.0	2,434.8	1.004	11.5	229.6	46	28,000	5,187	33,234
26.70	C	299.84	5.85	278.99	−3.0	2,438.8	1.003	0.0	211.2	0	0	4,417	4,417
22.25	A	295.39	5.85	278.99	−21.6	2,450.2	1.002	0.0	341.0	0	0	5,606	5,606
(12/06/96 data)										46	28,000	15,210	43,257
Totals 12/03/96–12/06/96								34.5	2,345.3	138	84,000	45,631	129,770
33.90	B	307.04	9.15	282.29	27.4	2,421.4	1.004	11.5	187.7	316	27,927	4,664	32,907
25.55	C	298.69	9.15	282.29	−7.8	2,441.5	1.003	0.0	43.4	0	0	714	714
32.80	A	305.94	9.15	282.29	22.8	2,424.0	1.004	0.0	305.8	0	0	7,261	7,261
(12/09/96 data)										316	28,243	12,639	40,882
Totals 12/06/96–12/09/96								34.6	1,610.6	949	83,781	37,916	122,646
41.95	B	315.09	12.20	285.34	61.5	2,401.8	1.005	11.5	82.0	707	27,620	2,452	30,780
28.85	C	301.99	12.20	285.34	6.1	2,433.6	1.003	0.0	148.6	0	0	2,481	2,481
32.80	A	305.94	12.20	285.34	22.8	2,460.0	1.004	0.0	183.4	0	0	3,793	3,793
(12/12/96 data)										707	27,620	8,727	37,054
Totals 12/09/96–12/12/96								34.5	1,241.8	2,122	82,861	26,180	111,163
50.89	B	324.03	6.35	279.49	99.2	2,416.6	1.007	11.5	21.9	1,143	27,839	981	29,963
33.40	C	306.54	6.35	279.49	25.3	2,422.6	1.004	0.0	56.8	0	0	1,542	1,542
38.31	A	311.45	6.35	279.49	46.1	2,410.7	1.005	0.0	66.5	0	0	2,138	2,138
(12/17/96 data)										1,143	27,839	4,661	33,643
Totals 12/12/96–12/17/96								57.6	725.9	5,716	139,194	23,305	168,215

Recovery of Air from SVE A, B, and C Wells (Airflow Measured Near SVE Wellheads)

70.60	B	343.74	5.57	278.71	182.2	2,331.2	1.010	11.5	34.5	2,096	26,808	2,265	31,169
41.90	C	315.04	5.57	278.71	61.3	2,401.9	1.005	0.0	86.8	0	0	3,170	3,170
58.57	A	331.71	5.57	278.71	131.6	2,361.1	1.008	0.0	50.6	0	0	2,703	2,703
(12/20/96 data)													
Totals 12/17/96–12/20/96								34.5	515.6	2,096	26,808	8,139	37,043
										6,287	**80,425**	**24,417**	**111,129**
85.17	B	358.31	5.88	279.02	243.3	2,294.4	1.013	18.9	37.1	4,592	43,296	2,978	50,866
44.01	C	317.15	5.88	279.02	70.2	2,396.8	1.006	0.0	62.2	0	0	2,385	2,385
68.95	A	342.09	5.88	279.02	175.3	2,335.3	1.010	0.0	24.0	0	0	1,528	1,528
(12/30/96 data)													
Totals 12/20/96–12/30/96								188.7	1,232.5	4,592	43,296	6,892	54,779
										45,915	**432,958**	**68,915**	**547,789**
89.03	B	362.17	11.65	284.79	259.5	2,284.5	1.013	46.7	−39.3	12,112	106,631	−3,080	115,663
55.70	C	328.84	11.65	284.79	119.5	2,368.2	1.008	0.0	39.4	0	0	1,749	1,749
70.40	A	343.54	11.65	284.79	181.4	2,331.7	1.010	0.0	2.1	0	0	123	123
(1/03/97 data)													
Totals 12/30/96–1/03/97								186.7	8.8	12,112	106,631	−1,207	117,536
										48,448	**426,524**	**−4,829**	**470,143**
90.88	B	364.02	−0.32	272.82	267.2	2,279.8	1.013	36.5	84.6	9,748	83,169	7,819	100,735
(1/08/97 data)								579.6			92,917		
92.40	B	365.54	−1.65	271.49	273.6	2,275.9	1.013	6.4	51.2	1,751	14,566	4,883	21,199
(1/10/97 data)											16,317		
Totals 1/03/97–1/10/97								195.2	525.5	52,240	444,976	48,859	546,075
								766.3			497,216		

Recovery of Air from Well B (Airflow Measured Before GAC Drums)

96.60	B	369.74	2.20	275.34	291.2	2,265.0	1.014	6.4	7.2	1,864	14,496	688	17,047
(1/11/97 data)											16,360		
96.60	B	369.74	−2.20	270.94	291.2	2,265.0	1.014	6.3	5.7	1,835	14,269	570	16,674
(1/14/97 data)											16,104		
94.90	B	368.04	−3.90	269.24	284.1	2,269.4	1.013	60.2	114.3	17,102	136,618	11,441	165,161
(1/15/97 data)											153,720		

Table C4 Actual Airflow Data (continued)

Temperature SVE Well Flow °C	K	Ambient Air Temperature °C	K	Water Enthalpy (kJ/kg) H_i	H_g	Heat Capacity of Air (kJ/kg-°C)	Component Mass (kg/day) Water	Air	Water (L)	Heat Content (kJ/day) Water (g)	Air	Total
98.20 (1/16/97 data)	371.34	−2.80	270.34	298.0	2,260.8	1.015	63.2	192.3	18,831	142,883 / 161,713	19,704	181,417
99.30 (1/17/97 data)	372.44	−3.30	269.84	302.6	2,257.9	1.015	152.2	440.3	46,052	343,657 / 389,709	45,841	435,550
Totals 1/10/97–1/17/97							300.9	771.1	89,353	680,461 / 769,814	79,383	849,198

Recovery of Air from Wells A and C (Airflow Measured Near A and C Wellheads)

Temperature SVE Well Flow °C	K	Ambient Air Temperature °C	K	Water Enthalpy (kJ/kg) H_i	H_g	Heat Capacity of Air (kJ/kg-°C)	Component Mass (kg/day) Water	Air	Water (L)	Heat Content (kJ/day) Water (g)	Air	Total
77.65 (1/18/97 data)	350.79	−1.10	272.04	211.8	2,313.4	1.011	123.8	435.9	26,224	286,402 / 312,626	34,704	347,330
79.90 (1/20/97 data)	353.04	4.40	277.54	221.2	2,307.8	1.012	82.5	466.2	18,251	190,394 / 208,645	35,614	244,259

Recovery of Air from Wells 14, 16, 18, & 21 (Airflow Measured Before GAC Drums)

Temperature SVE Well Flow °C	K	Ambient Air Temperature °C	K	Water Enthalpy (kJ/kg) H_i	H_g	Heat Capacity of Air (kJ/kg-°C)	Component Mass (kg/day) Water	Air	Water (L)	Heat Content (kJ/day) Water (g)	Air	Total
81.63 (1/21/97 data)	354.77	6.70	279.84	228.4	2,303.5	1.012	171.1	593.7	39,086	394,127 / 433,213	45,009	478,221
78.75 (1/22/97 data)	351.89	1.10	274.24	216.4	2,310.7	1.011	171.1	649.5	37,030	395,356 / 432,386	50,987	483,374
78.63 (1/23/97 data)	351.77	2.20	275.34	215.9	2,311.0	1.011	168.1	644.3	36,293	388,477 / 424,770	49,780	474,550
78.50 (1/24/97 data)	351.64	1.10	274.24	215.4	2,311.3	1.011	182.8	707.8	39,371	422,506 / 461,876	55,384	517,261
80.63 (1/25/97 data)	353.77	−0.60	272.54	224.3	2,306.0	1.012	186.6	796.2	41,847	430,297 / 472,144	65,434	537,578

Date													
82.75 (1/27/97 data)	355.89	7.80	280.94	233.2	2,300.6	1.012	209.4	845.5	48,812	481,635 / 530,447	64,113	594,560	
Totals 1/17/97–1/27/97							1,587.2 / 5,893.4	6,450.8 / 22,508.3	353,976	3,661,222 / 4,015,198	500,753		

Recovery of Air from Well B (Airflow Measured Before GAC Drums)

125.00 (1/28/97 data)	398.14	7.80	280.94	305.5	2,302.8	1.019	174.9	631.8	53,436	402,756 / 456,191	75,470	531,661

Recovery of Air from Well B (Airflow Measured Before GAC Drums)

128.00 (1/29/97 data)	401.14	7.80	280.94	305.5	2,308.4	1.019	158.6	890.6	48,456	366,114 / 414,570	109,113	523,683

Recovery of Air from Well B (Airflow Measured Before GAC Drums)

Date												
130.00 (1/30/97 data)	403.14	6.70	279.84	305.5	2,312.2	1.020	205.5	895.2	62,785	475,151 / 537,936	112,594	650,530
133.00 (1/31/97 data)	406.14	6.70	279.84	305.5	2,317.8	1.020	202.9	784.9	61,990	470,286 / 532,276	101,129	633,404
132.00 (2/01/97 data)	405.14	7.80	280.94	305.5	2,315.9	1.020	133.2	370.2	40,695	308,483 / 349,178	46,898	396,076
131.00 (2/03/97 data)	404.14	6.70	279.84	305.5	2,314.1	1.020	110.9	396.7	33,882	256,629 / 290,511	50,299	340,810
132.00 (2/04/97 data)	405.14	6.10	279.24	305.5	2,315.9	1.020	202.9	634.5	61,990	469,903 / 531,894	81,485	613,379
131.00 (2/05/97 data)	404.14	3.90	277.04	305.5	2,314.1	1.020	145.7	473.8	44,514	337,158 / 381,672	61,427	443,099
128.00 (2/06/97 data)	401.14	3.90	277.04	305.5	2,308.4	1.019	133.2	473.9	40,695	307,480 / 348,176	59,938	408,114

Table C4 Actual Airflow Data (continued)

Temperature SVE Well Flow °C	K	Ambient Air Temperature °C	K	Water Enthalpy (kJ/kg) H_f	H_{fg}	Heat Capacity of Air (kJ/kg-°C)	Component Mass (kg/day) Water	Air	Water (L)	Heat Content (kJ/day) Water (g)	Air	Total
						Recovery of Air from Well B (Airflow Measured Before GAC Drums)						
128.00 (2/07/97 data)	401.14	−1.10	272.04	305.5	2,308.4	1.019	133.2	475.5	40,695	307,480 / 348,176	62,568	410,744
132.50 (2/08/97 data)	405.64	−4.40	268.74	305.5	2,316.9	1.020	114.3	502.3	34,921	264,819 / 299,740	70,149	369,889
137.00 (2/10/97 data)	410.14	4.40	277.54	305.5	2,325.4	1.021	144.6	467.9	44,178	336,247 / 380,426	63,338	443,764
139.00 (2/11/97 data)	412.14	10.60	283.74	305.5	2,329.1	1.022	144.6	573.7	44,178	336,793 / 380,972	75,273	456,245
131.00 (2/12/97 data)	404.14	6.70	279.84	305.5	2,314.1	1.020	144.6	530.4	44,178	334,612 / 378,791	67,251	446,041
131.00 (2/13/97 data)	404.14	2.20	275.34	305.5	2,314.1	1.020	125.7	435.5	38,404	290,877 / 329,281	57,225	386,505
131.00 (2/14/97 data)	404.14	−1.70	271.44	305.5	2,314.1	1.020	106.4	354.3	32,507	246,215 / 278,723	47,955	326,678
Totals 1/27/97–2/14/97							2,637	9,756	805,568	6,103,897 / 9,909,447	1,255,750	8,165,197
134.50 (2/15/97 data)	407.64	1.10	274.24	305.5	2,320.6	1.021	118.1	485.7	36,082	274,068 / 310,150	66,149	376,299
						Recovery of Air from Well B (Airflow Measured Before GAC Drums)						
138.00 (2/17/97 data)	411.14	11.10	284.24	305.5	2,327.2	1.021	131.4	455.6	40,130	305,684 / 345,814	59,027	404,841

137.00 (2/18/97 data)	410.14	11.10	284.24	305.5	2,325.4	1.021	125.7	523.1	38,404	292,298 330,702	67,236	397,938
138.00 (2/19/97 data)	411.14	7.20	280.34	305.5	2,327.2	1.021	156.0	503.2	47,661	363,051 410,712	67,191	477,904(
137.00 (2/20/97 data)	410.14	6.70	279.84	305.5	2,325.4	1.021	102.6	409.6	31,346	238,582 269,929	54,485	324,414
136.00 (2/21/97 data)	409.14	4.40	277.54	305.5	2,323.5	1.021	125.7	497.0	38,404	292,061 330,465	66,774	397,238
136.00 (2/22/97 data)	409.14	1.10	274.24	305.5	2,323.5	1.021	104.9	368.1	32,049	243,733 275,782	50,702	326,484
136.00 (2/24/97 data)	409.14	-1.10	272.04	305.5	2,323.5	1.021	112.3	449.4	34,295	260,810 295,105	62,904	358,009
131.00 (2/25/97 data)	404.14	-1.10	272.04	305.5	2,314.1	1.020	110.5	438.6	33,760	255,703 289,463	59,103	348,566
Recovery of Air from Well B (Airflow Measured Before GAC Drums)												
130.00 (2/26/97 data)	403.14	6.10	279.24	305.5	2,312.2	1.020	114.3	552.1	34,921	264,281 299,202	69,785	368,987
133.00 (2/27/97 data)	406.14	7.20	280.34	305.5	2,317.8	1.020	125.7	354.3	38,404	291,350 329,754	45,462	375,216
134.00 (2/28/97 data)	407.14	3.30	276.44	305.5	2,319.7	1.020	98.8	360.4	30,185	229,187 259,372	48,047	307,419
Totals 2/14/97–2/28/97					1,669.5		6,302.0		510,068	3,877,302 4,387,370	838,796	5,226,166
131.00 (3/01/97 data)	404.14	2.20	275.34	305.5	2,314.1	1.020	114.3	489.1	34,921	264,496 299,417	64,269	363,686

Table C4   Actual Airflow Data (continued)

Temperature SVE Well Flow		Ambient Air Temperature		Water Enthalpy (kJ/kg)		Heat Capacity of Air (kJ/kg-°C)	Component Mass (kg/day)		Water (L)	Heat Content (kJ/day)		
°C	K	°C	K	H_l	H_g		Water	Air		Water (g)	Air	Total
133.00 (3/03/97 data)	406.14	10.00	283.14	305.5	2,317.8	1.020	64.8	435.5	19,782	150,079 / 169,861	54,649	224,510
125.00 (3/04/97 data)	398.14	12.80	285.94	305.5	2,302.8	1.019	87.4	484.2	26,703	201,263 / 227,965	55,374	283,339
122.00 (3/05/97 data)	395.14	0.00	273.14	305.5	2,297.1	1.018	57.2	439.7	17,476	131,397 / 148,873	54,634	203,507
120.00 (3/07/97 data)	393.14	6.70	279.84	305.5	2,293.4	1.018	59.1	499.3	18,041	135,425 / 153,466	57,610	211,076
Recovery of Air from Well B (Airflow Measured Before GAC Drums)												
101.00 (3/12/97 data)	374.14	20.00	293.14	305.5	2,258.0	1.015	35.0	517.8	10,687	78,983 / 89,670	42,569	132,240
Totals 2/28/97–3/12/97							681.4	5,871.6	208,182	1,563,081 / 1,771,263	611,643	2,382,906
89.00 (3/17/97 data)	362.14	14.40	287.54	259.4	2,284.6	1.013	35.7	469.7	9,270	81,652 / 90,922	35,510	126,432
Totals 3/12/97–3/17/97							178.7	2,348.6	46,348.8	408,261.9 / 454,610.7	177,551.6	632,162.2
Total Heat Utilized							7,820.7	39,705.6	2,175,310.6	18,068,926.5 / 20,244,237.1	3,734,272.4	23,978,509.5

Table C5 Thermodynamic Properties of Water

Water Temperature		Vapor Pressure	Specific Volume (m³/kg)		Enthalpy (kJ/kg)		
°C	K	(kPa)	V_l	V_g	H_l	H_{lg}	H_g
0.00	273.14	0.61	0.001000	206.100	0.0	2,500.9	2,500.9
1.86	275.00	0.70	0.001000	181.700	7.5	2,496.8	2,504.3
6.86	280.00	0.99	0.001000	130.300	28.1	2,485.4	2,513.5
11.86	285.00	1.39	0.001001	94.670	48.8	2,473.9	2,522.7
16.86	290.00	1.92	0.001001	69.670	69.7	2,462.2	2,531.9
21.86	295.00	2.62	0.001002	51.900	90.7	2,450.1	2,540.8
26.86	300.00	3.54	0.001004	39.100	111.7	2,438.4	2,550.1
31.86	305.00	4.72	0.001005	29.780	132.8	2,426.3	2,559.1
36.86	310.00	6.23	0.001007	22.910	153.9	2,414.3	2,568.2
41.86	315.00	8.14	0.001009	17.800	175.1	2,402.0	2,577.1
46.86	320.00	10.54	0.001011	13.960	196.2	2,389.8	2,586.0
51.86	325.00	13.53	0.001013	11.040	217.3	2,377.6	2,594.9
56.86	330.00	17.21	0.001015	8.809	238.4	2,365.3	2,603.7
61.86	335.00	24.71	0.001018	7.083	259.4	2,353.0	2,612.4
66.86	340.00	27.18	0.001021	5.737	280.5	2,340.5	2,621.0
71.86	345.00	33.77	0.001024	4.680	301.5	2,328.0	2,629.5
76.86	350.00	41.66	0.001027	3.844	322.5	2,315.4	2,637.9
81.86	355.00	51.05	0.001030	3.178	343.4	2,302.9	2,646.3
86.86	360.00	62.15	0.001034	2.643	364.4	2,290.1	2,654.5
91.86	365.00	75.21	0.001037	2.211	385.3	2,277.3	2,662.6
96.86	370.00	90.47	0.001041	1.860	406.3	2,264.3	2,670.6
100.00	373.14	101.33	0.001043	1.673	419.5	2,256.1	2,675.6
101.86	375.00	108.20	0.001045	1.573	427.3	2,251.2	2,678.5
106.86	380.00	128.80	0.001049	1.337	448.3	2,237.9	2,686.2
111.86	385.00	152.40	0.001053	1.142	469.3	2,224.5	2,693.8
116.86	390.00	179.50	0.001058	0.9800	490.4	2,210.9	2,701.3
121.86	395.00	210.40	0.001062	0.8445	511.5	2,197.0	2,708.5
126.86	400.00	245.60	0.001067	0.7308	532.7	2,182.9	2,715.6
131.86	405.00	285.40	0.001072	0.6349	554.0	2,168.6	2,722.6
136.86	410.00	330.20	0.001077	0.5537	575.3	2,154.0	2,729.3
141.86	415.00	380.60	0.001082	0.4846	596.7	2,139.1	2,735.8
146.86	420.00	437.00	0.001087	0.4256	618.2	2,123.9	2,742.1
151.86	425.00	499.90	0.001093	0.3750	639.8	2,108.4	2,748.2
156.86	430.00	569.90	0.001099	0.3314	661.4	2,092.7	2,754.1
94.00	367.14	81.74	0.001039	2.0608	394.3	2,282.9	2,677.2

Note: Conversion water vapor pressure from kPa to psia: 0.1451. Conversion of water-vapor specific volume from m³/kg to ft³/lb: 16.0186. Conversion of enthalpy from kJ/kg to Btu/lb: 0.4302. V_l = specific volume of liquid water; V_g = specific volume of gaseous water; H_l = enthalpy of liquid water; H_{lg} = enthalpy of vaporization of liquid water to gaseous water; H_g = enthalpy of gaseous water.

Table C5 Thermodynamic Properties of Water (continued)

Boiling Point vs. Elevation

Elevation		Boiling Point of Water			Std. Atmosphere Pressure	
m	ft	°F	°C	K	lb/in.²	kPa
0	0	212.0	100.0	373.1	14.70	101.309
250	820	210.6	99.2	372.4	14.26	98.277
500	1,640	208.9	98.3	371.4	13.85	95.451
750	2,461	207.5	97.5	370.6	13.44	92.626
1,000	3,281	206.1	96.7	369.9	13.03	89.800
1,250	4,101	204.4	95.8	368.9	12.64	87.112
1,500	4,921	203.0	95.0	368.1	12.26	84.493
1,750	5,741	201.6	94.2	367.4	11.89	81.943
2,000	6,562	199.9	93.3	366.4	11.53	79.462
2,500	8,202	197.1	91.7	364.9	10.83	74.638

Heat Content of Water

Water Temperature		Heat Capacity kJ/kg-K	Enthalpy kJ/kg	
°C	K		H_l (Calc)	H_l (Table)
0.00	273.14	4.116	0.0	0.0
1.86	275.00	4.119	7.7	7.5
6.86	280.00	4.128	28.3	28.1
11.86	285.00	4.137	49.1	48.8
16.86	290.00	4.145	69.9	69.7
21.86	295.00	4.152	90.8	90.7
26.86	300.00	4.159	111.7	111.7
31.86	305.00	4.165	132.7	132.8
36.86	310.00	4.171	153.7	153.9
41.86	315.00	4.177	174.8	175.1
46.86	320.00	4.182	196.0	196.2
51.86	325.00	4.187	217.2	217.3
56.86	330.00	4.192	238.4	238.4
61.86	335.00	4.197	259.6	259.4
66.86	340.00	4.202	281.0	280.5
71.86	345.00	4.207	302.3	301.5
76.86	350.00	4.212	323.7	322.5
81.86	355.00	4.217	345.2	343.4
86.86	360.00	4.222	366.7	364.4
91.86	365.00	4.228	388.3	385.3
96.86	370.00	4.233	410.0	406.3
100.00	373.14	4.237	423.7	419.5
101.86	375.00	4.239	431.8	427.3
106.86	380.00	4.246	453.7	448.3
111.86	385.00	4.253	475.7	469.3
116.86	390.00	4.260	497.8	490.4
121.86	395.00	4.268	520.1	511.5
126.86	400.00	4.276	542.5	532.7
131.86	405.00	4.286	565.1	554.0
136.86	410.00	4.296	587.9	575.3
141.86	415.00	4.306	610.9	532.7
146.86	420.00	4.318	634.1	554.0
151.86	425.00	4.330	657.6	575.3

Table C6 Thermodynamic Properties of Air

Properties of Dry Air						
Air Temperature		Pressure (kPa)	Specific Volume (Vg)		Density Air (kg/m³)	Heat Capacity kJ/kg-°C
°C	K		(m³/kg)	ft³/lb		
0.00	273.14	101.33	0.774	12.397	1.29199	0.999
0.00	273.14	101.33	0.774	12.397	1.29199	0.999
1.86	275.00	101.33	0.779	12.482	1.28325	1.000
6.86	280.00	101.33	0.793	12.709	1.26034	1.000
11.86	285.00	101.33	0.808	12.936	1.23823	1.001
16.86	290.00	101.33	0.822	13.163	1.21688	1.002
21.86	295.00	101.33	0.836	13.390	1.19625	1.002
26.86	300.00	101.33	0.850	13.616	1.17632	1.003
31.86	305.00	101.33	0.864	13.843	1.15703	1.004
36.86	310.00	101.33	0.878	14.070	1.13837	1.005
41.86	315.00	101.33	0.893	14.297	1.12030	1.005
46.86	320.00	101.33	0.907	14.524	1.10280	1.006
51.86	325.00	101.33	0.921	14.751	1.08583	1.007
56.86	330.00	101.33	0.935	14.978	1.06938	1.008
61.86	335.00	101.33	0.949	15.205	1.05342	1.009
66.86	340.00	101.33	0.963	15.432	1.03793	1.009
71.86	345.00	101.33	0.978	15.659	1.02288	1.010
76.86	350.00	101.33	0.992	15.886	1.00827	1.011
81.86	355.00	101.33	1.006	16.113	0.99407	1.012
86.86	360.00	101.33	1.020	16.340	0.98026	1.013
91.86	365.00	101.33	1.034	16.567	0.96683	1.013
96.86	370.00	101.33	1.048	16.794	0.95377	1.014
100.00	373.14	101.33	1.057	16.936	0.94574	1.015
101.86	375.00	101.33	1.063	17.021	0.94105	1.015
106.86	380.00	101.33	1.077	17.247	0.92867	1.016
111.86	385.00	101.33	1.091	17.474	0.91661	1.017
116.86	390.00	101.33	1.105	17.701	0.90486	1.018
121.86	395.00	101.33	1.119	17.928	0.89340	1.018
126.86	400.00	101.33	1.133	18.155	0.88224	1.019
132.00	405.14	101.33	1.148	18.389	0.87104	1.020
136.86	410.00	101.33	1.162	18.609	0.86072	1.021
141.86	415.00	101.33	1.176	18.836	0.85035	1.022
146.86	420.00	101.33	1.190	19.063	0.84023	1.023
151.86	425.00	101.33	1.204	19.290	0.83034	1.024
156.86	430.00	101.33	1.218	19.517	0.82069	1.024
161.86	435.00	101.33	1.233	19.744	0.81125	1.025
166.86	440.00	101.33	1.247	19.971	0.80203	1.026
171.86	445.00	101.33	1.261	20.198	0.79302	1.027
176.86	450.00	101.33	1.275	20.425	0.78421	1.028
176.86	450.00	101.33	1.275	20.425	0.78421	1.028
181.86	455.00	101.33	1.289	20.652	0.77559	1.029
186.86	460.00	101.33	1.304	20.879	0.76716	1.030
191.86	465.00	101.33	1.318	21.105	0.75891	1.031
196.86	470.00	101.33	1.332	21.332	0.75084	1.032
201.86	475.00	101.33	1.346	21.559	0.74294	1.033
206.86	480.00	101.33	1.360	21.786	0.73520	1.034
211.86	485.00	101.33	1.374	22.013	0.72762	1.035
216.86	490.00	101.33	1.389	22.240	0.72019	1.035
221.86	495.00	101.33	1.403	22.467	0.71292	1.036
226.86	500.00	101.33	1.417	22.694	0.70579	1.037
231.86	505.00	101.33	1.431	22.921	0.69880	1.038

Table C6 Thermodynamic Properties of Air (continued)

Air Temperature °C	Air Temperature K	Pressure (kPa)	Specific Volume (Vg) (m³/kg)	Specific Volume (Vg) ft³/lb	Density Air (kg/m³)	Heat Capacity kJ/kg-°C
236.86	510.00	101.33	1.445	23.148	0.69195	1.039
241.86	515.00	101.33	1.459	23.375	0.68523	1.040
246.86	520.00	101.33	1.474	23.602	0.67864	1.041
251.86	525.00	101.33	1.488	23.829	0.67218	1.042
256.86	530.00	101.33	1.502	24.056	0.66584	1.043
261.86	535.00	101.33	1.516	24.283	0.65962	1.044
266.86	540.00	101.33	1.530	24.510	0.65351	1.045
271.86	545.00	101.33	1.544	24.737	0.64751	1.046
276.86	550.00	101.33	1.559	24.963	0.64163	1.047
281.86	555.00	101.33	1.573	25.190	0.63585	1.048
286.86	560.00	101.33	1.587	25.417	0.63017	1.049
291.86	565.00	101.33	1.601	25.644	0.62459	1.050
296.86	570.00	101.33	1.615	25.871	0.61911	1.051
301.86	575.00	101.33	1.629	26.098	0.61373	1.052
306.86	580.00	101.33	1.644	26.325	0.60844	1.053

Humidity-Saturated Air Temperature

Air Temperature °C	Air Temperature K	Pressure (kPa)	Saturated Dry Air kg Water per kg Air	Saturated Dry Air lb Water per lb Air	Saturated Dry Air kg Water per m³ Air	Saturated Dry Air lb Water per ft³ Air
0.00	273.14	101.33	0.00378	0.00378	0.00488	0.00030
0.00	273.14	101.33	0.00378	0.00378	0.00488	0.00030
1.86	275.00	101.33	0.00432	0.00432	0.00554	0.00035
6.86	280.00	101.33	0.00615	0.00615	0.00775	0.00048
11.86	285.00	101.33	0.00864	0.00864	0.01070	0.00067
16.86	290.00	101.33	0.01201	0.01201	0.01462	0.00091
21.86	295.00	101.33	0.01652	0.01652	0.01976	0.00123
26.86	300.00	101.33	0.02250	0.02250	0.02647	0.00165
31.86	305.00	101.33	0.03039	0.03039	0.03516	0.00220
36.86	310.00	101.33	0.04076	0.04076	0.04641	0.00290
41.86	315.00	101.33	0.05438	0.05438	0.06092	0.00380
46.86	320.00	101.33	0.07224	0.07224	0.07967	0.00497
51.86	325.00	101.33	0.09589	0.09589	0.10412	0.00650
56.86	330.00	101.33	0.12731	0.12731	0.13614	0.00850
61.86	335.00	†01.33	0.20069	0.20069	0.21141	0.01320
66.86	340.00	101.33	0.22810	0.22810	0.23675	0.01478
71.86	345.00	101.33	0.31105	0.31105	0.31817	0.01986
76.86	350.00	101.33	0.43447	0.43447	0.43806	0.02735
81.86	355.00	101.33	0.63183	0.63183	0.62808	0.03921
86.86	360.00	101.33	0.98716	0.98716	0.96768	0.06041
91.86	365.00	101.33	1.79202	1.79202	1.73258	0.10817
96.86	370.00	101.33	5.18598	5.18598	4.94623	0.30881

Table C7 Heat Capacity (Cp) of Minerals

Mineral	Composition	Mineral Density (g/m²)	Formula Weight g/g mol	Heat Capacity (Cp) J per (g mol)-K(°C)			
				a	b*10⁻³	c*10⁵	d*10⁶
Quartz[a]	SiO_2	2.65		43.93	38.83	-9.69	0.00
Plagioclase[b]	$1/2(Na_2O)1/2(Al_2O_3)3(SiO_2)$	2.60		218.28	156.79	-49.70	-15.27
K–Feldspar[b]	$1/2(K_2O)1/2(Al_2O_3)3(SiO_2)$	2.60		238.36	119.21	-53.15	11.84
Calcite[a]	$CaCO_3$	2.72		104.52	21.92	-25.94	
Hematite[a]	Fe_2O_3	5.10		98.28	77.82	-14.85	
Magnetite[a]	Fe_3O_4	5.17		91.55	202.00	0.00	
Kaolinite[b]	$(Al_2O_3)2(SiO_2)2(H_2O)$	2.55		356.26	88.04	-56.49	
Chlorite[b]	$3(MgO)2(SiO_2)2(H_2O)$	2.60		385.74	87.95	-53.40	
Illite/Mica							
Illite[b]	$1/2(K_2O)2(Al_2O_3)7(SiO_2)1/2(Al_2O_3)2(H_2O)$			799.97	295.29	-166.13	11.84
Mica[b]	$1/2(K_2O)(Al_2O_3)3(SiO_2)1/2(Al_2O_3)(H_2O)$	2.76		431.31	129.59	-90.26	11.84
Illite/Smectite							
Illite[b]	$1/2(K_2O)2(Al_2O_3)7(SiO_2)1/2(Al_2O_3)2(H_2O)$			799.97	295.29	-166.13	11.84
Smectite[b]	$(Al_2O_3)4(SiO_2)n(H_2O)$, n = 2 or >2 (amorphous?)	2.20		444.12	165.70	-75.87	
CaO[a]	$Cp = a + bT + cT^{-2} + dT^2$		56.08	50.42	4.18	-8.49	
K₂O[a]	$Cp = a + bT + cT^{-2} + dT^2$		94.20	95.65	-4.94	-11.05	23.68
Na₂O[a]	$Cp = a + bT + cT^{-2} + dT^2$		61.98	55.48	70.21	-4.14	-30.54
MgO[a]	$Cp = a + bT + cT^{-2} + dT^2$		40.31	48.99	3.43	-11.34	
SiO₂[a]	$Cp = a + bT + cT^{-2} + dT^2$		60.08	43.93	38.83	-9.69	
Al₂O₃[a]	$Cp = a + bT + cT^{-2} + dT^2$		101.96	117.49	10.38	-37.11	
H₂O[a]	$Cp = a$		18.02	75.46	0.00	0.00	

[a] Literature values.
[b] Calculation: For example, the "a" value for illite, $1/2(K_2O)2(Al_2O_3)7(SiO_2)1/2(Al_2O_3)2(H_2O)$:

$a = (0.5)(a, K_2O) + (2)(a, Al_2O_3) + (7)(a, SiO_2) + (0.5)(a, Al_2O_3) + (2)(a, H_2O)$

$a = (0.5)(95.65) + (2)(117.46) + (7)(43.93) + (0.5)(117.46) + (2)(75.46) = 799.9$ J/(g mol)-K (°C)

Table C7 Heat Capacity (Cp) of Minerals (continued)

Mineral	Composition	Mineral Density (g/m³)	Formula Weight g/g mol	Heat Capacity (Cp) J per (g mol)-K(°C)			
				a	b^*10^{-3}	c^*10^{-5}	d^*10^6
Quartz[a]	SiO_2		60.08	0.731	0.646	−0.161	
Plagioclase[a]	$1/2(Na_2O)1/2(Al_2O_3)3(SiO_2)$		262.22	0.832	0.598	−0.190	−0.058
K-Feldspar[a]	$1/2(K_2O)1/2(Al_2O_3)3(SiO_2)$		278.34	0.856	0.428	−0.191	0.043
Calcite[a]	$CaCO_3$		100.09	1.044	0.219	−0.259	
Hematite[a]	Fe_2O_3		159.69	0.615	0.487	−0.093	
Magnetite[a]	Fe_3O_4		231.54	0.395	0.872	0.000	
Kaolinite[a]	$(Al_2O_3)2(SiO_2)2(H_2O)$		258.16	1.380	0.341	−0.219	
Chlorite[a]	$3(MgO)2(SiO_2)2(H_2O)$		277.13	1.392	0.317	−0.193	
Illite/Mica[a]							
Illite[a]	$1/2(K_2O)2(Al_2O_3)7(SiO_2)1/2(Al_2O_3)2(H_2O)$		758.63	1.054	0.389	−0.219	0.016
Mica[a]	$1/2(K_2O)(Al_2O_3)3(SiO_2)1/2(Al_2O_3)(H_2O)$		398.31	1.083	0.325	−0.227	0.030
Illite/Smectite							
Illite[a]	$1/2(K_2O)2(Al_2O_3)7(SiO_2)1/2(Al_2O_3)2(H_2O)$		758.63	1.054	0.389	−0.219	0.016
Smectite[a]	$(Al_2O_3)4(SiO_2)n(H_2O)$, n=2 or >2 (amorphous?)		378.33	1.174	0.438	−0.201	

[a] kJ/(kg)-K(°C) = [kJ/(g mol)-K(°C)]/(formula weight, g/g-mol)
Calculation: For example, for illite, a = (799.9)/(758.63) = 1.054 kJ/(kg)-K(°C)

Mineral	Composition	Kirtland Content (% dwb)	Heat Capacity (Cp) J per (kg)-K(°C)			
			a	$b*10^{-3}$	$c*10^{-6}$	$d*10^{6}$
Quartz	SiO_2	47.39	0.347	0.306	-0.076	
Plagioclase	$1/2(Na_2O)1/2(Al_2O_3)3(SiO_2)$	14.12	0.118	0.084	-0.027	-0.008
K-Feldspar	$1/2(K_2O)1/2(Al_2O_3)3(SiO_2)$	7.90	0.068	0.034	-0.015	0.003
Calcite	$CaCO_3$	5.55	0.058	0.012	-0.014	
Hematite	Fe_2O_3	0.50	0.003	0.002	0.000	
Magnetite	Fe_3O_4	1.51	0.006	0.013	0.000	
Kaolinite	$(Al_2O_3)2(SiO_2)2(H_2O)$	2.18	0.030	0.007	-0.005	
Chlorite	$3(MgO)2(SiO_2)2(H_2O)$	2.69	0.037	0.009	-0.005	
Illite	$1/2(K_2O)2(Al_2O_3)7(SiO_2)1/2(Al_2O_3)2(H_2O)$	3.53	0.037	0.014	-0.008	0.001
Mica	$1/2(K_2O)1/2(Al_2O_3)3(SiO_2)1/2(Al_2O_3)(H_2O)$	0.00	0.000	0.000	0.000	0.000
Illite(M.L.)	$1/2(K_2O)2(Al_2O_3)7(SiO_2)1/2(Al_2O_3)2(H_2O)$	3.41	0.036	0.013	-0.007	0.001
Smectite(M.L.)	$(Al_2O_3)4(SiO_2)n(H_2O)$, n=2 or >2 (amorphous?)	11.21	0.132	0.049	-0.022	
Totals		100.00	0.871	0.544	-0.181	-0.004

Temperature		Cp
°C	K	kJ/kg-K
0.00	273.14	0.777
25.00	298.14	0.830
50.00	323.14	0.873
75.00	348.14	0.911
100.00	373.14	0.944
125.00	398.14	0.973
150.00	423.14	1.000

Calculated Cp Values

kJ/(kg)-K(°C) = [J/(g mol)-K(°C)]/(formula weight, g/g-mol)

The "a" value for illite in the above table:
a = (% dwb)(a, illite)/100 = (3.53)(1.054)/100 = 0.037 kJ/(kg)-K(°C)

$Cp = 0.871 + 0.545*10^{-3}*T − 0.181*10^{+5}/T^2 − 0.004*10^{-6}*T^2$

T = K = °C + 273.14

Table C8 Mass Water Vapor in Exhaust Air

Date	Days	Flow	Wet Air Outflow (slpm)	Ambient Values Temp. °C	RH %	SVE Temp. °C	Before GAC °C	Humidity Dry Air kg H₂O/kg Air Inflow	Outflow	Dry Air Outflow Density kg/m³	Mass Total kg	Water Mass kg
Recovery of Air from SVE A, B, and C Wells (Airflow Measured Near SVE Wellheads)												
12/03/96–12/06/96	3	CC	470.0 / 429.6	6.1	56	26.1	6.1	0.00329	0.00587	1.264	2,345.3	9.4
12/06/96–12/09/96	3	CC	334.0 / 313.9	12.2	45	32.5	14.4	0.00887	0.01035	1.188	1,610.6	10.2
12/09/96–12/12/96	3	CC	261.0 / 233.3	12.2	38	37.3	13.3	0.00887	0.00961	1.232	1,241.8	7.8
12/12/96–12/17/96	5	CC	131.4 / 78.0	2.3	30	42.3	1.7	0.00448	0.00427	1.292	725.9	2.1
12/17/96–12/20/96	3	CC	122.7 / 95.5	5.6	21	54.1	9.3	0.00569	0.00736	1.250	515.6	3.2
12/20/96–1/03/97	14	CC	52.4 / 49.0	7.5	56	72.3	8.6	0.0064665	0.00702	1.25751	1,241.3	4.2
Recovery of Air from SVE A, B, and C Wells (Airflow Measured Near SVE Wellheads)												
1/03/97–1/09/97	6	CC	59.0 / 44.7	-2.0	55	92.9	14.4	0.00378	0.01035	1.227	474.2	3.9
Recovery of Air from SVE–B Well (Airflow Measured After Third GAC Drum)												
1/09/97–1/17/97	8	CC	82.8 / 58.1	-1.9	68	96.2	14.2	0.00378	0.01022	1.228	822.3	6.3
1/17/97–1/27/97	10	CoC	434.1 / 383.3	7.8	57	125.5	28.8	0.00662	0.02556	1.169	6,450.8	140.6
1/27/97–2/14/97	18	CC	327.2 / 325.9	4.5	55	131.7	32.4	0.00528	0.03151	1.155	9,755.5	279.0
2/14/97–2/28/97	14	CC	262.9 / 261.4	4.8	51	134.0	22.0	0.00539	0.01668	1.196	6,302.0	87.8
2/28/97–3/12/973	12	CC	277.3 / 283.7	12.4	30	115.8	21.5	0.00901	0.01619	1.198	5,871.6	79.2
3/12/97–3/17/97	5	CC	283.2 / 277.2	14.4	30	89.0	26.7	0.01035	0.02231	1.177	2,348.6	45.1
Totals												**678.9**

| Date | Heat Content | | |
	H_i kJ/kg	H_{fg} kJ/kg	Total kJ
12/03/96–12/06/96	–5.5	2,440.2	23,003
12/06/96–12/09/96	21.5	2,424.8	25,067
12/09/96–12/12/96	41.8	2,413.2	19,028
12/12/96–12/17/96	63.0	2,400.9	5,234
12/17/96–12/20/96	112.8	2,372.1	7,906
12/20/96–1/03/97	189.4	2,326.9	10,601
1/03/97–1/09/97	275.7	2,274.6	10,010
1/09/97–1/17/97	289.5	2,266.0	16,078
1/17/97–1/27/97	413.0	2,186.7	365,410
1/27/97–2/14/97	439.3	2,169.1	727,847
2/14/97–2/28/97	449.1	2,162.4	229,315
2/28/97–3/12/97	371.9	2,213.8	204,816
3/12/97–3/17/97	259.4	2,284.6	114,729
Totals			**1,759,045**

Appendix D

Demonstration RF Energy Data

Appendix D Listing and Description of Tables

Item	Title	Description
Table D1	Summary of KAI RF energy data	RF energy summary for each time period defined in Table 2.8 1. Dates included in the time period 2. Number of days in each time period 3. Average AC power output, kW 4. Average RF power output, kW 5. Average AC energy output, kW-h/day 6. Average RF energy output, kW-h/day 7. RF energy delivered to soil, kW-h/day (conversion efficiency below energy value = RF energy to soil/RF output) 8. RF energy delivered to soil, kJ/day (total energy delivered for the period below energy value)
Table D2	KAI energy data	Data recorded by KAI during the Kirtland AFB demonstration 1. Data and summary explanation of terms 2. Daily averages for the operation of the RF generator-antennae unit

Table D1 Summary of KAI RF Energy Data

RF Operating Time Period	Time (days)	Average AC Output (kW)	Average RF Output (kW)	Average AC Output (kWh/day)	Average RF Output (kWh/day)	RF Energy Delivered (kWh/day)	RF Energy Delivered (kJ/day)
12/03/96–12/06/96	3.00	8.99	1.63	685.27	36.78	29.83 81.10[a] Total kJ	107,386 322,157
12/06/96–12/09/96	3.00	6.99	1.93	251.11	68.12	55.92 82.08[a] Total kJ	201,298 603,893
12/09/96–1212/96	3.00	10.10	2.78	230.42	64.26	51.80 80.61[a] Total kJ	186,474 559,421
12/12/96–12/17/96	5.00	20.45	7.42	452.74	164.39	127.23 77.39[a] Total kJ	458,016 2,290,081
12/17/96–12/20/96	3.00	18.86	8.65	407.47	190.88	150.72 78.96[a] Total kJ	542,606 1,627,819
12/20/96–1/03/97	14.00	24.11	9.91	538.42	220.83	180.88 81.91[a] Total kJ	651,158 9,116,205
1/03/97–1/09/97	6.00	23.29	9.65	597.47	248.97	202.52 81.34[a] Total kJ	729,080 4,374,482
1/09/97–1/17/97	8.00	25.98	11.96	570.53	262.63	213.16 81.17[a] Total kJ	767,386 6,139,086
1/17/97–1/27/97	10.00	29.39	13.35	609.52	285.70	224.86 78.71[a] Total kJ	809,508 8,095,082
1/27/97–2/14/97	18.00	28.95	13.40	667.75	309.04	250.24 80.97[a] Total kJ	900,852 16,215,345
2/14/97–2/28/97	14.00	27.10	12.77	606.11	286.05	233.61 81.67[a] Total kJ	841,008 11,774,115
Total	87.00					Total kJ	61,117,686

[a] Efficiency = average RF output/RF energy delivered.

Table D2 KAI Energy Data

Data Summary

87	Days of operation
1,958.67	Hours of operation
81.61	Days of on-time operation
93.81%	% on time
47,829.22	kWh of AC energy
21,012.96	kWh of RF generator energy
16,977.14	kWh of RF energy delivered to soil
35.50%	Overall AC generator/RF delivered to soil efficiency (includes AC power for equipment)

Explanation of Terms

RF On Time	Time period over which the RF data were averaged
AC Power	Average power delivered by diesel generator
AC Energy	Energy delivered by diesel generator
RF Power	Average power delivered by RF generator
RF Energy	Energy delivered by RF generator
Cpl (Fwd)	Vector voltmeter fwd power measurement
Cpl (Rev)	Vector voltmeter rev power measurement
Return Loss % Loss	This value measures efficiency of RF delivery
RF power - delivered	Average RF power delivered to soil (includes 0.1537 dB cable insertion loss)
RF energy - delivered	RF energy delivered to soil

Date	Time (days)	RF On Time (h)	AC Power (kW)	AC Energy (kWh)	RF Power (kW)	RF Energy (kWh)	Cpl (fwd) (dB)
12/03/96	0	0.00	0.00	0.00	0.00	0.00	0.00
12/04/96	1	21.91	8.89	194.78	1.62	35.49	99.74
12/05/96	2	23.44	8.57	200.88	1.30	30.47	98.12
12/06/96	3	22.64	9.52	215.53	1.96	44.37	100.47
12/07/96	4	22.90	10.01	229.23	2.91	66.64	100.82
12/09/96	6	47.82	10.96	524.11	2.88	137.72	102.55
12/10/96	7	20.12	9.33	187.72	1.80	36.22	100.03
12/11/96	8	24.22	8.25	199.82	1.41	34.15	99.74
12/12/96	9	23.86	12.73	303.74	5.13	122.40	104.36
12/13/96	10	20.66	13.42	277.26	5.00	103.30	104.75
12/14/96	11	26.30	25.52	671.18	9.52	250.38	107.54
12/15/96	12	21.48	27.26	585.54	9.96	213.94	107.81
12/16/96	13	21.77	10.46	227.71	3.17	69.01	107.63
12/17/96	14	19.61	25.60	502.02	9.45	185.31	106.49
12/18/96	15	18.66	22.42	418.36	7.65	142.75	102.70
12/19/96	16	23.38	8.55	199.90	8.55	199.90	105.57
12/20/96	17	23.59	25.61	604.14	9.75	230.00	105.84
12/21/96	18	22.58	23.11	521.82	8.53	192.61	104.70
12/22/96	19	24.04	24.36	585.61	9.59	230.54	106.47
12/23/96	20	21.80	25.13	547.83	10.04	218.87	107.43
12/24/96	21	24.04	24.48	588.50	9.44	226.94	106.00
12/25/96	22	30.05	25.08	753.65	9.78	293.89	106.56
12/26/96	23	19.97	25.76	514.43	10.43	208.29	107.54
12/27/96	24	17.90	25.42	455.02	10.51	188.13	107.55
12/28/96	25	21.03	24.09	506.61	10.12	212.82	106.11
12/29/96	26	24.85	24.96	620.26	10.79	268.13	107.85
12/30/96	27	21.89	22.22	486.40	9.27	202.92	107.79
12/31/96	28	20.59	22.71	467.60	9.86	203.02	105.45
12/01/97	29	24.51	23.21	568.88	9.98	244.61	107.02
1/02/97	30	20.46	23.29	476.51	10.11	206.85	107.72
1/03/97	31	18.76	23.71	444.80	10.34	193.98	107.83

Date	Time (days)	RF On Time (h)	AC Power (kW)	AC Energy (kWh)	RF Power (kW)	RF Energy (kWh)	Cpl (fwd) (dB)
1/04/97	32	39.91	24.03	959.04	10.43	416.26	107.65
1/05/97	33	19.47	24.36	474.29	10.43	203.07	107.51
1/06/97	34	23.17	25.29	585.97	10.78	249.77	107.58
1/07/97	35	23.35	16.37	382.24	6.17	144.07	107.47
1/08/97	36	22.14	24.49	542.21	9.24	204.57	107.50
1/09/97	37	25.47	25.17	641.08	10.84	276.09	106.19
1/10/97	38	20.13	25.79	519.15	11.27	226.87	107.53
1/11/97	39	19.51	27.56	537.70	12.78	249.34	108.62
1/12/97	40	22.76	28.15	640.69	13.22	300.89	108.59
1/13/97	41	22.49	27.34	614.88	12.63	284.05	108.27
1/14/97	42	23.29	9.00	209.61	2.56	59.62	99.53
1/15/97	43	21.96	30.52	670.22	14.98	328.96	108.98
1/16/97	44	23.04	29.29	674.84	13.84	318.87	107.71
1/17/97	45	23.10	30.18	697.16	14.39	332.41	108.81
1/18/97	46	11.80	26.34	310.81	8.26	97.47	94.06
1/19/97	47	10.58	31.22	330.31	14.55	153.94	108.90
1/20/97	48	23.45	31.41	736.56	14.86	348.47	109.13
1/21/97	49	23.05	27.29	629.03	11.94	325.84	105.69
1/22/97	50	23.97	30.15	722.70	14.03	336.30	108.12
1/23/97	51	21.92	24.41	535.07	11	241.12	106.31
1/24/97	52	22.32	29.88	666.92	14.36	320.52	108.94
1/25/97	53	24.06	29.99	721.56	14.1	339.25	107.03
1/26/97	54	20.33	31.03	630.84	14.88	302.51	108.98
1/27/97	55	25.20	32.2	811.44	15.54	391.61	109.07
1/28/97	56	23.26	30.41	707.34	14.22	330.76	107.88
1/29/97	57	23.42	28.64	670.75	13.35	312.66	106.99
1/30/97	58	21.95	29.35	644.23	13.42	294.57	107.22
1/31/97	59	22.15	29.09	644.34	13.4	296.81	106.72
2/01/97	60	24.25	18.81	456.14	7.28	176.54	93.79
2/02/97	61	22.36	29.62	662.30	15.17	339.20	108.14
2/03/97	62	24.75	29.99	742.25	15.46	382.64	109.09
2/04/97	63	23.28	30.5	710.04	15.68	365.03	107.94
2/05/97	64	23.76	29.24	694.74	15.52	368.76	108.12
2/06/97	65	23.28	25.5	593.64	11.03	256.78	106.09
2/07/97	66	23.74	32.02	760.15	13.91	330.22	108.3
2/08/97	67	22.22	29.00	644.38	12.07	268.20	102.84
2/09/97	68	25.10	33.06	829.81	14.57	365.71	108.83
2/10/97	69	22.74	32.71	743.83	14.53	330.41	108.81
2/11/97	70	23.78	28.91	687.48	12.12	288.21	106.99
2/12/97	71	22.88	30.38	695.09	14.91	341.14	107.72
2/13/97	72	23.67	26.93	637.43	12.28	290.67	106.28
2/14/97	73	18.34	27.02	495.55	12.24	224.48	107.79
2/15/97	74	23.27	30.12	700.89	14.86	345.79	108.49
2/16/97	75	22.83	27.52	628.28	12.99	296.56	108.41
2/17/97	76	22.04	28.79	634.53	14.39	317.16	108.79
2/18/97	77	24.55	29.32	719.81	14.95	367.02	108.26
2/19/97	78	20.03	27.03	541.41	12.86	257.59	107.64
2/20/97	79	21.41	26.4	565.22	12.36	264.63	106.65
2/21/97	80	22.64	28.94	655.20	14.25	322.62	107.51
2/22/97	81	23.19	29.62	686.89	14.35	332.78	107.15
2/23/97	82	24.65	28.26	696.61	13.41	330.56	108.21
2/24/97	83	22.82	18.79	428.79	6.865	156.66	97.275
2/25/97	84	20.99	9.32	195.63	0.32	6.72	86.34
2/26/97	85	22.27	31.24	695.71	15.4	342.96	108.24
2/27/97	86	21.89	32.1	702.67	16	350.24	109.06
2/28/97	87	19.86	31.92	633.93	15.78	313.39	108.98

Table D2 KAI Energy Data (continued)

Date	Time (days)	RF On Time (h)	AC Power (kW)	AC Energy (kWh)	RF Power (kW)	RF Energy (kWh)	Cpl (fwd) (dB)
Totals		1,958.67		47,829.22		21,012.96	
Hours (87 days)		2,088.00			43.93%	AC to RF	
% Operational		93.81			10.73	Average kW	

Date	Time (days)	Cpl (Rev) (dB)	Return Loss (dB)	Loss %	RF Power Delivered (kW)	RF Energy Delivered (kWh)	Notes Ave. RF Power/ Total RF Energy
12/03/96	0	0.00	0.00	0.00	0.00	0.00	
12/04/96	1	91.60	8.14	15.35	1.37	29.00	
12/05/96	2	89.75	8.37	14.55	1.11	25.13	
12/06/96	3	92.89	7.58	17.46	1.62	35.35	1.37/89.49
12/07/96	4	92.44	8.38	14.52	2.49	54.98	
12/09/96	6	94.36	8.19	15.17	2.44	112.77	
12/10/96	7	91.99	8.04	15.70	1.52	29.47	
12/11/96	8	91.65	8.09	15.52	1.19	27.85	
12/12/96	9	96.66	7.70	16.98	4.26	98.08	2.38/323.14
12/13/96	10	97.13	7.62	17.30	4.14	82.46	
12/14/96	11	100.34	7.20	19.05	7.71	195.62	
12/15/96	12	101.37	6.44	22.70	7.70	159.63	
12/16/96	13	100.85	6.78	20.99	2.50	52.63	
12/17/96	14	99.16	7.33	18.49	7.70	145.79	
12/18/96	15	95.37	7.33	18.49	6.24	112.31	
12/19/96	16	98.22	7.35	18.41	6.98	157.43	
12/20/96	17	98.35	7.49	17.82	8.01	182.44	6.37/1,088.31
12/21/96	18	96.77	7.93	16.11	7.16	155.97	
12/22/96	19	98.39	8.08	15.56	8.10	187.90	
12/23/96	20	99.28	8.15	15.31	8.50	178.92	
12/24/96	21	97.83	8.17	15.24	8.00	185.66	
12/25/96	22	98.34	8.22	15.07	8.31	240.93	
12/26/96	23	99.13	8.41	14.42	8.93	172.05	
12/27/96	24	99.07	8.48	14.19	9.02	155.82	
12/28/96	25	97.79	8.32	14.72	8.63	175.18	
12/29/96	26	99.45	8.40	14.45	9.23	221.40	
12/30/96	27	99.59	8.20	15.14	7.87	166.22	8.37/1840.05
12/31/96	28	97.27	8.18	15.21	8.36	166.16	
1/01/97	29	98.85	8.17	15.24	8.46	200.12	
1/02/97	30	99.64	8.08	15.56	8.54	168.59	
1/03/97	31	99.86	7.97	15.96	8.69	157.35	8.51/692.23
1/04/97	32	99.37	8.28	14.86	8.88	342.08	
1/05/97	33	99.05	8.46	14.26	8.94	168.07	
1/06/97	34	98.92	8.66	13.61	9.31	208.27	
1/07/97	35	98.84	8.63	13.71	5.32	120.00	
1/08/97	36	101.39	6.11	24.49	6.98	149.10	
1/09/97	37	97.83	8.36	14.59	9.26	227.62	9.00/1215.13
1/10/97	38	99.05	8.48	14.19	9.67	187.90	
1/11/97	39	101.78	6.84	20.70	10.13	190.85	
1/12/97	40	100.61	7.98	15.92	11.12	244.18	
1/13/97	41	99.77	8.50	14.13	10.85	235.44	
1/14/97	42	92.05	7.48	17.86	2.10	47.27	
1/15/97	43	100.81	8.17	15.24	12.70	269.13	
1/16/97	44	99.54	8.17	15.24	11.73	260.88	
1/17/97	45	100.84	7.97	15.96	12.09	269.65	10.05/1705.30
1/18/97	46	86.85	7.21	19.01	6.69	76.19	
1/19/97	47	101.27	7.63	17.26	12.04	122.94	
1/20/97	48	102.15	6.98	20.04	11.88	268.93	

Date	Time (days)	Cpl (Rev) (dB)	Return Loss (dB)	Loss %	RF Power Delivered (kW)	RF Energy Delivered (kWh)	Notes Ave. RF Power/ Total RF Energy
1/21/97	49	97.90	7.79	16.63	9.95	221.46	
1/22/97	50	100.27	7.85	16.41	11.73	271.35	
1/23/97	51	98.29	8.02	15.78	9.26	196.02	
1/24/97	52	101.05	7.89	16.26	12.03	259.08	
1/25/97	53	99.28	7.75	16.79	11.73	272.48	
1/26/97	54	101.12	7.86	16.37	12.44	244.20	
1/27/97	55	101.22	7.85	16.41	12.99	315.98	10.07/2248.63
1/28/97	56	99.94	7.94	16.07	11.93	267.95	
1/29/97	57	99.11	7.88	16.29	11.17	252.62	
1/30/97	58	99.25	7.97	15.96	11.28	238.95	
1/31/97	59	98.76	7.96	16.00	11.26	240.66	
2/01/97	60	86.00	7.79	16.63	6.07	142.06	
2/02/97	61	100.29	7.85	16.41	12.68	273.69	
2/03/97	62	101.19	7.90	16.22	12.95	309.43	
2/04/97	63	100.08	7.86	16.37	13.11	294.67	11.31/2020.04
2/05/97	64	100.16	7.96	16.00	13.04	299.00	
2/06/97	65	98.11	7.98	15.92	9.27	208.39	
2/07/97	66	100.38	7.92	16.14	11.66	267.29	
2/08/97	67	95.06	7.78	16.67	10.06	215.71	
2/09/97	68	100.86	7.97	15.96	12.24	296.66	
2/10/97	69	100.82	7.99	15.89	12.22	268.26	
2/11/97	70	98.93	8.06	15.63	10.23	234.71	
2/12/97	71	99.76	7.96	16.00	12.53	276.61	
2/13/97	72	98.29	7.99	15.89	10.33	235.99	
2/14/97	73	99.88	7.91	16.18	10.26	181.62	11.18/2484.23
2/15/97	74	100.64	7.85	16.41	12.42	279.01	
2/16/97	75	100.45	7.96	16.00	10.91	240.46	
2/17/97	76	100.81	7.98	15.92	12.10	257.39	
2/18/97	77	100.32	7.94	16.07	12.55	297.33	
2/19/97	78	99.54	8.10	15.49	10.87	210.12	
2/20/97	79	98.62	8.03	15.74	10.41	215.22	
2/21/97	80	99.43	8.08	15.56	12.03	262.95	11.61/1762.49
2/22/97	81	99.06	8.09	15.52	12.12	271.34	
2/23/97*	82	100.09	8.12	15.42	11.34	269.87	
2/24/97	83	89.54	7.74	16.85	5.71	125.74	
2/25/97	84	78.99	7.35	18.41	0.26	5.29	
2/26/97	85	99.71	8.53	14.03	13.24	284.60	
2/27/97	86	100.46	8.60	13.80	13.79	291.40	
2/28/97	87	100.47	8.51	14.09	13.56	259.86	10.00/1,508.10

Totals   16,977.14

[RF Energy (kWh) / RF Energy Delivered (kWh)][100] = 80.79% RF to Heat

* Data not available for 2/23/97. Used average of 2/22 and 2/24 data.

Appendix E

**Idealized Scenarios for Kirtland AFB
RF-SVE Demonstration**

Appendix E Listing and Description of Tables

Item	Title	Description
Calculations	Example calculations for Table E2, Case 1	Example calculations for Case 1
Calculations	Example calculations for Table E3, Case 1	Example calculations for Case 1
Table E1	RF-SVE idealized case studies Kirtland AFB demonstration	1. Description of three idealized cases for the RF-SVE demonstration at Kirtland AFB 2. Conceptual area and volume of Kirtland AFB demonstration site 3. General site conditions
Table E2	RF-SVE design basis for idealized Cases 1, 2, and 3 Kirtland AFB demonstration	For each case: 1. Operating assumptions for SVE system 2. Operating assumptions for RF system 3. Operating assumptions for SVOC contaminant removal
Table E3	Heat balance for idealized Cases 1, 2, and 3 Kirtland AFB demonstration	For each case: 1. Heat requirements for wet soil 2. Heat requirements for water vapor 3. Heat SVE air to 100°C 4. Heat SVE air to 131°C 5. Heat loss to the surroundings 6. Total heat requirements

EXAMPLE CALCULATIONS FOR TABLE E2, CASE 1

The calculated data in Appendix E tables were obtained using MS Excel. In most cases, the values within the example calculations have been rounded to an appropriate significant digit.

Case 1 — SVE System

Air extraction wells = 1
Maximum dry-airflow per extraction well = 280 slpm
AIR_{total} = maximum volumetric airflow
AIR_{total} = (280 slpm/well)(1 well)(60 min/h)(24 h/day) /(1,000 L/m³) = 403 m³/day
AIR_{mass} = maximum mass airflow = (403 m³/day)(1.204 kg/m³) = 485 kg/day

Case 1 — RF System

Data from Table E1:
  Soil density = 1,788 kg/m³
  Initial soil moisture = 7.0% dwb
  Final soil moisture = 0.5% dwb
Data from Table E2:
  Total RF antenna (RFA) = 1
  kW per RF antenna (kW/RFA) = 13.20
  Antenna efficiency (RF energy to heat) = 80.8%
  Antenna on time = 93.8%
  Efficiency of AC power to RF energy = 43.9%

1. Heat wet soil to 100°C

Total heat content and mass of soil moisture from Table E3, heat balances:
$H_{wet\ soil}$, total heat content of wet soil at 100°C = 5.116 million kJ = 1,421 kWh
Mass of soil (dry basis) = (28.32 m³)(1,778 kg/m³) = 50,347 kg
Total mass of moisture in soil = (50,347 kg)(7.0%/100%) = 3,524 kg
Total mass of soil moisture to be removed = 3,524 kg − (50,347 kg)(0.5%/100%) = 3,273 kg

Operating from ambient to 100°C:
Air extraction wells = 1 (SVE B)
Density of air at 20°C = 1.204 kg/m³
Set airflow per well = 190 slpm
Set humidity of air at 100°C = 0.2500 kg water/kg dry air

Change the value (**9.95 days**) until agreement with "days" value in 4.
T_{total}, Total time to treat site at 100°C = **9.95 days (iterative value with "days" in 4).**

AIR_{mass} = (190 Lpm)(60 min/h)(24 h/day)(1.204 kg/m³)(1 well)/(1,000 L/m³) = 329 kg/day
Mass of dry air to remove water = (329 kg/day)(9.95 day) = 3,277 kg
Mass of soil moisture removed at 100°C = (3,277 kg)(0.25 kg/kg) = 819 kg
Check: Mass of dry air = (819 kg)/(0.2500 kg/kg) = *3,277 kg*

2. Remove water and DRO from wet soil

Operating from 100°C–131.5°C:
Airflow per well = 280 liters per minute
Air extraction wells = 1 (SVE B)
Density of air at 20°C = 1.204 kg/m³

Change the value (**13.78 days**) until agreement with "days" value in 4.
T_{total}, total time to treat site up to 131.5°C = **13.78 days (iterative value with "days" in 4).**

Total mass of soil moisture to be removed = 3,273 kg
Mass of soil moisture removed = (3,273 kg – 819 kg) = 2,453 kg
AIR_{mass} = (280 Lpm)(60 min/h)(24 h/day)(1.204 kg/m³)(1 well)/(1,000 L/m³) = 485 kg/day
Mass of dry air to remove water = (13.78 days)/(485 kg/day) = 6,689 kg dry air
Adjust humidity until the *Check* value for mass of dry air equals calculated value
Humidity of air at 131.5°C = *0.3668* kg water/kg dry air
Check: mass of dry air = (2,453 kg)/(0.3668 kg/kg) = *6,689 kg*

3. Heat dry soil from 100°C to 131.5°C
Total heat content from Table E3, heat balances
$H_{wet\ soil}$, total heat content of dry soil at 131.5°C = 1,574 kWh
$H_{wet\ soil}$, total heat content of dry soil at 100°C = 1,092 kWh
$H_{dry\ soil}$, heat dry soil from 100°C–131.5°C = (1,574 – 1,092 kWh) = 483 kWh
$T_{dry\ soil}$, time to heat dry soil from 100°C–131.5°C
$T_{dry\ soil}$ = (483 kWh)/[(1 RFA)(13.2 kW/RFA)(80.8%/100%)(93.8%/100%)]/(24 h/day)
$T_{dry\ soil}$ = 2.01 days

4. Heat wet soil, heat dry air, and remove water-DRO

Operating from ambient to 100°C
Total heat content from Table E3, heat balances
H_{total}, total heat requirement from RF system = 8.597 million kJ = 2,388 kWh
T_{total}, total time to treat site at 100°C
T_{total} = (2,388 kWh)/[(1 RFA)(13.2 kW/RFA)(80.8%/100%)(93.8%/100%)]/(24 h/day)
T_{total} = **9.95 days (match # days in 1).**

Operating from 100°C to 131.5°C
Total heat content from Table E3, heat balances
H_{total}, total heat requirement from RF system = 11.884 million kJ = 3,301 kWh
T_{total}, total time to treat site at 131.5°C
T_{total} = (3,301 kWh)/[(1 RFA)(13.2 kW/RFA)(80.8/100%)(93.8/100%)]/(24 h/day)
T_{total} = **13.78 days (match # days in 2).**

Case 1 — DRO Contaminant Removal

Mass of DRO removed in air
Mass of soil (dry basis) = 50,374 kg
DRO concentration = 920 mg/kg
Final DRO concentration = 250 mg/kg
Mass of DRO removed in air = (50,374 kg)(920 – 250 mg/kg)/1,000,000 mg/kg
Mass of DRO removed in air = 33.7 kg

DRO removal rates per extraction well
Operating at 100°C:

Rate = (33.7 kg DRO)[(819 kg moisture removed)/3,273 kg total moisture)]/[(1 well)(9.95 days)]
Rate = 0.85 kg/day

Operating at 131.5°C:
Rate = [33.7 kg DRO – (0.85 kg/day)(9.95 day)(1 wells)]/[(1 wells)(13.75 day)]
Rate = 1.84 kg/day

DRO removal rate: mg/kg of soil per day:
DRO mg/kg-day = (33.7 kg DRO/50,375 kg soil)/(9.95 + 13.75 day)
DRO mg/kg-day = 28.27

EXAMPLE CALCULATIONS FOR TABLE E3, CASE 1

The calculated data in Appendix E tables were obtained using MS Excel. In most cases, the values within the example calculations have been rounded to an appropriate significant digit.

Case 1 — Heat Requirement for Wet Soil

C_p, heat capacity of dry soil at 20°C
  Use soil heat capacity (C_p) in Table C1
  C_p = 0.777 kJ/kg-°C + (20 – 0°C)/(25 – 0°C)(830 – 0.777 kJ/kg-°C)
  C_p = 0.820 kJ/kg-°C

C_p, heat capacity of dry soil at 100°C
  Use soil heat capacity (C_p) in Table C1
  C_p = 0.944 kJ/kg-°C

C_p, heat capacity of dry soil at 131.5°C
  Use soil heat capacity (C_p) in Table C1
  C_p = 0.974 kJ/kg-°C + (131.5 – 125°C)/(150 – 125°C)(1.000 – 0.974 kJ/kg-°C)
  C_p = 0.981 kJ/kg-°C

H_i, heat content of moisture at 100 and 20°C
  Use enthalpy table for liquid water (H_i) in Table C5
  H_i = 419.5 kJ/kg at 100°C
  H_i = 69.7 kJ/kg + (20.0 – 16.86°C)/(21.86 – 16.86°C)(90.7 – 69.7 kJ/kg)
  H_i = 82.9 kJ/kg at 20°C

H_{soil}, total heat content of dry soil at 100°C
  H_{soil} = [(50,374 kg)(0.944 kJ/kg-°C)(100°C) – (50,374 kg)(0.820 kJ/kg-°C)(20°C)]
  H_{soil} = 3,930,000 kJ
  H_{soil} = 3,930,000 kJ/(1 million kJ/1.0 million kJ) = 3.930 million kJ

H_{water}, total heat content of moisture at 100°C
  Mass of soil moisture = (7.0% H_2O/100)(50,347 kg) = 3,524 kg
  H_{water} = (3,524 kg)(419.5 kJ/kg – 82.9 kJ/kg)/1,000,000 = 1.186 million kJ

$H_{wet\ soil}$, Total heat content of wet soil at 100°C
  $H_{wet\ soil}$ = (3.930 million kJ + 1.186 million kJ) = 5.116 million kJ
  $H_{wet\ soil}$ = (5,116,000 kJ)(1,000 J/kJ)/(3,600,000 J/kWh) = 1,421 kWh

$H_{dry\ soil}$, total heat content of dry soil at 131.5°C
  $H_{dry\ soil}$ = [(50,347 kg)(0.9807 kJ/kg-°C)(131.5°C) – (50,347 kg)(0.8195 kJ/kg-°C)(20°C)]/1,000,000 = 5.668 million kJ
  $H_{dry\ soil}$ = (5,668,000 kJ)(1,000 J/kJ)/(3,600,000 J/kWh) = 1,574 kWh

Case 1 — Heat Requirements for Water Vapor

Mass of soil moisture to be removed = ((7.0 – 0.5%)/100))(50,347) = 3,273 kg

Heat wet soil from ambient to 100°C
  Mass of soil moisture removed heating to 100°C: Table E2
  Mass of soil moisture removed heating to 100°C = 819 kg
  H_{lg} = heat of vaporization of water, Table C5 = 2,256 kJ/kg
  $H_{water\ vapor}$, heat content of water vapor at 100°C

$H_{water\ vapor}$ = (819 kg)(2,256 kJ/kg)/(1,000,000) = 1.849 million kJ
$H_{water\ vapor}$ = (1,849,000 kJ)(1,000 J/kJ)/(3,600,000 J/kWh) = 513 kWh

Remove water and DRO from wet soil
Mass of soil moisture removed heating to 131.5°C: Table E2
Mass of soil moisture removed heating to 131.5°C = 2,453 kg
$H_{water\ vapor}$, heat content of water vapor at 131.5°C
$H_{water\ vapor}$ = (2,453 kg)(2,256 + 46 kJ/kg)/(1,000,000) = 5.649 million kJ
$H_{water\ vapor}$ = (5,649,000 kJ)(1,000 J/kJ)/(3,600,000 J/kWh) = 1,569 kWh
$H_{total\ vapor}$, total content of water vapor
$H_{total\ vapor}$ = (1.849 million kJ + 5.649 million kJ) = 7.497 million kJ
$H_{total\ vapor}$ = (7,497,000 kJ)(1,000 J/kJ)/(3,600,000 J/kWh) = 2,083 kWh

Case 1 — Heat SVE Air to 100°C

Heat Wet Soil to 100°C
Ambient air at 20°C
Mass of SVE air at 100°C = 3,277 kg
Cp_{air}, heat capacity of air at 100 and 20°C from Table C6
Cp_{air} (20°C) = 1.0210 kJ/kg-°C; Cp_{air} (100°C) = 1.0147 kJ/kg-°C
H_{air}, heat content of dry air at 100°C
H_{air} = (3,277 kg)[(100°C)(1.0147 kJ/kg-°C) − (20°C)(1.0210 kJ/kg-°C)]/10^6 = 0.2669 million kJ
H_{air} = (266,900 kJ)(1,000)/(3,600,000 J/kWh) = 74 kWh

Remove water and DRO from wet soil
Mass of SVE air at 131.5°C = 6,689 kg
Cp_{air}, heat capacity of air at 131.5 and 20°C from Table C5
Cp_{air} (20°C) = 1.0210 kJ/kg-°C; Cp_{air} (131.5°C) = 1.0025 kJ/kg-°C
H_{air}, heat content of dry air at 131.5°C
H_{air} = (6,689 kg)[(131.5°C)(1.0025 kJ/kg-°C) − (20°C)(1.0210 kJ/kg-°C)]/106
H_{air} = 0.7632 million kJ
H_{air} = (763,200 kJ)(1,000)/(3,600,000 J/kWh) = 212 kWh

Case 1 — Heat Loss to the Surroundings

RF-SVE heat loss to the surroundings = 25%

Heat wet soil to 100°C
Heat content of wet and dry soil from Table E3
$H_{wet\ soil}$ (100°C) = 5.116 million kJ
$H_{dry\ soil}$ (131.5°C) = 5.668 million kJ
H_{total} = Total heat requirement for RF system = 8.598 million kJ = 3,308 kWh
$T_{100°C}$, total time to treat site at 100°C = 9.95 day
$T_{131°C}$, total time to treat site at 131.5°C = 13.78 day
H_{sur}, heat loss to surroundings = (25/100%)(8.598 million kJ) − (25/100%)[(5.116 + 5.668 million kJ)/2](13.78 day/(9.95 + 13.78 days)
H_{sur}, heat loss to surroundings = 1.367 million kJ = 391 kWh

Remove water and DRO from wet soil
$H_{wet\ soil}$ (100°C) =5.116 million kJ
$H_{dry\ soil}$ (131.5°C) = 5.668 million kJ
H_{total}, total heat requirement from RF system = 11.910 million kJ = 3,308 kWh
$T_{100°C}$, total time to treat site at 100°C = 9.95 day
$T_{131°C}$, total time to treat site at 131.5°C = 13.78 day
H_{sur}, heat loss to surroundings = 25/100%)(11.910 million kJ) + (25%/100)[(5.116 + 5.668 million kJ)/2] [13.78 day/(9.95 + 13.78 days)]
H_{sur}, heat loss to surroundings = 3.760 million kJ = 1,033 kWh

Case 1 — Total Heat Requirements

Operating at 100°C
$H_{wet\ soil}$, total heat content of wet soil = 5.116 million kJ
H_{water} vapor, total heat content of water vapor = 1.849 million kJ
H_{air}, heat content of dry air at 100°C = 0.267 million kJ
H_{sur}, heat loss to surroundings = 1.367 million kJ (iterative value with H_{sur} above)
Change the value (1.367 million kJ) until agreement with H_{sur} above
H_{total}, total heat requirement from RF system = (5.111 + 1.849 + 0.267 + 1.367) million kJ
H_{total}, total heat requirement from RF system = 8.598 million kJ

Operating at 131.5°C

H_{dry}, heat dry from 100 to 131.5°C = 1.738 million kJ

$H_{water\ vapor}$, total heat content of water vapor = 5.649 million kJ

H_{air}, heat content of dry air at 100°C = 0.763 million kJ

H_{sur}, heat loss to surroundings = **3.760 million kJ (iterative value with H_{sur} above)**

Change the value (3.760 million kJ) until agreement with H_{sur} above

H_{total}, total heat requirement from RF system = (1.738 + 5.649 +0.763 + 3.760) million kJ

H_{total}, **total heat requirement from RF system = 11.910 million kJ**

Table E1 RF-SVE Idealized Case Studies, Kirkland AFB Demonstration

Cost Study Cases

Case 1: 9.290 m² Site 1
RF power to applicator at 13.20 kW
Heat wet soil to 100°C
Remove water & light-end DRO compounds at 100°C
Heat dry soil areas to 131.5°C
Remove heavy-end DRO compounds
Final DRO content <500 mg/kg

Case 2: 9.290 m² Site 1
RF power to applicator at 7.11 kW
Heat wet soil to 100°C
Remove water & light-end DRO compounds at 100°C
Heat dry soil areas to 131.5°C
Remove heavy-end DRO compounds
Final DRO content <500 mg/kg

Case 3: 9.290 m² Site 1
RF power to applicator at 7.11 kW
Heat wet soil to 100°C
Remove water & light-end DRO compounds at 100°C
Heat dry soil areas to 131.5°C
Remove heavy-end DRO compounds
Final DRO content <500 mg/kg

Area and Volume of Sites

Site 1

Depth of design volume	3.048
Length of design volume	3.048 m
Width of design volume	3.048 m
Thickness of contamination	3.048 m
Area	9.290 m²
Volume	28.317 m³

Site 2

Depth of heated volume	2.286 m
Radius of RF zone	3.048 m
Thickness of contamination	4.572 m
Number of RF zones	1
Area	29.19 m²
Volume	133.44 m³

Site Conditions

Soil has similar physical properties to the Kirtland AFB site

Initial DRO	*920* mg/kg (av value) (10,000 mg/kg max value)
Initial moisture	7.0%
Final moisture	0.5%
Soil density	1,778 kg/m³ (dry basis) (111 lb/ft³)
Soil porosity	34%
Ambient air temp	15–30 °C
Initial soil temp	20 °C
Water table depth	80 m
Depth to contaminated soil	3.048 m
SSTL for DRO	250 mg/kg (less than value)
No free NAPL	
No smear zone	
Include impermeable barrier at site (concrete?)	

Note: 3.28084 ft = 1.0 m; 4,046.86 m² = 1.0 acre; 16.018 lb/ft³ = 1.0 kg/m³; 0.02832 m³ = 28.32 L = 1.0 ft³; 3.6 × 10⁶ J = 1.0 kWh.

Table E2   RF-SVE Design Basis for Idealized Cases 1, 2, and 3
Kirtland AFB Demonstration

Case 1 — SVE System

Time to heat soil and remove water-DRO	23.73 days
Air injection wells	4
Air extraction wells	1
Standard (std) air temperature	20 °C
Standard (std) air pressure	101.3 kPa
Maximum dry-airflow per injection well	0 std L/min (Lpm)
Maximum dry-airflow per extraction well	280 std L/min (Lpm)
AIR_{total}, maximum total volumetric dry-airflow	403 std m³/day
Density of dry air at 20°C and 101.3 kPa	1.204 kg/m³
AIR_{mass}, maximum total mass dry-airflow	485 kg/day

Case 1 — RF System

Number of RF antenna wells	2
Total RF antenna (RFA)	1
KW per RF antenna (kW/RFA)	13.20
Antenna efficiency (RF energy to heat)	80.8%
Antenna on time	93.8%
Efficiency of AC power to RF energy	43.9%

1. Heat wet soil to 100°C

$H_{wet\ soil}$, total heat content of wet soil at 100°C	5.116 million kJ	1,421 kWh
Mass of soil moisture to be removed	3,273 kg	

Operating from ambient to 100°C:

$T_{100°C}$, total time to treat site up to 100°C	*9.95 days (change #)*
Humidity of air (RF-SVE demonstration data)	0.2500 kg water/kg dry air
Mass of soil moisture removed at 100°C	819 kg of water
Mass of dry air to remove water at 100°C	3,277 kg dry air — airflow = *190* Lpm
Check	*3,277*

2. Remove water and DRO from wet soil
Operating from 100–131.5°C:

$T_{131°C}$, total time to treat site up to 131.5°C	*13.78 days (change #)*
Humidity of air (RF-SVE demonstration)	0.3668 kg water/kg dry air
Mass of soil moisture removed	2,453 kg of water
Mass of dry air to remove water	6,689 kg dry air - airflow = *280*Lpm
Check	*6,689*

3. Heat dry soil from 100°C to 131.5°C

$H_{dry\ soil}$, total heat content of dry soil at 131.5°C	5.668 million kJ	1,574 kWh
$H_{dry\ soil}$, total heat content of dry soil at 100°C	3.930 million kJ	1,092 kWh
$H_{dry\ soil}$, heat dry soil from 100°C to 131.5°C	1.738 million kJ	483 kWh
$T_{dry\ soil}$, time to heat soil from 100°C to 131.5°C	2.01 days	

4. Heat wet soil, heat dry air, and remove water-DRO
Operating from ambient to 100°C

H_{total}, total heat requirement from RF system	8.598 million kJ	2,388 kWh
$T_{100\ c}$, total time to treat site up to 100°C	9.95 days (match above #)	

Operating from 100-131.5°C:

H_{total}, total heat requirement from RF system	11.910 million kJ	3,308 kWh
$T_{131\ c}$, total time to treat site up to 131.5°C	13.78 days (match above #)	

**Table E2   RF-SVE Design Basis for Idealized Cases 1, 2, and 3 (continued)
Kirtland AFB Demonstration**

Case 1 — DRO Contaminant Removal

Mass of DRO removed in air

Mass of soil (dry basis)	50,347 kg
Initial DRO concentration	920 mg/kg
Final DRO concentration	250 mg/kg
Mass of DRO removed in air	33.7 kg

DRO removal rates per extraction well

Operating at 100°C	0.85 kg/day
Operating at 131.5°C	1.84 kg/day

DRO rate of removal

mg per kg of soil per day	28.23 mg/kg-day

DRO removal rate assumptions
1. Demonstration rate of DRO removal for one extraction well
   Airflow at ~280 std Lpm
   DRO removal rate = ~2.0 kg/day
2. Demonstration DRO averaged ~900 mg/kg and site 1 contains 8,000 mg/kg
3. Demonstration DRO was mixed with high concentrations of MRO that averaged ~6,650 mg/kg
4. Final average DRO (mg/kg) is similar to site-1's SSTL value of 500 mg/kg

Case 2 — SVE System

Time to heat soil and remove water-DRO	**45.38 days**

Air injection wells	4
Air extraction wells	1
Standard (std) air temperature	20 °C
Standard (std) air pressure	101.3 kPa
Dry-airflow per injection well	0 std L/min (Lpm)
Dry-airflow per extraction well	280 std L/min (Lpm)
AIR_{total}, total volumetric dry-airflow	403 std m³/day
Density of dry air at 20°C and 101.3 kPa	1.204 kg/m³
AIR_{mass}, total mass dry-airflow	485 kg/day

Case 2 — RF System

RF applicator wells	2
Total RF applicators (RFA)	1
kW per RF applicator (kW/RFA)	7.11
Applicator efficiency (RF energy to heat)	80.8%
Applicator on time	93.8%
Efficiency of AC power to RF energy	43.9%

1. **Heat wet soil to 100°C**

$H_{wet\ soil}$, total heat content of wet soil at 100°C	5.116 million kJ	1,421 kWh
Mass of soil moisture to be removed	3,273 kg	

   Operating from ambient to 100°C:

$T_{100°C}$, total time to treat site up to 100°C	**18.72 days (change #)**
Humidity of air (RF-SVE demonstration data)	0.1000 kg water/kg dry air
Mass of soil moisture removed at 100°C	617 kg of water
Mass of dry air to remove water at 100°C	6,166 kg dry air — airflow = *190* Lpm
Check	*6,166*

2. **Remove water and DRO from wet soil**
   Operating from 100–131.5°C:

$T_{131°C}$, total time to treat site up to 131.5°C	**26.66 days (change #)**
Humidity of air (RF-SVE demonstration)	0.2053 kg water/kg dry air
Mass of soil moisture removed	2,656 kg of water
Mass of dry air to remove water	12,941 kg dry air — airflow = *280* Lpm
Check	*12,940*

3. Heat dry soil from 100–131.5°C

$H_{wet\ soil}$, total heat content of dry soil at 131.5°C	5.668 million kJ	1,574 kWh
$H_{wet\ soil}$, total heat content of dry soil at 100°C	3.930 million kJ	1,421 kWh
$H_{dry\ soil}$, heat dry soil from 100–131.5°C	1.738 million kJ	153 kWh

$T_{dry\ soil}$, time to heat soil from 100–131.5°C   **1.19 days**

4. Heat wet soil, heat dry air, and remove water-DRO
Operating from ambient to 100°C
H_{total}, total heat requirement from RF system   8.714 million kJ   2,421 kWh

$T_{100°C}$, total time to treat site up to 100°C   **18.72 days (match # above)**

Operating from 100–131.5°C
H_{total}, total heat requirement from RF system   12.412 million kJ   3,448 kWh

$T_{131°C}$, total time to treat site up to 131.5°C   **26.66 days (match # above)**

Case 2- DRO Contaminant Removal

Mass of DRO removed in air
Mass of soil (dry basis)	50,347 kg
Initial DRO concentration	920 mg/kg
Final DRO concentration	250 mg/kg
Mass of DRO removed in air	33.7 kg

DRO removal rates per extraction well
Operating at 100°C	0.34 kg/day
Operating at 131.5°C	1.03 kg/day
14.76 mg/kg-day	

DRO removal rate assumptions
1. Demonstration rate of DRO removal for 1 extraction well
   Airflow at ~280 std Lpm
   DRO removal rate = ~2.0 kg/day
2. Demonstration DRO averaged ~900 mg/kg and Site 1 contains 8,000 mg/kg
3. Demonstration DRO was mixed with high concentrations of MRO that averaged ~6,650 mg/kg
4. Final average DRO (mg/kg) is similar to Site-1's SSTL value of 500 mg/kg

Case 3 — SVE System

Time to heat soil and remove water-DRO	**45.59 days**
Air injection wells	4
Air extraction wells	1
Standard (std) air temperature	20 °C
Standard (std) air pressure	101.3 kPa
Airflow per injection well	0 std L/min (Lpm)
Airflow per extraction well	280 std L/min (Lpm)
AIR_{total}, total volumetric airflow	403 std m³/day
Density of air at 20°C and 101.3 kPa	1.204 kg/m³
AIR_{mass}, total mass airflow	485 kg/day

Case 3 — RF System

RF applicator wells	2
Total RF applicators (RFA)	1
kW per RF applicator (kW/RFA)	7.11
Applicator efficiency (RF Energy to Heat)	80.8%
Applicator on time	93.8%
Efficiency of AC power to RF energy	43.9%

1. Heat wet soil to 100°C

$H_{wet\ soil}$, total heat content of wet soil at 100°C	5.668 million kJ	1,574 kWh
Mass of soil moisture to be removed	3,273 kg	

**Table E2   RF-SVE Design Basis for Idealized Cases 1, 2, and 3 (continued)
Kirtland AFB Demonstration**

Operating from ambient to 100°C:	
$T_{100°C}$, total time to treat site up to 100°C	**18.43 days (change #)**
Humidity of air (RF-SVE demonstration data)	0.0800 kg water/kg dry air
Mass of soil moisture removed at 100°C	521 kg of water
Mass of dry air to remove water at 100°C	6,518 kg dry air — airflow = *204* Lpm
Check	*6,518*

2. Remove water and DRO from wet soil
Operating from 100°C to 131.5°C:

$T_{131°C}$, total time to treat site up to 131.5°C	**27.16 days (change #)**
Humidity of air (RF-SVE demonstration data)	0.2087 kg water/kg dry air
Mass of soil moisture removed	2,751 kg of water
Mass of dry air to remove water	13,184 kg dry air — airflow = *280* Lpm
Check	*13,182*

3. Heat dry soil from 100°C to 131.5°C

$H_{wet\ soil}$, total heat content of dry soil at 131.5°C	5.668 million kJ	1,574 kWh
$H_{wet\ soil}$, total heat content of dry soil at 100°C	3.930 million kJ	1,421 kWh
$H_{dry\ soil}$, heat dry soil from 100–131.5°C	1.738 million kJ	153 kWh
$T_{dry\ soil}$, time to heat soil from 100–131.5°C	**1.19 days**	

4. Heat wet soil, heat dry air, and remove water-DRO
Operating from ambient to 100°C

H_{total}, total heat requirement from RF system	8.027 million kJ	2,383 kWh
$T_{100°C}$, total time to treat site up to 100°C	**18.43 days (match above #)**	
Operating at 131.5°C		
H_{total}, total heat requirement from RF system	13.832 million kJ	3,513 kWh
$T_{130°C}$, total time to treat site at 131.5°C	**27.16 days (match # above)**	

Case 3 — DRO contaminant removal

Mass of DRO removed in air	
Mass of soil (dry basis)	50,347 kg
Initial DRO concentration	920 mg/kg
Final DRO concentration	250 mg/kg
Mass of DRO removed in air	34 kg
DRO removal rates per extraction well	
Operating at 100°C	0.29 kg/day
Operating at 131.5°C	1.04 kg/day
DRO rate of removal	
mg/kg of soil per day	14.70 mg/kg-day

DRO removal rate assumptions
1. Demonstration rate of removal for 1 extraction well
   Airflow at ~280 std. Lpm
   DRO removal rate = ~2.0 kg/day
2. Demonstration DRO averaged ~900 mg/kg and site 1 contains 8,000 mg/kg
3. Demonstration DRO was mixed with high concentrations of MRO that averaged ~6,650 mg/kg
4. It was estimated that only ~50% of the DRO compounds can be removed by RF-SVE operating at 100°C.

**Table E3   Heat Balances for Idealized Cases 1, 2, and 3
Kirtland AFB Demonstration**

Case 1 — Heat Requirement for Wet Soil

Average soil temperature	20.0 °C	
Boiling point of water	100.0 °C	
Average final temperature	131.5 °C	
Volume of soil	28.32 m³	
Mass of soil (dry basis)	50,347 kg	
Mass of soil (dry basis)	50.35 t	
Total mass of moisture in the soil	3,524	
Mass of soil moisture to be removed	3,273 kg	
C_p, heat capacity of dry soil at 131.5°C	0.9807 kJ/kg-°C	
C_p, heat capacity of dry soil at 100°C	0.9444 kJ/kg-°C	
C_p, heat capacity of dry soil at 20°C	0.8195 kJ/kg-°C	
H_l, heat content of soil moisture at 100°C	419.5 kJ/kg	
H_l, heat content of soil moisture at 20°C	82.9 kJ/kg	
H_{soil}, total heat content of dry soil at 100°C	3.930 million kJ	1,092 kWh
H_{water}, total heat content of moisture at 100°C	1.186 million kJ	330 kWh
$H_{wet\ soil}$, total heat content of wet soil at 100°C	**5.116 million kJ**	**1,421 kWh**
H_{soil}, total heat content of dry soil at 131.5°C	5.668 million kJ	1,574 kWh
$H_{dry\ soil}$, total heat content of dry soil at 131.5°C	**5.668 million kJ**	**1,574 kWh**

Case 1 — Heat Requirements for Water Vapor

Mass of soil moisture to be removed	3,273 kg of water	
Heat wet soil from ambient to 100°C		
Mass of soil moisture removed up to 100°C	819 kg of water	
H_{lg}, heat of vaporization for water at 100°C	2,256 kJ/kg	
$H_{water\ vapor}$, heat content of water vapor at 100°C	1.849 million kJ	513 kWh
Remove water and DRO from wet soil		
Mass of soil moisture removed heating to 131.5°C	2,453 kg of water	
H_{lg}, heat of vaporization for water at 100°C	2,256 kJ/kg	
H_g, heat content of water vapor: 100–131.5°C	46 kJ/kg	
$H_{water\ vapor}$, heat content of water vapor at 131.5°C	5.649 million kJ	1,569 kWh
$H_{total\ vapor}$, total heat content of water vapor	7.497 million kJ	2,083 kWh

Case 1 — Heat SVE Air to 100°C

Heat wet soil to 100°C		
Heat air from 20–100°C		
$T_{100°C}$, total time to treat site at 100°C	9.95 days	
Humidity of SVE air at 100°C	0.2500 kg water/kg dry air	
Mass of dry SVE air	3,277 kg	
$C_{p\ air}$, heat capacity of air at 100°C	1.0147 kJ/kg-°C	
$C_{p\ air}$, heat capacity of air at 20°C	1.0025 kJ/kg-°C	
H_{air}, heat content of dry air at 100°C	**0.2669 million kJ**	**74 kWh**

Case 1 — Heat SVE Air to 131.5°C

Remove water and DRO from wet soil		
Heat air from 20–131.5°C		
$T_{131°C}$, total time to treat site at 131.5°C	13.78 days	
Humidity of SVE air at 100°C	0.3668 kg water/kg dry air	
Mass of dry SVE air	6,689 kg	
$C_{p\ air}$, heat capacity of air at 131.5°C	1.0201 kJ/kg-°C	
$C_{p\ air}$, heat capacity of air at 20°C	1.0025 kJ/kg-°C	
H_{air}, heat content of dry air at 131.5°C	**0.7632 million kJ**	**212 kWh**

**Table E3   Heat Balances for Idealized Cases 1, 2, and 3 (continued)
Kirtland AFB Demonstration**

Case 1 — Heat Loss to the Surroundings

The heat loss to the surroundings is based on the heat balance for the RF-SVE demonstration.

RF-SVE heat loss to the surroundings, Phase 1	25% of total RF heat	
RF-SVE heat loss to the surroundings, Phase 2	25% of total RF heat	
$H_{wet\ soil}$, total heat content of wet soil at 100°C	5.116 million kJ	1,421 kWh
$H_{dry\ soil}$, total heat content of dry soil at 131°C	5.668 million kJ	1,574 kWh

Heat wet soil to 100°C
  Operating from ambient to 100°C
  H_{sur} includes heat loss by wet soil:

H_{sur}, **heat loss to surroundings**	1.367 million kJ	391 kWh

Remove water and DRO from wet soil
  Operating from 100°C to 131.5°C
  H_{sur} includes heat loss by wet and dry soils

H_{sur}, **heat loss to surroundings**	3.760 million kJ	1,033 kWh

Case 1 — Total Heat Requirements

Heat wet soil to 100°C

$H_{wet\ soil}$, total heat content of wet soil	5.116 million kJ	1,421 kWh
$H_{water\ vapor}$, total heat content of water vapor	1.849 million kJ	513 kWh
H_{air}, heat content of dry air at 100°C	0.267 million kJ	74 kWh
H_{sur}, heat loss to surroundings	*1.367 million kJ*	380 kWh
H_{total}, **total heat requirement from RF system**	**8.598 million kJ**	**2,388 kWh**
Check	*8.598 million kJ*	

Remove water and DRO from wet soil

H_{dry}, heat dry soil from 100°C to 131.5°C	1.738 million kJ	483 kWh
$H_{water\ vapor}$, total heat content of water vapor	5.649 million kJ	1,569 kWh
H_{air}, heat content of dry air at 131.5°C	0.763 million kJ	212 kWh
H_{sur}, heat loss to surroundings	*3.760 million kJ*	1,044 kWh
H_{total}, **total heat requirement from RF system**	**11.910 million kJ**	**3,308 kWh**
Check	*11.910 million kJ*	
H_{total}, **project heat requirement from RF system**	**20.509 million kJ**	**5,697 kWh**
Check	*25.00%*	

Case 2 — Heat Requirement for Wet Soil

Average soil temperature	20.0 °C
Boiling point of water	100.0 °C
Average final temperature	131.5 °C
Vol of soil	28.32 m³
Mass of soil (dry basis)	50,347 kg
Mass of soil (dry basis)	50.35 t
Total mass of moisture in the soil	3,524 kg
Mass of soil moisture to be removed	3,273 kg
Cp, heat capacity of dry soil at 131.5°C	0.9807 kJ/kg-°C
Cp, heat capacity of dry soil at 100°C	0.9444 kJ/kg-°C
Cp, heat capacity of dry soil at 15°C	0.8195 kJ/kg-°C
H_l, heat content of soil moisture at 100°C	419.5 kJ/kg
H_l, heat content of soil moisture at 20°C	82.9 kJ/kg

H_{soil}, total heat content of dry soil at 100°C	3.930 million kJ	1,092 kWh
H_{water}, total heat content of moisture at 100°C	1.186 million kJ	330 kWh
$H_{wet\ soil}$, total heat content of wet soil at 100°C	5.116 million kJ	1,421 kWh
H_{soil}, total heat content of dry soil at 131.5°C	5.668 million kJ	1,574 kWh
$H_{dry\ soil}$, total heat content of dry soil at 131.5°C	5.668 million kJ	1,574 kWh

Case 2 — Heat Requirements for Water Vapor in Air

Mass of soil moisture to be removed   3,273 kg of water

Heat wet soil to 100°C
   Operating at 100°C
   Mass of soil moisture removed at 100°C   617 kg of water
   H_{lg}, heat of vaporization for water at 100°C   2,256 kJ/kg
   $H_{water\ vapor}$, heat content of water vapor at 100°C   1.391 million kJ   386 kWh

Remove water and DRO from wet soil
   Mass of soil moisture removed at 100°C   2,656 kg of water
   H_{lg}, heat of vaporization for water at 100°C   2,256 kJ/kg
   H_g, heat content of water vapor: 100–131.5°C   46 kJ/kg
   $H_{water\ vapor}$, heat content of water vapor at 131.5°C 6.116 million kJ   1,699 kWh

   $H_{total\ vapor}$, total heat content of water vapor   7.507 million kJ   2,085 kWh

Case 2 — Heat SVE Air to 100°C

Heat wet soil to 100°C
   T_{total}, total time to treat site at 100°C   18.72 days
   Humidity of SVE air at 100°C   0.1000 kg water/kg dry air
   Mass of dry SVE air   3,273 kg
   Cp_{air}, heat capacity of air at 100°C   1.0147 kJ/kg-°C
   Cp_{air}, heat capacity of air at 20°C   1.0025 kJ/kg-°C
   H_{air}, heat content of dry air at 100°C   0.266 million kJ   74 kWh

Case 2 — Heat SVE Air to 131.5°C

Remove water and DRO from wet soil
   T_{total}, total time to treat site at 131.5°C   26.66 days
   Humidity of SVE air at 100°C   0.2053 kg water/kg dry air
   Mass of dry SVE air   12,941 kg
   Cp_{air}, heat capacity of air at 131.5°C   1.0201 kJ/kg-°C
   Cp_{air}, heat capacity of air at 20°C   1.0025 kJ/kg-°C
   H_{air}, Heat Content of Dry Air at 131.5°C   1.442 million kJ   401 kWh

Case 2 — Heat Loss to the Surroundings

The heat loss to the surroundings is based on the heat balance for the RF-SVE demonstration.

RF-SVE heat loss to the surroundings   35% of total RF heat
RF-SVE heat loss to the surroundings   20% of total RF heat

$H_{wet\ soil}$, total heat content of wet soil at 100°C   5.116 million kJ   1,421 kWh

$H_{dry\ soil}$, total heat content of dry soil at 131°C   5.668 million kJ   1,574 kWh

Heat wet soil to 100°C
   Operating at 100°C
   H_{sur} includes heat loss by wet soil
   H_{sur}, heat loss to surroundings   1.941 million kJ   555 kWh

Remove water and DRO from wet soil
   Operating at 131.5°C
   H_{sur} includes heat loss by wet soil
   H_{sur}, heat loss to surroundings   3.116 million kJ   1,499 kWh

Case 2 — Total Heat Requirements

Heat wet soil to 100°C

$H_{wet\ soil}$, total heat content of wet soil	5.116 million kJ	1,421 kWh
$H_{water\ vapor}$, total heat content of water vapor	1.391 million kJ	386 kWh
H_{air}, heat content of dry air at 100°C	0.266 million kJ	74 kWh
H_{sur}, heat loss to surroundings	1.941 million kJ	539 kWh
H_{total}, total heat requirement from RF system	8.714 million kJ	2,421 kWh
Check	*8.714 million kJ*	

Table E3   Heat Balances for Idealized Cases 1, 2, and 3 (continued)
Kirtland AFB Demonstration

Remove water and DRO from wet soil

H_{dry}, heat dry soil from 100–131.5°C	1.738 million kJ	483 kWh
$H_{water\ vapor}$, total heat content of water vapor	6.116 million kJ	1,699 kWh
H_{air}, heat content of dry air at 131.5°C	1.442 million kJ	401 kWh
H_{sur}, heat loss to surroundings	*3.116 million kJ*	866 kWh

H_{total}, **total heat requirement from RF system**	**12.412 million kJ**	**3,448 kWh**
Check	*12.412 million kJ*	
H_{total}, **project heat requirement from RF system**	**21.126 million kJ**	**5,868 kWh**
Check	*23.94%*	

Case 3 — Heat Requirement for Wet Soil

Average soil temperature	20.0 °C
Boiling point of water	100.0 °C
Average final temperature	131.5 °C
Volume of soil	28.32 m³
Mass of soil (dry basis)	50,347 kg
Mass of soil (dry basis)	50.35 t
Total mass of moisture in the soil	3,524
Mass of soil moisture to be removed	3,273 kg

Cp, heat capacity of dry soil at 131.5°C	0.9807 kJ/kg-°C
Cp, heat capacity of dry soil at 100°C	0.9444 kJ/kg-°C
Cp, heat capacity of dry soil at 20°C	0.8195 kJ/kg-°C
H_l, heat content of soil moisture at 100°C	419.5 kJ/kg
H_l, heat content of soil moisture at 20°C	82.9 kJ/kg

H_{soil}, total heat content of dry soil at 100°C	3.930 million kJ	1,092 kWh
H_{water}, total heat content of moisture at 100°C	1.186 million kJ	330 kWh

$H_{wet\ soil}$, **total heat content of wet soil at 100°C**	**5.116 million kJ**	**1,421 kWh**

H_{soil}, total heat content of dry soil at 131.5°C	5.668 million kJ	1,574 kWh

$H_{wet\ soil}$, **total heat content of dry soil at 131.5°C**	**5.668 million kJ**	**1,574 kWh**

Case 3 — Heat Requirements for Water Vapor in Air

Mass of soil moisture to be removed	3,273 kg of water

Heat wet soil from ambient to 100°C

Mass of soil moisture removed at 100°C	521 kg of water	
H_{lg}, heat of vaporization for water at 100°C	2,256 kJ/kg	
$H_{water\ vapor}$, **heat content of water vapor at 100°C**	**1.176 million kJ**	327 kWh

Remove water and DRO from wet soil

Mass of soil moisture removed at 131°C	2,751 kg of water	
H_{lg}, heat of vaporization for water at 100°C	2,256 kJ/kg	
H_g, heat content of water vapor: 100–131.5°C	46 kJ/kg	
$H_{water\ vapor}$, **heat content of water vapor at 131.5°C**	**6.335 million kJ**	1,760 kWh

$H_{total\ vapor}$, **total heat content of water vapor**	**7.511 million kJ**	**2,086 kWh**

Case 3 — Heat SVE Air to 100°C

Heat wet soil to 100°C

T_{total}, total time to treat site at 100°C	18.43 days	
Humidity of SVE air at 100°C	0.0800 kg water/kg dry air	
Mass of dry SVE air	6,518 kg	
Cp_{air}, heat capacity of air at 100°C	1.0147 kJ/kg-°C	
Cp_{air}, heat capacity of air at 20°C	1.0025 kJ/kg-°C	
H_{air}, **heat content of dry air at 100°C**	**0.531 million kJ**	147 kWh

Case 3 — Heat SVE Air to 131°C

Remove water and DRO from wet soil

T_{total}, total time to treat site at 131.5°C	27.16 days

Humidity of SVE air at 100°C	0.2087 kg water/kg dry air	
Mass of dry SVE air	13,184 kg	
Cp_{air}, heat capacity of air at 131.5°C	1.0201 kJ/kg-°C	
Cp_{air}, heat capacity of air at 20°C	1.0025 kJ/kg-°C	
H_{air}, heat content of dry air at 131.5°C	1.498 million kJ	416 kWh

Case 3 - Heat Loss to the Surroundings

The heat loss to the surroundings is based on the heat balance for the RF-SVE demonstration.

RF-SVE heat loss to the surroundings	25% of total RF heat	
RF-SVE heat loss to the surroundings	25% of total RF heat	
$H_{wet\ soil}$, total heat content of wet soil at 100°C	5.116 million kJ	1,574 kWh
$H_{wet\ soil}$, total heat content of dry soil at 131°C	5.668 million kJ	1,574 kWh

Heat wet soil to 100°C
  Operating at 100°C
  H_{sur} includes heat loss by wet soil

H_{sur}, heat loss to surroundings	1.204 million kJ	334 kWh

Remove water and DRO from wet soil
  Operating at 131.5°C
  H_{sur} includes heat loss by wet soil

H_{sur}, heat loss to surroundings	4.261 million kJ	1,184 kWh

Case 3 — Total Heat Requirements

Heat wet soil to 100°C

$H_{wet\ soil}$, total heat content of wet soil	5.116 million kJ	1,574 kWh
$H_{water\ vapor}$, total heat content of water vapor	1.176 million kJ	327 kWh
H_{air}, heat content of dry air at 100°C	0.531 million kJ	147 kWh
H_{sur}, heat loss to surroundings	1.204 million kJ	334 kWh
H_{total}, total heat requirement from RF system	8.027 million kJ	2,383 kWh
Check	*8.579 million kJ*	

Remove water and DRO from wet soil

H_{dry}, heat dry soil from 100–131.5°C	1.738 million kJ	153 kWh
$H_{water\ vapor}$, total heat content of water vapor	6.335 million kJ	1,760 kWh
H_{air}, heat content of dry air at 131.5°C	1.498 million kJ	416 kWh
H_{sur}, heat loss to surroundings	4.261 million kJ	1,184 kWh
H_{total}, total heat requirement from RF system	13.832 million kJ	3,513 kWh
Check	*12.645 million kJ*	

H_{total}, project heat requirement from RF system
  Check   25.00%

Appendix F

Organic Compound Vapor Pressures

Appendix F Listing and Description of Tables

Item	Title	Description
Table F1	Organic Vapor Pressure Data Table 3-7 (Perry and Green, 1984)	1. Air temperature, °C and K 2. Vapor pressure at the specified temperature, mm Hg and kPa 3. Mass organic component, kg, per kg of dry air at the specified temperature
Table F2	Mass removal rates from organic vapor pressure data	1. Air temperature, °C and K 2. Mass removal rates, kg per day per m^3/min, of normal alkanes at specified temperature

Table F1   Organic Vapor Pressure Data

Dodecane ($C_{12}H_{26}$)

| Air Temperature | | Vapor Pressure | | kg($C_{12}H_{26}$) |
°C	K	mm Hg	kPa	per kg Air
17.9	291.0	0.2	0.03	0.0025
32.9	306.0	0.5	0.07	0.0062
47.8	320.9	1.0	0.13	0.0124
75.8	348.9	5.0	0.67	0.0621
90.0	363.1	10.0	1.33	0.1250
104.6	377.7	20.0	2.67	0.2535
121.7	394.8	40.0	5.33	0.5210
132.1	405.2	60.0	8.00	0.8038
146.2	419.3	100.0	13.33	1.4209
167.2	440.3	200.0	26.66	3.3492
191.0	464.1	400.0	53.32	10.4183

Octadecane ($C_{18}H_{38}$)

| Air Temperature | | Vapor Pressure | | kg($C_{18}H_{38}$) |
°C	K	mm Hg	kPa	per kg Air
84.9	358.0	0.2	0.03	0.0037
101.9	375.0	0.5	0.07	0.0092
119.6	392.7	1.0	0.13	0.0185
152.1	425.2	5.0	0.67	0.0928
169.6	442.7	10.0	1.33	0.1868
187.5	460.6	20.0	2.67	0.3787
207.4	480.5	40.0	5.33	0.7784
219.7	492.8	60.0	8.00	1.2010
236.0	509.1	100.0	13.33	2.1230
260.6	533.7	200.0	26.66	5.0039
288.0	561.1	400.0	53.32	15.5658

Tetradecane ($C_{14}H_{30}$)

| Air Temperature | | Vapor Pressure | | kg($C_{14}H_{30}$) |
°C	K	mm Hg	kPa	per kg Air
44.9	318.0	0.2	0.03	0.0029
61.9	335.0	0.5	0.07	0.0072
76.4	349.5	1.0	0.13	0.0144
106.0	379.1	5.0	0.67	0.0723
120.7	393.8	10.0	1.33	0.1455
135.6	408.7	20.0	2.67	0.2950
152.7	425.8	40.0	5.33	0.6064
164.0	437.1	60.0	8.00	0.9356
178.5	451.6	100.0	13.33	1.6538
201.8	474.9	200.0	26.66	3.8980
226.8	499.9	400.0	53.32	12.1256

Nonodecane ($C_{19}H_{40}$)

| Air Temperature | | Vapor Pressure | | kg($C_{19}H_{40}$) |
°C	K	mm Hg	kPa	per kg Air
97.9	371.0	0.2	0.03	0.0039
115.9	389.0	0.5	0.07	0.0097
133.2	406.3	1.0	0.13	0.0195
166.3	439.4	5.0	0.67	0.0978
183.5	456.6	10.0	1.33	0.1970
200.8	473.9	20.0	2.67	0.3993
220.0	493.1	40.0	5.33	0.8208
232.8	505.9	60.0	8.00	1.2663
248.0	521.1	100.0	13.33	2.2384
271.8	544.9	200.0	26.66	5.2760
299.8	572.9	400.0	53.32	16.4121

Hexadecane ($C_{16}H_{34}$)

| Air Temperature | | Vapor Pressure | | kg($C_{16}H_{34}$) |
°C	K	mm Hg	kPa	per kg Air
73.9	347.0	0.2	0.03	0.0033
89.9	363.0	0.5	0.07	0.0082
105.3	378.4	1.0	0.13	0.0164
135.2	408.3	5.0	0.67	0.0825
149.8	422.9	10.0	1.33	0.1661
164.7	437.8	20.0	2.67	0.3367
181.3	454.4	40.0	5.33	0.6921
193.2	466.3	60.0	8.00	1.0679
208.5	481.6	100.0	13.33	1.8876
231.7	504.8	200.0	26.66	4.4492
258.3	531.4	400.0	53.32	13.8402

Heneicosane ($C_{21}H_{44}$)

| Air Temperature | | Vapor Pressure | | kg($C_{21}H_{44}$) |
°C	K	mm Hg	kPa	per kg Air
115.9	389.0	0.2	0.03	0.0043
135.9	409.0	0.5	0.07	0.0107
152.6	425.7	1.0	0.13	0.0215
188.0	461.1	5.0	0.67	0.1081
205.4	478.5	10.0	1.33	0.2176
223.2	496.3	20.0	2.67	0.4410
243.4	516.5	40.0	5.33	0.9065
255.3	528.4	60.0	8.00	1.3986
272.0	545.1	100.0	13.33	2.4722
296.0	569.1	200.0	26.66	5.8272
323.8	596.9	400.0	53.32	18.1268

Note: Italic values were extrapolated using the plots of the original data.

Example calculation for mass organic vapor per unit mass dry air:
  VP = Vapor Pressure n-Alkane, kPa
  AP = Atmospheric Pressure, kPa = 101.33 kPa; kg(Organic) per kg dry air = [VP/(AP − VP)]*[mol weight organic/mol weight air].

  For dodecane, VP = 8.0 kPa at 132.1°C (405.2 K); kg($C_{12}H_{26}$ vapor) per kg dry air = [8/(101.3 − 8.0)]*[170.34/18.16] = 0.8038 kg/kg.

Table F2   Mass Removal Rates from Organic Vapor Pressure

($C_{12}H_{26}$)		($C_{14}H_{30}$)		($C_{16}H_{34}$)	
Air Temp (K)	kg/day/ m³/min	Air Temp (K)	kg/day/ m³/min	Air Temp (K)	kg/day/ m³/min
291.0	4.3	318.0	4.6	347.0	4.8
306.0	10.3	335.0	10.9	363.0	11.4
320.9	19.6	349.5	20.9	378.4	22.0
348.9	90.2	379.1	96.7	408.3	102.3
363.1	174.1	393.8	187.2	422.9	201.0
377.7	339.0	408.7	365.6	437.8	388.9
394.8	670.3	425.8	725.1	454.4	773.0
405.2	1008.3	437.1	1092.9	466.3	1167.0
419.3	1719.2	451.6	1867.5	481.6	1998.4
440.3	3868.1	474.9	4170.2	504.8	4477.1

($C_{18}H_{38}$)		($C_{19}H_{40}$)		($C_{21}H_{44}$)	
Air Temp (K)	kg/day/ m³/min	Air Temp (K)	kg/day/ m³/min	Air Temp (K)	kg/day/ m³/min
358.0	5.2	371.0	5.3	389.0	5.6
375.0	12.7	389.0	12.7	409.0	13.3
392.7	23.8	406.3	24.4	425.7	25.7
425.2	111.0	439.4	113.0	461.1	119.4
442.7	213.3	456.6	220.0	478.5	230.3
460.6	418.4	473.9	427.2	496.3	452.7
480.5	824.1	493.1	842.6	516.5	894.5
492.8	1232.9	505.9	1274.2	528.4	1341.0
509.1	2115.3	521.1	2187.4	545.1	2305.2
533.7	4753.0	544.9	4919.4	569.1	5195.1

Example calculation for organic removal rate:
  Assume constant airflow = 1.0 m³/min; da = density of air at the specified temp.
  V mass (Table F1) = vapor mass in dry air at the specified temp, kg/kg dry air;
    kg(organic)/day/(m³/min) = [V mass]*[da]*[24h/day]*[60 min/h].
  For dodecane, V mass = 0.8083 kg/kg at 132.1°C (405.2 K); da = 0.87104 kg/m³ at 132.1°C
    (405.2 K); kg($C_{12}H_{26}$ vapor)/day/(m³/min) = [0.8038]*[0.87104]*24*60 = 1008 kg/day per m³/min.

Appendix G

Partitioning Interwell Tracer Tests (PITT)

G1 PARTITIONING INTERWELL TRACER TESTS

G1.1   Introduction

Background on the Tracer Test

The partitioning tracer test has been used since the 1970s for characterizing oil in reservoirs (Allison et al., 1991 and Tang, 1992). Recently, the partitioning interwell tracer test (PITT) has been adapted for detection of non-aqueous phase liquids (NAPL) in the subsurface. This method was first applied to the detection of NAPL contamination in the saturated zone. It was demonstrated to be effective at both laboratory and field scales (Jin et al., 1995 and Jackson et al., 1997). The PITT technology has also been applied to the detection of NAPL in the vadose zone (Whitley et al., 1995 and INTERA, 1997).

A PITT involves several steps. First, a hydraulic flow field is set up between injection and extraction wells that bracket the zone of possible contamination. A mix of tracers is then introduced in the injection well and detected at the extraction site. The injected mixture includes a nonpartitioning tracer and at least one partitioning tracer. The nonpartitioning tracer exists solely in the mobile phase, so its travel time is dependent only on the flow rate and swept volume. The partitioning tracer is distributed between the mobile and resident immobile phases as it moves along the stream paths, and lags behind relative to the fraction of time spent in the immobile phases. A comparison of the partitioning and nonpartitioning tracer response curves can thus provide quantitative information about immobile phases present in the subsurface. Method of moments analyses are used to delineate the NAPL saturation from the tracer response curves (Appendix G2).

Purpose of the Study

The purpose of the PITT was primarily to help characterize the hydrocarbon contamination beneath a fire-training pit at Kirtland Air Force Base in Albuquerque, NM (Section 3.4). The second purpose of the PITT was to evaluate the remedial effectiveness of RF-SVE (Daniel et al., 1995).

Originally, the PITT was to provide information about the total petroleum hydrocarbons (TPH) existing in the subsurface. It was later determined that removal of the diesel range organic (DRO) hydrocarbons was the most important factor concerning the remediation effort. Thermodynamic modeling was employed concurrent with the PITT results to help delineate the removal of the DRO fraction.

G1.2   Laboratory Studies

The purpose of the laboratory studies was to gather information about the thermodynamic interaction of the tracers and the site hydrocarbons. Specifically, this included two tasks:

1. Determine the equilibrium partition coefficients, KN, between each tracer and the site hydrocarbons.

$$K_N = \frac{C_N}{C_A}$$

where C_N is the concentration of the tracer in the hydrocarbon NAPL
C_A is the concentration of the tracer in the air phase.
A complete theoretical explanation can be found in Appendix G2.

2. Develop a thermodynamic model to predict how the composition of the hydrocarbon mixture affects these partition coefficients.

Laboratory Materials and Equipment

Two composites were made from pre-demonstration soil samples. The first composite consisted of samples from boreholes 1 and 2, at depths ranging from 3 to 4 m below ground surface (bgs). The second composite consisted of samples from boreholes 3 through 6, at depths ranging from 3 to 4 m bgs. Sections 3.4.6 and 3.4.7 and Appendix A give more information about these samples. The soil was sieved to a maximum 2-mm grain size. A subsample was taken from each soil composite and extracted for TPH by the University of Texas Petroleum and Geosystems Engineering (PGE) personnel. The analytical procedure for TPH involved sonication and a methylene chloride solvent (EPA, 1992).

The first soil composite contained 10,840 mg TPH per kg bulk soil and was used for the first study (KIRT11). Bulk soil includes the soil solids plus resident hydrocarbons and water. The second soil composite contained 13,550 mg TPH per kg bulk soil and it was used for the next two studies (KIRT12) and (KIRT14). The complete extracted TPH had a density in both cases of 0.88 g/cm³. A true boiling point (TBP) analysis of the TPH was completed by Southern Petroleum Labs of Houston, TX. The TBP analysis was necessary for compositional modeling and is further discussed in Section G1.3.

Table G1.1 summarizes the characteristics of the chemical compounds used as gas tracers in these experiments.

Table G1.1 Summary of Tracers Used in Experiments

Tracer Name	Formula	M.W. (g/mol)	B.P. (C)
Methane	CH_4	16.0	−161.0
Difluoromethane	CF_2H_2	52.0	−51.7
Perfluoromethylcyclohexane	C_7F_{14}	350	76.0
Perfluoro-1,3-dimethylcyclohexane	C_8F_{16}	400	101.5
Perfluoro-1,3,5-dimethylcyclohexane	C_9F_{18}	450	125.0
Perfluorodecalin	$C_{10}F_{18}$	462	141.5

The perfluorocarbon series had previously been established as a viable set of air/NAPL partitioning gas tracers for characterizing tetrachloroethylene contamination (Whitley et al., 1995 and INTERA, 1997). A gas tracer mixture containing all of the compounds was made in a 22.4-L vessel and pressurized with nitrogen. The mixture was verified to be single phase through gas chromatographic analysis at elevated temperatures. Table G1.2 includes the relative mass of each tracer in the mixture.

Figure G1.1 shows a schematic of the experimental setup. The injected air was a compressed source throttled to 138 kPa (20 psi) before reaching the Porter Model 3VCD-1000 flow controller that was fitted with the low-flow adapter. The tracers were introduced at the tracer injection valve, a ten-port, two-position Valco manual switching type plumbed with a single 1-milliliter (mL) injection loop. Two types of columns were used for the soil packs. A 30-cm long by 2.5-cm diameter Kontes glass chromatography column was used in the first experiment. The next 2 experiments used a 30-cm long by 2.2-cm diameter stainless steel column with Swagelok end fittings. The autosampling system consisted of a ten-port, two-position Valco valve with actuator wired to a Buck Model 610 gas chromatograph. The autosampling valve was fitted with a 0.1-mL sample loop. Sampling and analysis were controlled using a Dell Latitude 433cx PC installed with Peaksimple for Windows software. A bubble meter from Fisher Scientific was used to monitor flow rates.

Table G1.2 Summary of Column Data and Results

Test Name	Tracer	Q (mL/min)	V_P (mL)	S_W	V_N (mL)	S_N	t_1 (min)	K_N	Mass Inj. (μg)	Mass Rec. (μg)	% Recovery
KIRT11	CH_4	0.156	58.3	0.28	3.76	0.065	246	0.0	39.5	39.7	100.4
	CF_2H_2						422	0.0	128.4	130.0	101.3
	C_7F_{14}						445	8.3	1.38	1.38	100.3
	C_8F_{16}						622	15.6	1.52	1.56	102.5
	C_9F_{18}						850	25.0	1.24	1.24	99.5
KIRT12	CH_4	0.104	44.1	0.32	3.30	0.075	369	0.0	85.5	79.3	92.8
	CF_2H_2						619	0.0	68.1	69.0	101.2
	C_7F_{14}						650	8.8	16.8	15.7	93.6
	C_8F_{16}						836	14.7	16.8	16.8	100.3
	C_9F_{18}						1078	22.4	16.8	16.9	100.9
	$C_{10}F_{18}$						2543	68.6	16.8	18.8	112.3
KIRT14	CH_4	0.100	46.7	0.29	3.18	0.068	430	0.0	73.9	76.4	103.5
	CF_2H_2						670	0.0	58.8	61.7	104.8
	C_7F_{14}						729	9.4	14.5	13.9	96.4
	C_8F_{16}						951	16.4	14.5	15.2	108.8
	C_9F_{18}						1222	24.9	14.5	14.0	96.9
	$C_{10}F_{18}$						2750	73.2	14.5	13.8	95.4

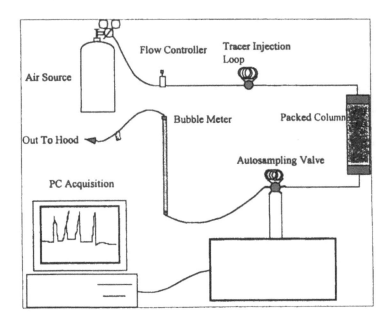

Figure G1.1 Schematic of laboratory tracer partitioning study.

Laboratory Procedures

The columns were packed by attaching the sleeve to a vibrating jig and adding soil in incremental 1-cm lifts. Each lift was lightly tamped and the column rotated to help prevent preferential pathways in the longitudinal direction. Important pack properties include

1. Total pore volume (V_P): pack volume not occupied by mineral solids
2. Volume of TPH (V_N): volume of TPH extracted from pack
3. TPH saturation (S_N): V_N/V_P
4. Water saturation (S_W): V_W/V_P where V_W is the total volume of water in the pack

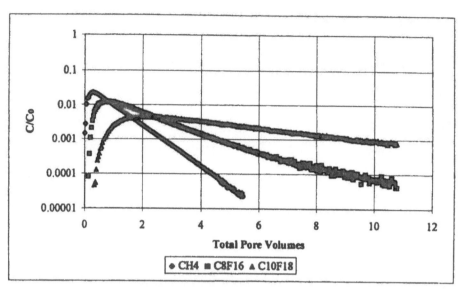

Figure G1.2 Tracer response curves, experiment KIRT12.

These properties for each experiment are included in Table G1.2.

After a column was put online, the flow rate was allowed to reach steady state overnight to help ensure that it remained constant over the course of the experiment. The flow rate (Q) for each experiment can be found in Table G1.2. The 1-cm³ sample loop was then filled with the tracer mixture at atmospheric pressure, and the valve was switched to inject the mixture on the column.

Effluent sampling occurred every 15 min over the course of a given experiment. Each experiment would typically last 5 to 7 days, depending on the mean residence time (ti) of the heaviest tracer, $C_{10}F_{18}$. The mean residence time (t_i) refers to the average amount of time that a particular tracer i spends in the column. $C_{10}F_{18}$ has the largest partition coefficient (K_N) and correspondingly has the longest mean residence time. Appendix G2 contains an explanation of the mathematical relationship between t_i and K_N.

Methane and difluoromethane were detected by flame ionization (FID), while the perfluorocarbons were analyzed using an electron capture detector (ECD). Detection limits were <10 parts tracer per million parts mixture by weight (ppm-w) for the FID detector and <1 ppm-w for the ECD detector.

Laboratory Results

Figure G1.2 shows the tracer response curves for the experiment (KIRT12), which are typical of all three experiments. Table G1.2 gives a summary of the experimental results for the three experiments. The partition coefficient values (K_N) for a given tracer compound were consistent among the 3 laboratory experiments, varying 9.8% at most. For example, the K_N values for C_7F_{14} were 8.3, 8.8, and 9.4 for the 3 experiments. This range of variation is typical for this type of experiment (Whitley, 1997).

Compositional Analysis

The necessary preparatory, thermodynamic, compositional work was completed by Hansen (1997). The following section provides a summary of this work.

Because of the compositional complexity of the TPH, a relationship between the tracer partition coefficients and the TPH composition was required. Hydrocarbon mixtures are suitable for thermodynamic analysis using the Peng-Robinson equation of state (EOS) (Khan, 1992). Useful

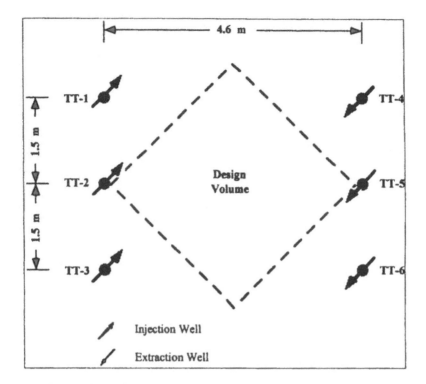

Figure G1.3 Plan view of PITT well layout.

routines of the compositional simulator UTCOMP (Chang et al., 1990) were arranged in a single program that includes this EOS.

The EOS requires the PVT properties of the components as input; so, this information was needed for both the tracers and the TPH. The thermodynamic properties of the tracers were available from the literature, but the TPH properties were unknown. The TBP analysis by Southern Petroleum Labs was used with carbon number correlations to develop a pseudofractional representation of the site hydrocarbons for input into the EOS (Hansen, 1997).

The EOS was tuned to match the tracer gas/TPH partition coefficients determined in the laboratory experiments described earlier. The tuned EOS was then used as a predictive tool to determine tracer gas/TPH partition coefficients for different mixture compositions. This was necessary due to the change in TPH composition that was caused by the RF-SVE remediation of the site soils.

G1.3 Field Test

Field Tracer Test Design

The well layout called for three well-pair groupings at three depths per grouping as shown in Figure G1.3, plan view. Figure G1.4 shows the various depths of each well grouping. Designated TT-numbers are consistent with those set out in Section 3.5.2. The letter designations in Figure G1.4 indicate the different depths for each numbered well. Computer simulations with this layout were run to help determine the PITT operational parameters. The purpose of these simulations was to design a PITT likely to be effective given variable subsurface parameters, since the exact soil permeability (k) distribution and TPH saturation (S_N) distribution were unknown. UTCHEM, the three-dimensional UT multicomponent multiphase flow and transport simulator (Delshad et al., 1996), was used for the simulations. The concentration breakthrough curves were analyzed with the method of moments. A base case scenario was set with the following parameters:

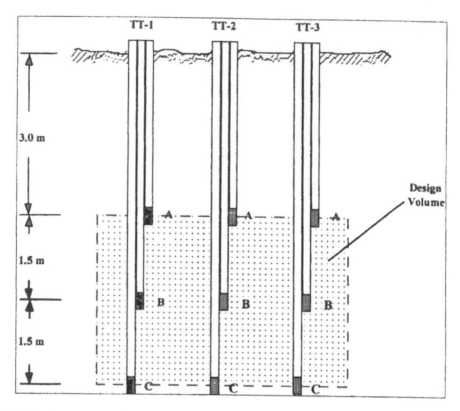

Figure G1.4 Profile view of PITT well layout.

- A total TPH volume of 5.7 m³
- A stochastic permeability field assumed to be log-normal with an average k of 10 D
- A variance ln(k) of 2.2
- A correlation length of 2 m in the horizontal plane

These parameters are statistical inputs used by UTCHEM to generate various possible k and S_N scenarios to simulate variable field conditions.

The flow rate and mass of tracers injected were the two main characteristics that needed to be determined with the simulations. The flow rate was adjusted to achieve a mean residence time of 1 day for the conservative tracer methane (Appendix G2). After the flow rate was set, the mass of tracer injected was adjusted according to the concentration envelope. This envelope was defined as the two curves limiting the concentrations of the nine wells at any given time-step.

To define this envelope, the simulated response data (tracer concentration at each time-step) were compiled from all nine wells for a single tracer. Here, time was defined by volume of air produced at the extraction wells. At each time-step, the tracer concentrations from all nine wells were compared and the minimum and maximum concentrations were extracted. The curves defined by these minimum and maximum concentrations are shown in Figure G1.5 which define the concentration envelope for C_9F_{18}. This process was completed for each tracer in the simulation.

The peak tracer concentration needed to fall below the maximum detection limit and approximately two orders of magnitude above the minimum detection limit to ensure that the breakthrough curve data were complete.

Over 50 simulations were completed and the results suggested that the following test characteristics were appropriate for the design volume:

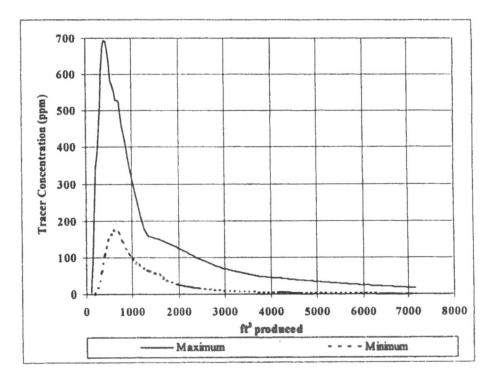

Figure G1.5 Actual concentration envelope for C_9F_{18}.

- A total airflow rate of 23.85 m³/day (6300 ft³/day) approximately evenly distributed over the 9 screens
- A total mass of 900 gm of the conservative tracer, methane
- A total mass of 200 gm of each of the partitioning tracers, the perfluorocarbon series (Table G1.1).

The simulations indicated that the PITT should last a minimum of 5 days, but 7 days would increase the accuracy of the results. The tracer injection time has a minor effect on the results. The expected pressure drops were between 690 and 1380 Pa. These were the design parameters indicated by the simulations and were followed whenever possible in the actual test. The actual PITT test parameters are described later in this section.

Materials and Equipment

The tracer compounds used in the field tests were the same as those studied in the laboratory. The tracers are listed in Table G1.4. The table also includes the total injected mass of each tracer. As with the laboratory experiments, methane was used as the conservative tracer, and the perfluorcarbons (C_xF_{2x}) were used for the partitioning tracers. The tracer mixture (Table G1.4) was contained in a 454-L (120-gal) tank pressurized to 2400 Pa (350 psi) with nitrogen. Figure G1.6 shows a schematic of the field test setup.

The gas flow for both injection and extraction was provided by two 1.5-kW Rotron blowers. Main flow lines consisted of 2.54 cm (1 in.) ID flexible PVC tubing. Gas flow rates were measured with an in-line turbine flow meter wired to a totalizer readout. On the injection side, the main flow was connected to the nine well points with a manifold. Cole-Parmer acrylic block-type rotameters with built-in throttling valves were connected to each single well line. The wells were 1.27-cm (0.5 in.) ID PVC tubes with perforated 1-ft intervals at the terminal ends. On the extraction side,

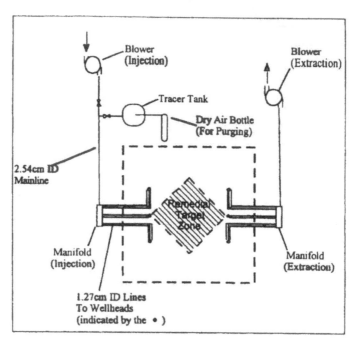

Figure G1.6 Schematic of field equipment setup.

a similar manifold system was used with 1-L (0.26-gal) knock-out bottles connected before the rotameters, to prevent water condensation in the meters.

Sampling occurred at each of the extraction wellheads. Flexible PVC 0.635 cm (0.25 in.) ID lines connected the 9 wellhead extraction points to the 10-position Valco dead-end switching valve. The tenth valve position was connected to a gas chromatograph (GC) calibration standard. The sample was drawn from the wellhead to the GC sample loop under vacuum from a Cole-Parmer 93.2 W high-capacity vacuum pump. The GC sample loop was plumbed to a ten-port Valco valve with a two-position electric actuator. A Buck Model 610 gas chromatograph provided the chemical analysis. A Dell Latitude 433cx PC installed with Peaksimple for Windows software controlled the GC and switching valves. The analytical system was completely automated.

Field Test Parameters

The actual parameters (Tables G1.3 and G1.4) followed the design parameters as closely as possible. However, several of the wells had very low permeabilities that prevented the design flow of 850 L/h (30 ft³/h) air under the pressure applied by the blowers. Table G1.3 shows the flow rates for each of the wells for PITT 1 and 2.

During the 9-month interlude between the 2 tests, a faulty tank allowed about 2/3 of the tracer mixture to leak from the second tank that had been slated for PITT2. The necessary refill was completed in the field. Therefore, the injected masses for the two tests were not identical. Table G1.4 shows the injected mass and injection time for each test, along with the mass recoveries.

Before PITT2 was completed, the site was resaturated with water. This was necessary because the heating process had dried out the soil to the point where the tracer compounds might sorb to the soil.

Data Analysis

Changing TPH Composition — The data from PITT1 were analyzed for average saturations using method of moments theory, with exponential extrapolation of the tracer response curves (see

Table G1.3 Injection and Extraction Flow Rates Used in the PITT

	Injection			Extraction	
Well	PITT1 (L/h)	PITT2 (L/h)	Well	PITT1 (L/h)	PITT2 (L/h)
TT1A	850	850	TT1A	850	850
TT1B	850	850	TT1B	850	850
TT1C	850	850	TT1C	850	850
TT2A	850	340	TT2A	850	566
TT2B	0	0	TT2B	566	850
TT2C	850	850	TT2C	28	28
TT3A	850	850	TT3A	850	283
TT3B	850	850	TT3B	850	850
TT3C	850	850	TT3C	850	850

Table G1.4 Tracer Injection Mass and Duration

	PITT1: $t_s = 2.1$ h			PITT2: $t_s = 2.0$ h		
Tracer	Injected (g)	Recovered (g)	Recovery (%)	Injected (g)	Recovered (g)	Recovery (%)
CH_4	500	285.1	57.0	497	296.9	59.8
CF_2H_2	400	229.6	57.4	290	173.3	59.8
C_7F_{14}	100	58.0	58.0	126	73.5	58.3
C_8F_{16}	100	57.4	57.4	150	88.7	59.1
C_9F_{18}	59	33.2	56.3	65	37.1	57.1
$C_{10}F_{18}$	42	22.9	54.5	70	43.3	61.8

Note: t_s is the duration of the tracer mixture injection.

Table G1.5 K values for the Two PITT Tests

	KPITT1 (L/L)	KPITT2 (L/L)
C_7F_{14}	8.80	7.43
C_8F_{16}	15.56	12.70
C_9F_{18}	24.10	19.35
$C_{10}F_{18}$	70.90	55.28

Appendix G2.2 for an explanation). The partition coefficient values (K) that were used were determined from the laboratory experiments, since the TPH used in those experiments was similar in composition to the site TPH (Section G1.2).

During the remediation effort, the composition of the TPH changed. Therefore, the laboratory-determined K values were not directly valid for analysis of the PITT2 data. Instead, the tuned EOS was used to predict the new K values through the following iterative process:

1. The initial volume of TPH was known from PITT1. The initial composition was also known from the true boiling point analysis (Hansen, 1997).
2. A small mass increment was removed from the lightest fraction (C_{13}–C_{20}). This removal did two things: first, it decreased the total volume of TPH. Second, it changed the overall TPH composition and therefore the K values predicted by the EOS.
3. The new K values were used to analyze the PITT2 data using method of moments (Appendix G2).
4. The current TPH volume predicted by the PITT2 data was compared to the volume left after subtracting the mass increment in step #2.
5. Steps 2 through 4 were repeated. When these two volumes were equal, the convergence criterion was reached. The new K values at convergence are given in Table G1.5.

Inverse Modeling to Determine Spatial TPH Distribution — More than the average saturation and total mass can be extracted from the PITT test breakthrough curves. The method of inverse

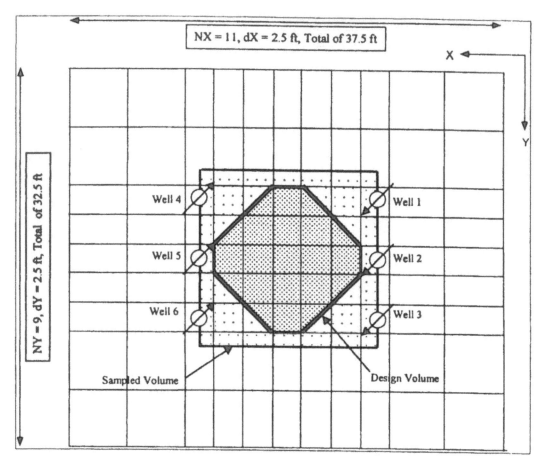

Figure G1.7 Plan view of grid used in tracer simulations.

modeling (also called parameter estimation or non-linear regression) allows for the calculation of the saturation or mass distribution in addition to their average and total. This procedure was done in two steps for this dataset. First, the permeability distribution was solved using the conservative tracer data. Next, using the permeability distribution obtained in the first step, the saturation distribution was determined from the partitioning tracer data. In each case, the distributions were determined by fitting simulated tracer response curves to the actual field data. The fitting was accomplished through minimization of the least square differences between the field and model data. Figure G1.7 shows the grid used in the simulations.

PITTs Tests Overall Remedial Performance — Using the method outlined earlier, the data from PITT1 and PITT2 yielded an estimate of the amount of TPH and DRO hydrocarbons removed by the RF-SVE process, as shown in Table G1.6.

As outlined in Section G1.2, the values in Table G1.6 were determined based on pseudofractional modeling of the site TPH. The TBP analysis was used to define fractions of the TPH by carbon number, including the DRO fraction (C_{13}–C_{20}). The EOS was used to correlate the tracer–air–TPH K values based on the composition of the site hydrocarbon. Then the previously outlined iterative procedure was used with the field data and thermodynamic data to converge to the values given in Table G1.6. For a complete understanding of the thermodynamic modeling, refer to Hansen (1997). For a more complete understanding of the method of moments analysis of the field data, refer to Appendix G2.

Table G1.6 TPH and DRO Mass Removal Calculated by Method of Moments

		MoleWt_av (g/mole)	Density, ρ (g/ml)	Mole Fraction	Volume (L)	Mass (kg)	% Removal by Mass
PITT1	TPH	366	0.905	1.000	2280	2,100	—
	DRO	253	0.843	0.320	490	410	—
PITT2	TPH	403	0.914	1.000	1990	1,800	—
	DRO	253	0.843	0.150	180	150	—
Decrease	TPH					300	13
	DRO					260	63

Note: These values should be considered with a variability of approximately 19% (Dwarakanath, 1997).

Table G1.7 Summary of Field Results for PITT1

Well	V_P (L)	S_W	V_N (L)	S_N
TT4A	14,500	0.14	320	0.022
TT4B	29,800	0.06	420	0.014
TT4C	40,000	0.10	280	0.007
TT5A	16,200	0.26	430	0.027
TT5B	9,700	0.09	160	0.017
TT5C	700	0.16	20	0.023
TT6A	7,000	0.16	60	0.008
TT6B	10,900	0.18	210	0.020
TT6C	37,300	0.15	380	0.010
Totals	166,400	0.131	2,280	0.014

Note: V_P = swept pore volume; S_W = saturation of tracers in water phase; V_N = vol of NAPL detected; S_N = saturation of tracers in NAPL phase.

Results by Wellpoint — Examples of the field-test response curves can be found in Appendix G4, Figures G4.1 through G4.36. The shape of the response curves indicated a complete sweep, showing no evidence of tracer bypassing or the presence of extremely low permeability layers. The shape of the curves and the mean residence times (Appendix G2) were also consistent with the streamline modeling completed before the tests, another indicator of complete sweep. It is important to note that normal heterogeneities in the flow field which cause uneven distribution of tracer do not invalidate the PITT results, as long as there are not totally impenetrable zones.

A complete set of results from the tests can be found in Appendix G5, Tables G5.1 and G5.2. Because each of the nine extraction wells gave information about streamlines captured by that particular well, a separate value was determined for that swept volume, V_P (volume contacted by captured tracers). The corresponding TPH volume, V_N (TPH volume detected by captured tracers), was also determined. Appendix G3 gives details on how these values were calculated. Tables G1.7 and G1.8 summarize the averaged field tracer test results at each well for PITT 1 and 2.

Note that the total volume of TPH detected in each case from Tables G1.7 and G1.8 corresponds to the TPH volume in Table G1.6. Appendix G3 provides a detailed explanation of the inverse modeling process.

Figure G1.8 shows a side-by-side comparison of average saturations for each wellpoint between the two tests. Note that for wells TT4C, TT5C, and TT6A, it seems that the TPH saturation has increased for that zone. This was due to the difference in swept volume between the two PITTs.

One trend shown in Figure G1.8 was the consistently high TPH saturation (S_N) near the middle zone (well 5, all depths) for both tests. Note that the saturation appeared to decrease on the outsides (wells 4 and 6, all depths) from PITT1 to PITT2, while the decrease was less evident in the middle zone (well 5, all depths). This may be a result of the remedial process or may represent a change in the flow paths.

Table G1.8 Summary of Field Results for PITT2

Well	V_P (L)	S_W	V_N (L)	S_N
TT4A	12,600	0.10	110	0.009
TT4B	26,900	0.08	360	0.014
TT4C	35,200	0.15	470	0.013
TT5A	8,000	0.30	190	0.023
TT5B	22,700	0.10	440	0.019
TT5C	800	0.25	20	0.026
TT6A	2,600	0.18	30	0.010
TT6B	17,700	0.10	180	0.010
TT6C	28,000	0.13	180	0.007
Totals	154,500	0.125	1,990	0.013

Table G1.9 Summary of Mass Comprised
in the Different Volumes (kg)

	Mass in Tracer Volumes (kg)
PITT1	2500
PITT2	2200
Reduction	300

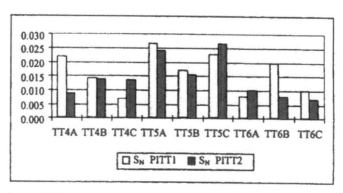

Figure G1.8 Comparison of TPH saturation pre- and post-remediation.

Because of the likely difference in the stream paths between the two PITTs due to the heating, SVE, and water resaturation, it was difficult to make definite conclusions about the spatial distribution of the TPH from this averaged data. This was a problem much better suited to analysis by inverse modeling.

What could be determined by the averaged PITT data was the total volume (V_N) and saturation S_N of TPH present in the subsurface before and after remediation. The TPH volume decreased from 2280 to 1990 L while the DRO hydrocarbon volume decreased from 490 to 180 L. Since the density of the hydrocarbon fractions was known, then these data could be stated on a mass basis, as in Table G1.6 and Section G1.4.

Remedial Performance by Inverse Modeling — Table G1.9 shows the inverse modeling results for hydrocarbon mass determination in the tracer volume. The TPH reduction determined from this method was 288 kg, which is 10% higher than that determined through the method of moments analysis. Given the range of uncertainty for both methods, the reductions determined from the two methods can be considered to be in good agreement.

G1.4　Conclusions

Laboratory Experiments

The laboratory work done for this study showed that partition coefficients can be reliably determined for hydrocarbon mixtures derived from field samples, with less than 10% variation among the experiments. The column test was adequate for both screening candidate tracer compounds and for quantifying their interaction with the site hydrocarbon mixture (Section G1.2).

The perfluorocarbon family of compounds was suitable for use as gas/hydrocarbon partitioning tracers. These compounds demonstrated a spectrum of partition coefficient values from 8.8 to 70.9 which were adequate for detection of a wide range of TPH saturations of about $0.002 < S_N < 0.23$ (Appendix G2, Equation 23). These compounds should be considered for any future gas/hydrocarbon PITT application.

Field Test

The PITT swept a much larger volume, 480 m^3, than the design volume of 28.3 m^3. Therefore, the PITT did not provide specific information within the imaginary boundaries of the design volume. However, when combined with compositional modeling, the PITT provided information about the amount and composition of the TPH in the tracer volume before and after remediation.

The PITT results from method of moments analysis indicated that 410 kg of DRO range hydrocarbons and 2100 kg of TPH existed in the tracer volume before remediation. The remedial effort reduced the DRO by 63% and the TPH by 13%, leaving 150 and 1800 kg, respectively.

The PITT results from inverse modeling showed that 288 kg of TPH were removed from the tracer volume. This result agreed closely (within 10%) with the method of moments result. This agreement provided additional confidence to the overall PITT results.

G2　PITT THEORY

G2.1　Method of Moments

The partition coefficient, K, of a partitioning tracer quantifies the fraction of tracer spent in the air and liquid phases. The partition coefficient is defined as the ratio of the concentration [M/L^3] of tracer i in the liquid phase j to the concentration of tracer i in the air phase, or

$$K_{i,j} = \frac{C_{i,j}}{C_{i,G}}$$

(G1)

where　$C_{i,j}$ is the concentration of i in liquid phase j
　　　$C_{i,G}$ is the concentration of i in the gas phase.
The overall flux F_i in a permeable medium flowing N phase is (Jin, 1995)

$$F_i = \sum_{j=1}^{N} f_j C_{ij}$$

(G2)

The total fluid-phase concentration C_i of tracer i is

$$C_i = \sum_{j=1}^{N} S_j C_{ij} \tag{G3}$$

where C_{ij} = Concentration of tracer i in phase j
f_j = Fractional flow of phase j
S_j = Saturation of phase j

Combining Equation (G1) with Equations (G2) and (G3) and using subscripts N, W, and G for NAPL, water, and gas, respectively, yields

$$F_i = \left(f_N K_{i,N} + f_G + f_W K_{i,W} \right) C_{i,G} \tag{G4}$$

$$C_i = \left(S_N K_{i,N} + S_G + S_W K_{i,W} \right) C_{i,G} \tag{G5}$$

The dimensionless residence time for any component i can be expressed as (Jin, 1995)

$$\bar{t}_D = \frac{\partial C_i}{\partial F_i}$$

For steady-state flow,

$$\bar{t}_D = \frac{C_i}{F_i} \tag{G6}$$

Combining Equation (G6) with Equations (G4) and (G5) yields

$$\bar{t}_D = \frac{S_N K_{i,N} + S_G + S_W K_{i,W}}{f_N K_{i,N} + f_G + f_W K_{i,W}} \tag{G7}$$

Assuming the NAPL and water phases are immobile, then

$$fN = 0$$

$$fW = 0$$

$$fG = 1$$

and from Equation (G7),

$$\bar{t}_D = S_N K_{i,N} + S_G + S_W K_{i,W} \tag{G8}$$

Since

$$S_N + S_G + S_W = 1$$

then

$$\bar{t}_D = S_W\left(K_{i,W} - 1\right) + S_N\left(K_{i,N} - 1\right) + 1 \tag{G9}$$

For tracers 1 and 2 which have partition coefficients $K_{1,W}$, $K_{1,N}$ and $K_{2,W}$ $K_{2,N}$

$$\bar{t}_{D,1} = S_W\left(K_{1,W} - 1\right) + S_N\left(K_{1,N} - 1\right) + 1 \tag{G10}$$

$$\bar{t}_{D,1} = S_W\left(K_{2,W} - 1\right) + S_N\left(K_{2,N} - 1\right) + 1 \tag{G11}$$

The swept pore volume, V_P, is inherent in $\bar{t}_D$, and allows the dimensionless residence times to be related to the experimentally determined residence times, $\bar{t}$, as follows:

$$V_P = \frac{Q\bar{t}_1}{\bar{t}_{D,1}} = \frac{Q\bar{t}_2}{\bar{t}_{D,2}} \tag{G12}$$

where

$$\bar{t}_i = \frac{\displaystyle\int_0^{t_f} t C_i dt}{\displaystyle\int_0^{t_f} C_i dt}$$

and Q is the volumetric flow rate.

Substituting (G10) and (G11) into (G12),

$$\frac{\bar{t}_1}{S_W\left(K_{1,W} - 1\right) + S_N\left(K_{1,N} - 1\right) + 1} = \frac{\bar{t}_2}{S_W\left(K_{2,W} - 1\right) + S_N\left(K_{2,N} - 1\right) + 1} \tag{G13}$$

Rearranging,

$$S_N = \frac{\bar{t}_2\left[S_W\left(K_{1,W} - 1\right) + 1\right] - \bar{t}_1\left[S_W\left(K_{2,W} - 1\right) + 1\right]}{\bar{t}_1\left(K_{2,N} - 1\right) - \bar{t}_2\left(K_{1,N} - 1\right)} \tag{G14}$$

A nonpartitioning tracer is generally included as a reference tracer in a PITT. If tracer 1 is a nonpartitioning tracer ($K_{N,1} = 0$, $K_{W,1} = 0$), then Equation (G14) simplifies to

$$S_N = \frac{\bar{t}_2\left(1 - S_W\right) - \bar{t}_1\left[S_W\left(K_{2,W} - 1\right) + 1\right]}{\bar{t}_1\left(K_{2,N} - 1\right) + \bar{t}_2} \tag{G15}$$

For a NAPL partitioning tracer where $K_{W,2} = 0$, Equation (G15) further reduces to

$$S_N = \frac{(\bar{t}_2 - \bar{t}_1)(1 - S_W)}{\bar{t}_1(K_{2,N} - 1) + \bar{t}_2} \tag{G16}$$

For a water partitioning tracer where $K_{N,2} = 0$, Equation (G15) can be solved for S_W, yielding

$$S_W = \frac{(\bar{t}_2 - \bar{t}_1)(1 - S_N)}{\bar{t}_1(K_{2,W} - 1) + \bar{t}_2} \tag{G17}$$

The volume of NAPL detected can be determined by multiplying the NAPL saturation by the swept pore volume, or

$$V_N = S_N \cdot V_P \tag{G18}$$

From Equation (G16)

$$S_N = \frac{(\bar{t}_2 - \bar{t}_1)S_G}{\bar{t}_1 K_{2,N}} \tag{G19}$$

and recognizing that

$$V_P = \frac{Q\bar{t}_1}{S_G} \tag{G20}$$

substituting Equations (G19) and (G20) in Equation (G18) yields

$$V_N = \frac{(\bar{t}_2 - \bar{t}_1)Q}{K_{2,N}} \tag{G21}$$

For the case of multiple injection wells and producing wells, the equations apply to each well pair where the flow rate in Equations (G20) and (G21) can be calculated as

$$Q = \frac{m}{M} Q_T \tag{G22}$$

where M is the total mass of tracer injected, Q_T is the total rate of all of the extraction wells, and m is the mass recovered from the individual producer. If a retardation factor, $R_F = t_2/t_1$, is defined, then Equation G19 can be rearranged to

$$R_F = 1 + \frac{K_N S_N}{(1 - S_N - S_W)} \tag{G23}$$

Jin (1995) gives criteria for a range of retardation factors that are viable in a field test, where $1.2 < R_F < 3.0$. Values lower than 1.2 are in a noise region, whereas values higher than 3.0 may cause too long a test. Therefore, given a range of K values for a tracer mixture, a range of likely detectable saturations can be determined.

G2.2   Extrapolation of Experimental Tracer Response Curves

To ensure the accuracy of the method of moments the tracer data should be complete, since a substantial amount of information is contained in the tail of a response curve (Jin, 1995). Extended tails may be incomplete because of concentrations falling below detection limits or premature termination of field tests due to time constraints.

An extrapolation technique has been developed to improve the accuracy of the response data (Jin, 1995).

It can be shown that for an assumed exponential decline of the tracer data,

$$\bar{t} = \frac{\int_0^{t_b} tC\,dt + m(m + t_b)C_b}{\int_0^{t_b} C\,dt + mC_b} - \frac{t_s}{2} \tag{G24}$$

where m = inverse slope of the straight line tail when the tracer reponse curves are plotted on a semi-log scale and C_b = tracer concentration at time t_b.

G3 INVERSE MODELING OF PITT

G3.1   Introduction

More than the average saturation and total mass can be extracted from the PITT breakthrough curves. The method of inverse modeling (also called parameter estimation or nonlinear regression) allows for the calculation of the saturation or mass distribution in addition to their average and total. On this dataset, the procedure was done in two steps. First, the permeability distribution was determined using the conservative tracer data; then, using the permeability distribution obtained in the first step, the saturation distribution was determined from the partitioning tracer data. The two steps involved similar objective functions:

$$\min f(k(x,y,z)) = \sum_{i=1}^{nwell} \sum_{j=1}^{ntime} \left(C_{i,j,\text{model}} - C_{i,j,\text{field}}\right)^2 \tag{G25}$$

$$\min f(sat(x,y,z)) = \sum_{i=1}^{nwell} \left(R_{i,j,\text{model}} - R_{i,j,\text{field}}\right)^2 \tag{G26}$$

where C is the tracer concentration and R is the retardation factor (Appendix G1).

Equation (G25) was used for determination of the permeability field. Although Equation (G25) could also be used for saturation determination, Equation (G26) has been proven to perform better in this type of problem (Harneshaug, 1996). This equation inherently involves integrating all the concentration data for one well in a single value called the residence time, avoiding the choice of the dispersivity coefficient to which the results are very sensitive. An accurate determination of the permeability field is an important factor in the success of the whole procedure as it regulates the tracer flow paths.

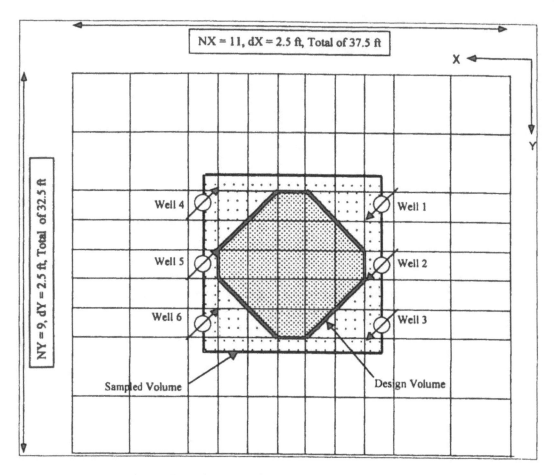

Figure G3.1 Map view of the grid used in the simulations.

The CONJUGATE code used in the permeability distribution calculation uses a conjugate gradient approach (Datta-Gupta and King, 1995). The second step of the procedure was executed with UTSTREAM which uses a Newton's method approach (Kurihara, 1995). Both simulators use the same forward model simulation. The grid used was similar to the one in the pretest simulation and has 9 layers of 11 × 9 blocks (Figures G3.1 and G3.2).

At the end of these simulations, each grid block will have a contaminant mass associated to it. By summing the mass contained in each grid block part of a given volume, the total mass within the volume can be calculated. The discrete nature of numerical modeling leads to an accurate approximation of the previously defined volumes.

Earlier numerical simulations showed that the tracer swept volume was smaller than the simulation volume (480 m³ compared to 1000 m³) but with similar proportions. The *sampled* volume is shown on Figures G3.1 and G3.2 and has a volume of 324 m³. The approximated *design* volume has a volume of 28.3 m³.

A common pitfall of inverse modeling is to mistake a local minimum for a global minimum. One of the ways to get around it is to choose a starting point as close as possible to the expected solution. The initial estimates used in the simulations were carefully constructed from the first moment analysis and a loose interpolation of the coring data. Because this inverse modeling method uses saturation and partition coefficients, it is at the present stage of development unable to distinguish DRO from TPH. The problem was solved in terms of TPH. However, because most of

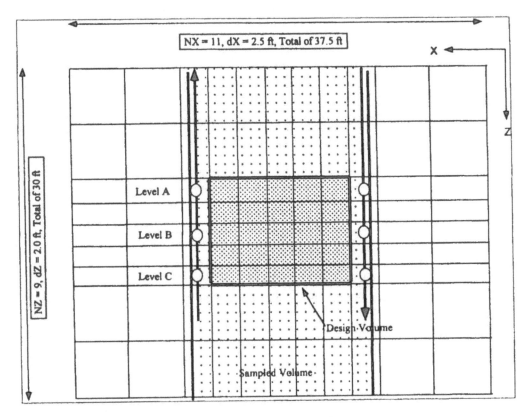

Figure G3.2 Cross-sectional view of the grid used in the simulations.

the DRO is extracted before a substantial part of heavier compounds is removed, most of the TPH removed was likely to be DRO (Hansen, 1997).

G3.2 Results and Discussion

Figures G4.39 and G4.40 show a comparison of the field and simulated data for each PITT. The simulation curves do not exactly fit all the data points since these data are noisy by nature. The more critical match (and the one upon which the fit was based) was the fit of the center of mass (rather than the highest value) of the actual and simulated response curves.

Another feature of importance was the presence of multiple peaks, which is indicative of the heterogeneity of the subsurface. If the permeability or TPH mass distribution model succeeded in capturing most of the double peaks, this strongly suggested that the model is accurate. Spurious double peaks also indicate that the model representation was not totally accurate.

The curves are plotted on a log scale to give details of the tail behavior. The tails also contain information on the heterogeneity of the subsurface; in particular, its slope is a function of the standard deviation of the distribution. A model slope steeper than the field data slope means that model permeability or TPH mass distribution failed to account for all the permeability variation in the field.

The maximum value of the simulated and field peaks is of lesser importance than factors such as dispersivity and porosity. After several trials, a value of 0.15-m (0.5-ft) has been retained for the dispersivity that controls how much spreading there is around a peak (tall and thin or flat and broader). The porosity field used in the inverse modeling originated partially from the interpretation

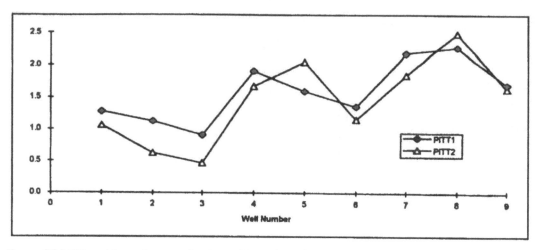

Figure G3.3 CH_4 residence time as a function of the well number.

of first moment analysis of the mixed tracer CH_2F_2 and partially from the core water-content analysis. From the dry weight analysis, the total porosity is about 0.32. However, the porosity open to gas flow is reduced by the presence of water which, as far as the gas is concerned, acts as part of the solid matrix. An average of 7% w/w gives a volumetric water content of about 0.12, leaving a volumetric gas content of 0.2, the value used in the simulations. It should be noted that the porosity is variable throughout the porous media as demonstrated by core data and other studies; however, as stated previously, this has only a minor impact on the results.

An overall comparison of the PITT1 and PITT2 (Figure G3.3) shows that the conservative-tracer residence times are very similar, suggesting that the heating process did not dramatically alter the permeability field.

Nevertheless, it should be noted that the conservative tracer data of a PITT truly yield the relative permeability rather than absolute permeability because of the presence of water. That is the reason why the permeability field is expected to change slightly when taken at two different times. This effect has been minimized for PITT2 because the soil had to be wetted to avoid tracer sorption. In addition, given the lack of pressure data for the tracer tests, the permeability field can only be determined to a multiplicative constant. Figures G4.41 to G4.44 show selected map views and cross-sections of the PITT1 and PITT2 permeability fields. The best data fit occurred when a horizontal lower permeability zone was inserted at the level of the design volume in the middle zone of the simulation grid. The same feature was also found by means of other techniques such as a higher blow count or a higher moisture content indicating a higher clay content zone. A geometric average for the PITT1 and PITT2 permeability fields of this middle zone was arbitrarily set to 5 D to fit earlier air pump tests. The upper and lower zones were defined as encompassing the volumes, respectively, from $Z = 1$ to $Z = 3$ and from $Z = 7$ to $Z = 9$. Other results and standard deviations (σ_{lnk}) are shown in Table G3.1.

The choice of the model permeability field clearly influences the mass distribution since the general shape of the conservative and partitioning tracer curves is the same. This means that, if the model permeability field is inaccurate, the inverse modeling procedure will try to account for it by adding to or removing mass from the "true" model mass distribution. So, although the fit may be correct, the contaminant mass distribution may be incorrect due to the inaccuracies in the perme-ability field.

It is believed that this work used a reasonably accurate model permeability field. The calculations of the TPH saturations were done mainly with C_9F_{18} [determined to be the most reliable tracer both in terms of residence time (>1 day) and retardation factor (>1.2)]. The same partition coefficients

Table G3.1 Summary of Permeability Field Results

		PITT1	PITT2
Upper zone	Geometric av. (D)	6.50	17.00
	σ_{lnk}	1.39	1.34
Middle zone	Geometric av. (D)	5.00	5.00
	σ_{lnk}	0.99	0.97
Lower zone	Geometric av. (D)	12.70	42.80
	σ_{lnk}	1.03	1.77
Total	Geometric av. (D)	6.90	10.30
	σ_{lnk}	1.25	1.50

Table G3.2 Summary of Mass Comprised in the Different Volumes (kg)

	Tracer Vol (kg)
PITT1	2470
PITT2	2180
Reduction	290

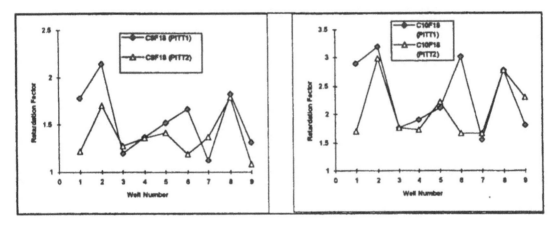

Figure G3.4 Comparison of retardation factors for the C_9 and C_{10} tracers.

defined earlier were used in this work. Figure G3.4 indicates that most of the TPH is still in the subsurface.

Most (five) wells show no decrease of the retardation factors, and some show a slight increase which is likely due to the slightly different swept volumes because of changes in the permeability field. The saturation calculations yield the following results in terms of mass (Table G3.2): A mass of 290 kg was removed from the tracer swept volume in close agreement with the mass determined from the first moment analysis. The mass distribution plots are shown in Figure G4.45.

The inverse modeling method has its own limitations. The local vs. global minimum problem is always present. It was tentatively overcome by a lot of manual adjustment and input at every stage of the process. However, the lack of uniqueness of the models is the biggest limitation to be overcome. One can find several solutions all as good as the other, with low objective function value and good model to field matches. Basically, one single value in a grid block cannot be trusted as such but only when averaged with the surroundings grid blocks. The design volume contained more than 100 blocks and was large enough to eliminate this problem.

G4 PITT FIGURES

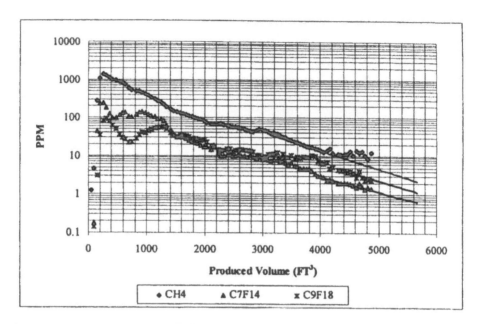

Figure G4.1 Tracer response curves for PITT1 well TT4A.

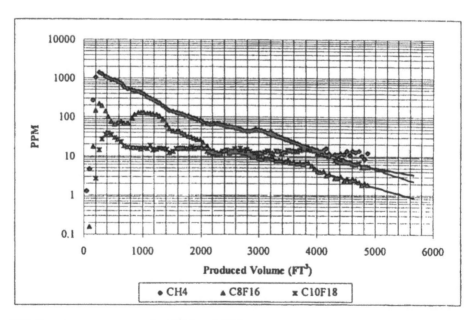

Figure G4.2 Tracer response curves for PITT1 well TT4A.

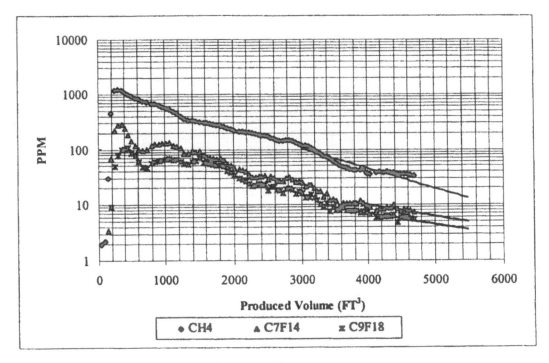

Figure G4.3 Tracer response curves for PITT1 well TT4B.

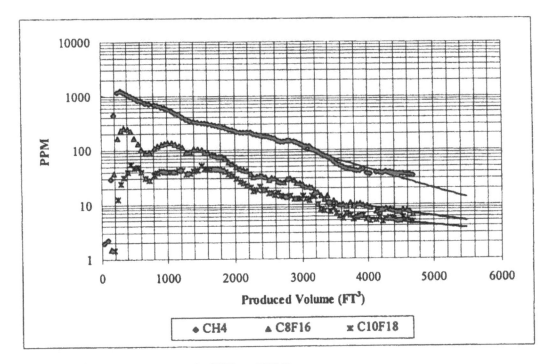

Figure G4.4 Tracer response curves for PITT1 well TT4B.

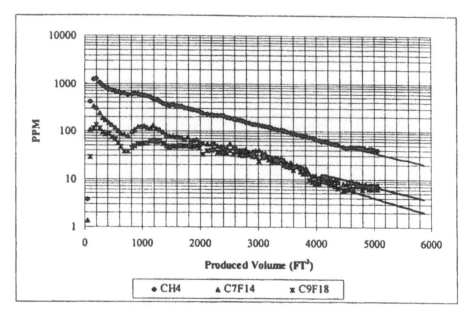

Figure G4.5 Tracer response curves for PITT1 well TT4C.

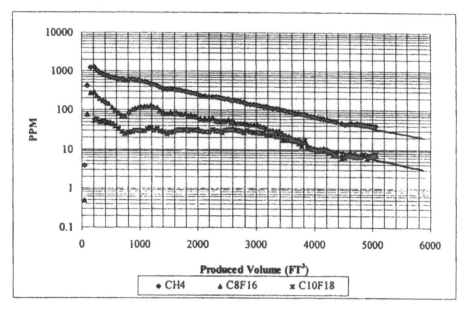

Figure G4.6 Tracer response curves for PITT1 well TT4C.

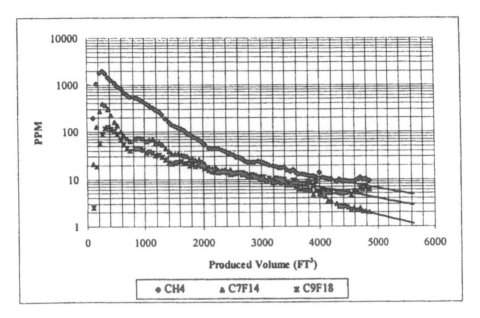

Figure G4.7 Tracer response curves for PITT1 well TT5A.

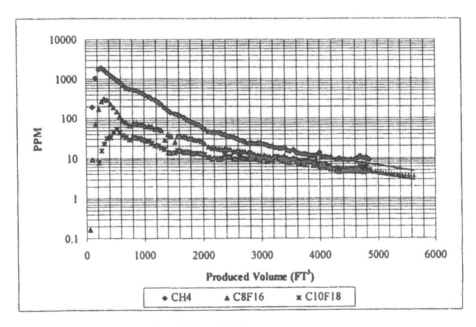

Figure G4.8 Tracer response curves for PITT1 well TT5A.

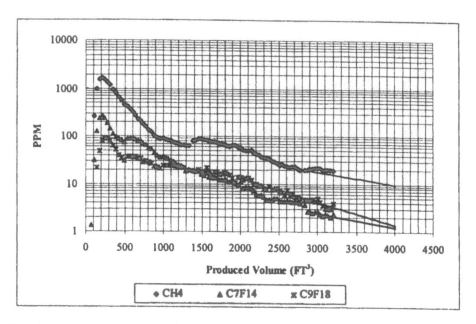

Figure G4.9 Tracer response curves for PITT1 well TT5B.

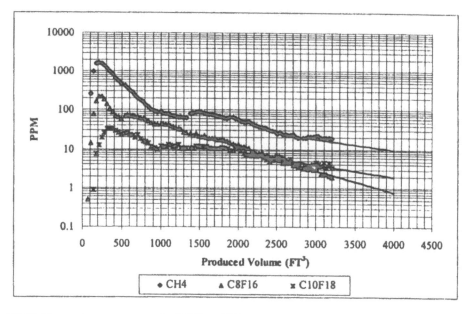

Figure G4.10 Tracer response curves for PITT1 well TT5B.

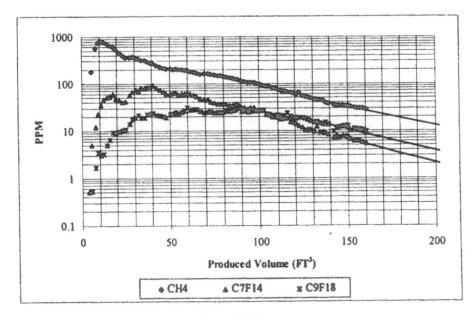

Figure G4.11 Tracer response curves for PITT1 well TT5C.

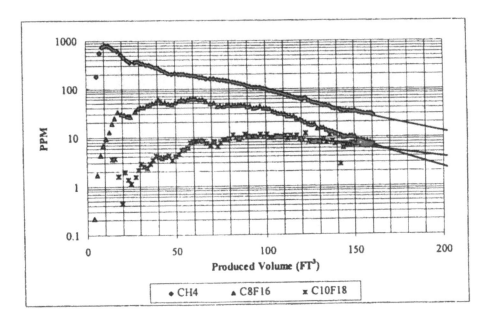

Figure G4.12 Tracer response curves for PITT1 well TT5C.

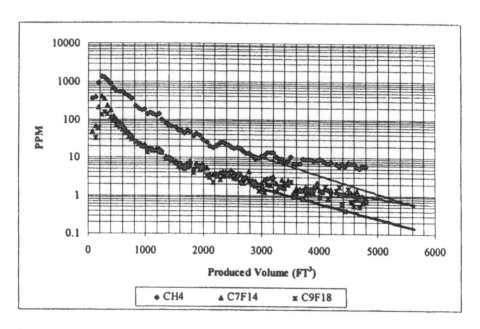

Figure G4.13 Tracer response curves for PITT1 well TT6A.

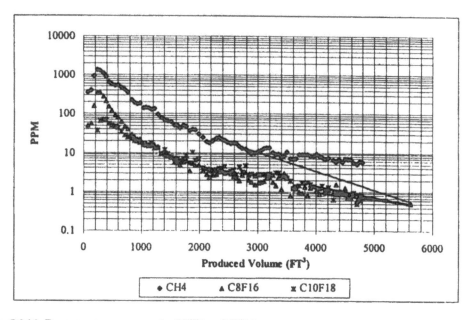

Figure G4.14 Tracer response curves for PITT1 well TT6A.

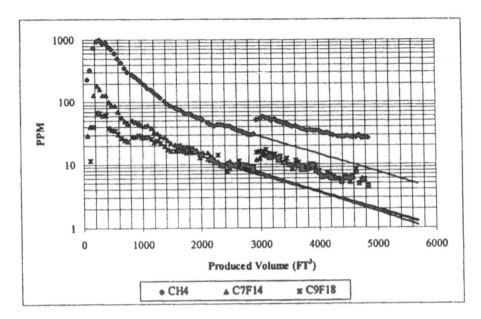

Figure G4.15 Tracer response curves for PITT1 well TT6B.

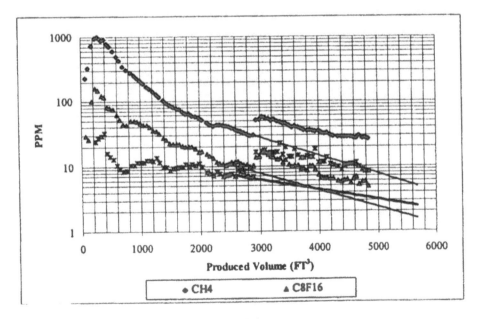

Figure G4.16 Tracer response curves for PITT1 well TT6B.

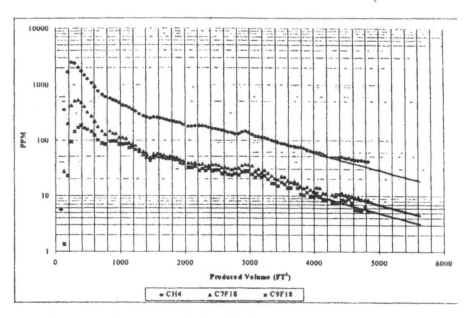

Figure G4.17 Tracer response curves for PITT1 well TT6C.

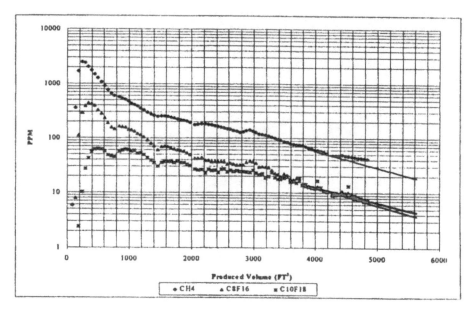

Figure G4.18 Tracer response curves for PITT1 well TT6C.

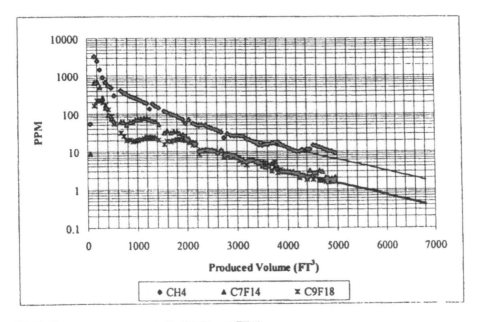

Figure G4.19 Tracer response curves for PITT2 well TT4A.

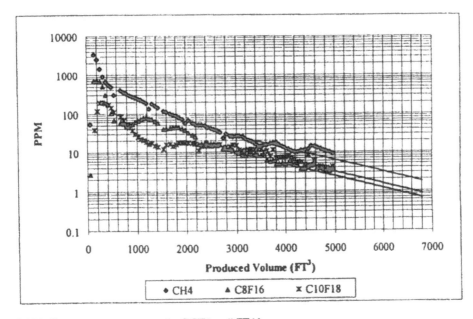

Figure G4.20 Tracer response curves for PITT2 well TT4A.

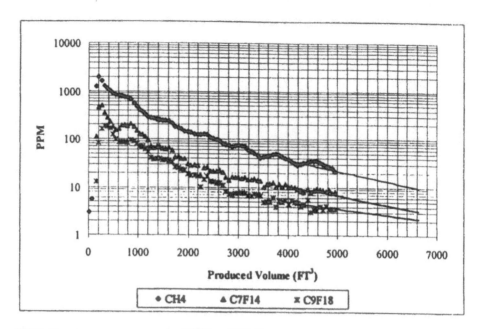

Figure G4.21 Tracer response curves for PITT2 well TT4B.

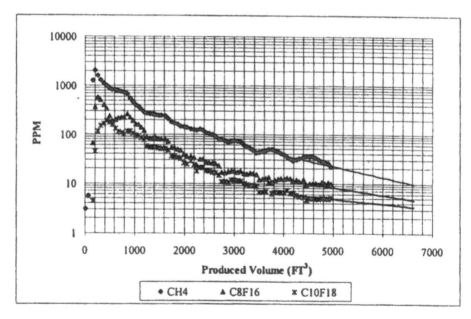

Figure G4.22 Tracer response curves for PITT2 well TT4B.

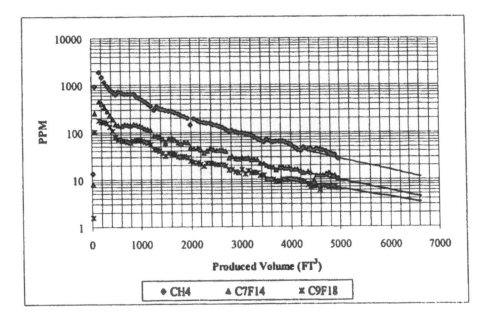

Figure G4.23 Tracer response curves for PITT2 well TT4C.

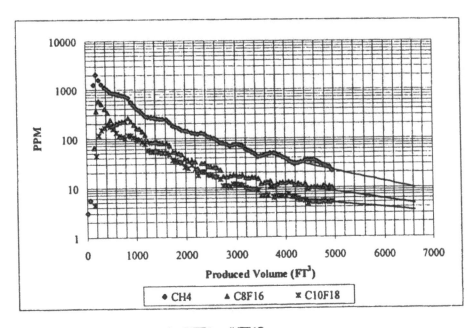

Figure G4.24 Tracer response curves for PITT2 well TT4C.

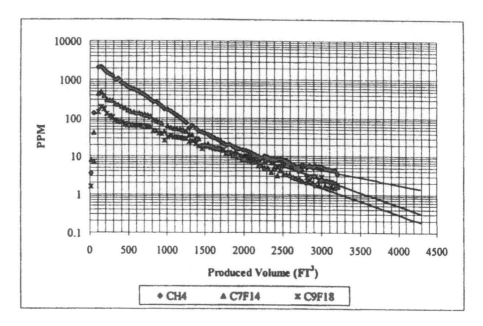

Figure G4.25 Tracer response curves for PITT2 well TT5A.

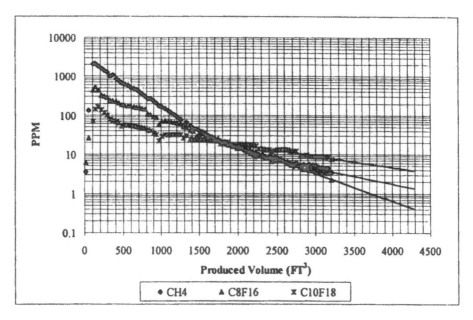

Figure G4.26 Tracer response curves for PITT2 well TT5A.

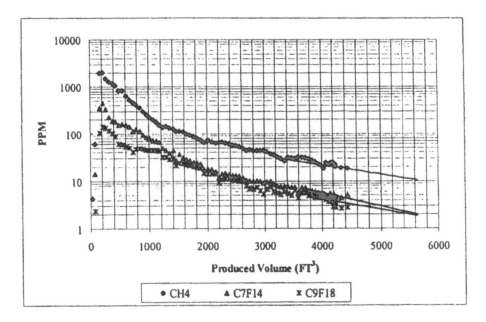

Figure G4.27 Tracer response curves for PITT2 well TT5B.

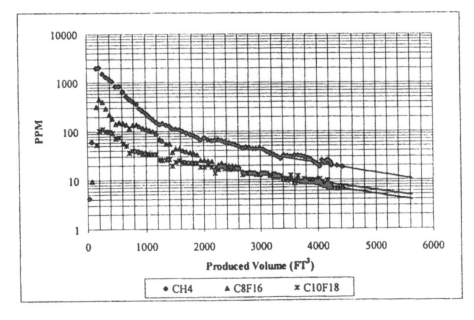

Figure G4.28 Tracer response curves for PITT2 well TT5B.

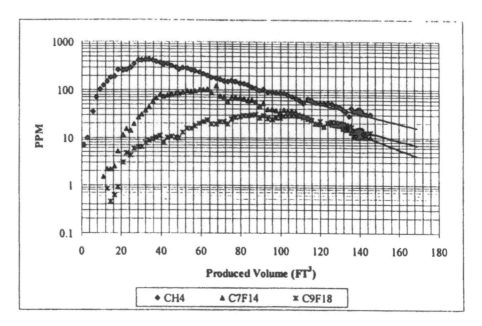

Figure G4.29 Tracer response curves for PITT2 well TT5C.

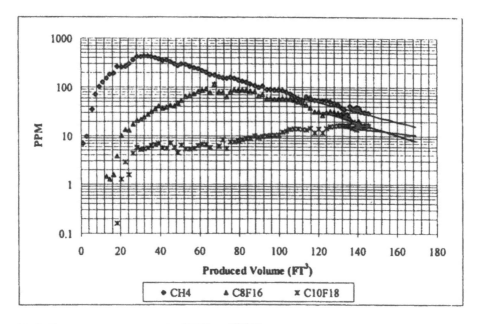

Figure G4.30 Tracer response curves for PITT2 well TT5C.

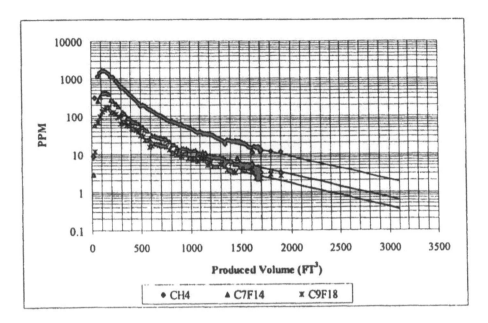

Figure G4.31 Tracer response curves for PITT2 well TT6A.

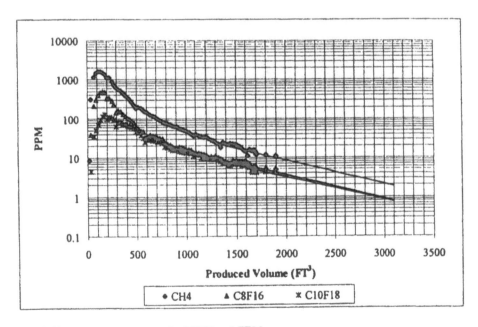

Figure G4.32 Tracer response curves for PITT2 well TT6A.

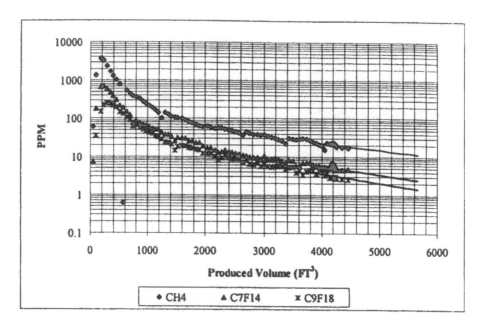

Figure G4.33 Tracer response curves for PITT2 well TT6B.

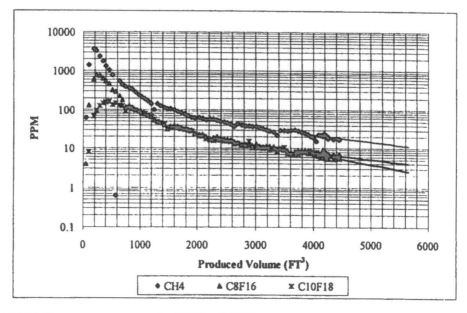

Figure G4.34 Tracer response curves for PITT2 well TT6B.

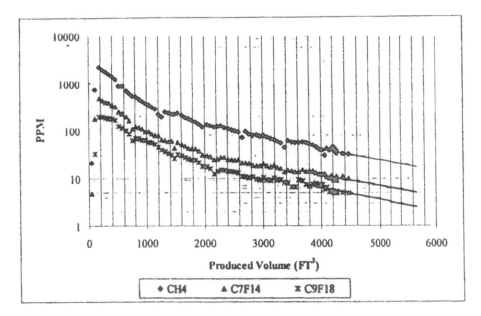

Figure G4.35 Tracer response curves for PITT2 well TT6C.

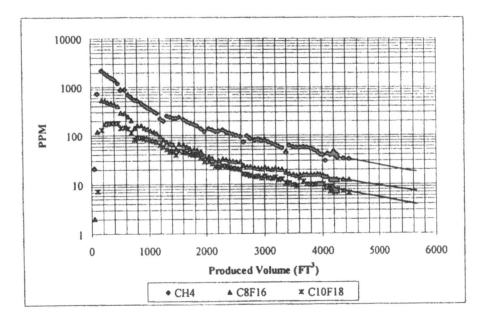

Figure G4.36 Tracer response curves for PITT2 well TT6C.

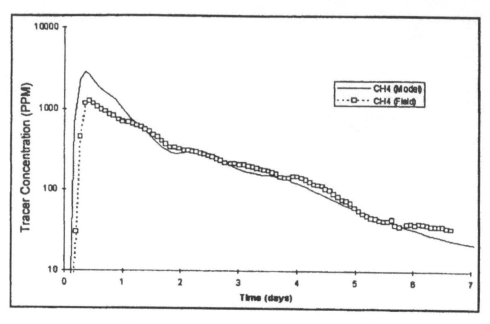

Figure G4.37 Comparison of field data and model-predicted conservative tracer concentrations for extraction well TT4B (PITT1).

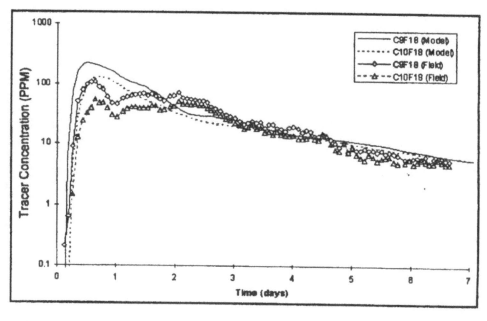

Figure G4.38 Comparison of field data and model-predicted partitioning tracer concentrations for extraction well TT4B (PITT1).

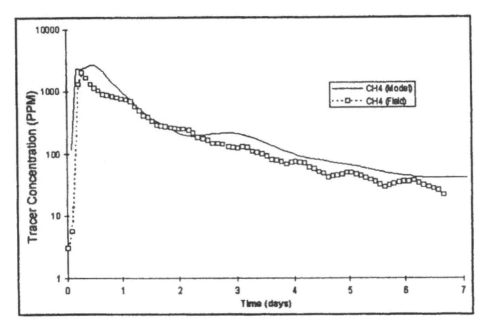

Figure G4.39 Comparison of field data and model-predicted partitioning tracer concentrations for extraction well TT4A (PITT2).

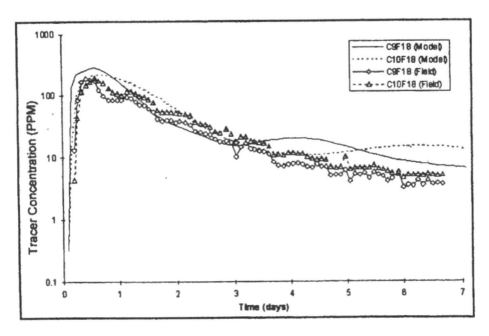

Figure G4.40 Comparison of field data and model-predicted conservative tracer concentrations for extraction well TT4B (PITT2).

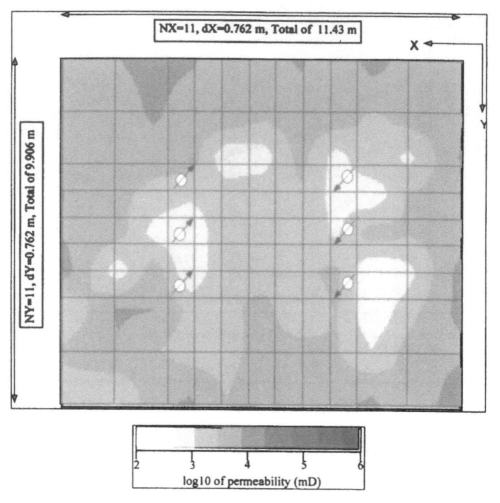

Figure G4.41 Map view of the pre-test permeability field at a 5 m depth.

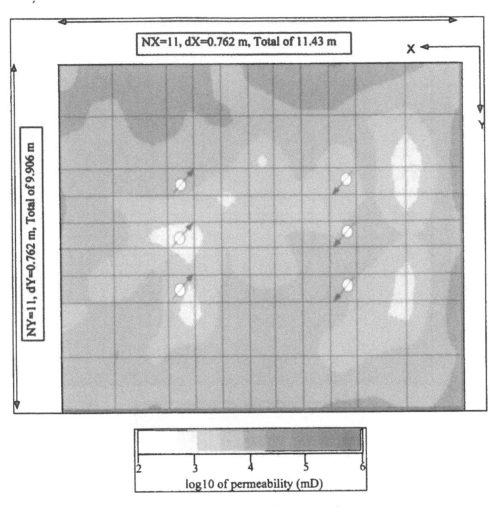

Figure G4.42 Map view of the post-test permeability field at a 5 m depth.

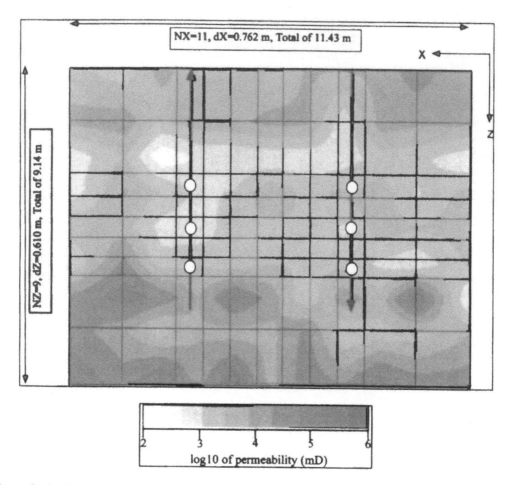

Figure G4.43 Cross-sectional view of the pretest permeability field at the TT2-TT5 level.

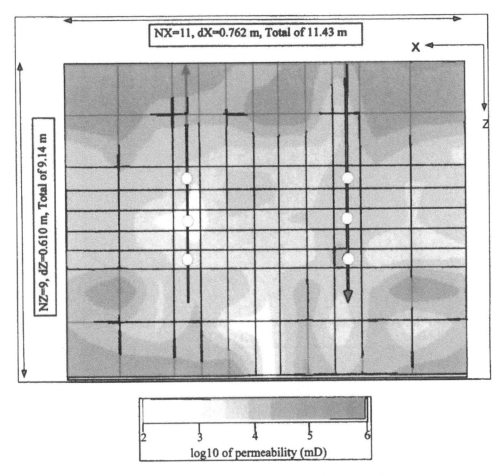

Figure G4.44 Cross-sectional view of the post-test permeability field at the TT2-TT5 level.

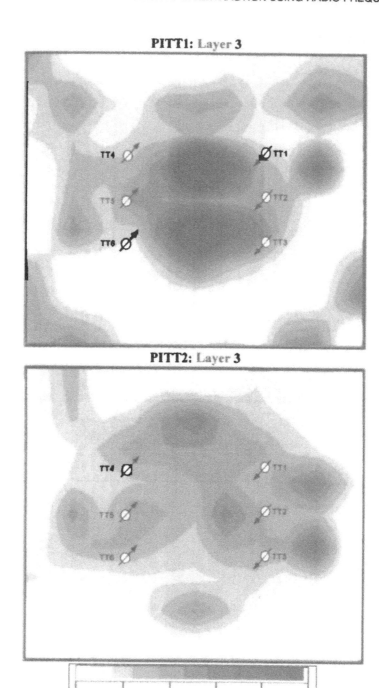

Figure G4.45 Map view of the saturation field at the third layer.

Table G5.1 Complete Results for PITT1 Using Extrapolated Response Curves

Well	Tracer	Residence Time (ft³)	Average Flow (ft³/h)	Mass Recovered (g)	Rf	Swept Volume (ft³)	Kw (LG/LL)	Sw	Kn (LG/LL)	Vn (ft³)	Sn	mg/kg (approx.)
TT4A	CH_4	916	30	32.2	1.00				0			
	CF_2H_2	1,176	30	25.4	1.28		1.7	0.140	0		0.018	
	C_7F_{14}	1,037	30	6.4	1.13	530		0.140	8.83	6.68	0.013	2,185
	C_8F_{16}	1,188	30	6.3	1.30	523		0.140	15.56	8.39	0.016	2,781
	C_9F_{18}	1,635	30	3.5	1.78	508		0.140	24.10	13.77	0.027	4,688
	$C_{10}F_{18}$	2,661	30	2.5	2.90	508		0.140	70.90	11.41	0.022	3,892
	average					513				11.19	**0.020**	**3,387**
TT4B	CH_4	1,368	30	44.6	1.00				0			
	CF_2H_2	1,520	30	36.0	1.11		1.7	0.060	0		0.013	
	C_7F_{14}	1,592	30	9.3	1.16	1,067		0.060	8.83	18.25	0.017	3,011
	C_8F_{16}	1,671	30	9.4	1.22	1,074		0.060	15.56	14.19	0.013	2,328
	C_9F_{18}	1,876	30	5.3	1.37	1,028		0.060	24.1	14.64	0.014	2,511
	$C_{10}F_{18}$	2,602	30	3.9	1.90	1,043		0.060	70.9	12.32	0.012	2,082
	average					1,053				14.85	**0.014**	**2,483**
TT4C	CH_4	1,562	30	49.1	1.00				0			
	CF_2H_2	1,866	30	41.3	1.19		1.7	0.102	0		0.004	
	C_7F_{14}	1,562	30	10.5	1.00	1,409		0.102	8.83	0.02	0.000	2
	C_8F_{16}	1,563	30	10.5	1.00	1,409		0.102	15.56	0.04	0.000	5
	C_9F_{18}	1,750	30	6.0	1.12	1,376		0.102	24.1	6.14	0.004	780
	$C_{10}F_{18}$	2,431	30	4.4	1.56	1,414		0.102	70.9	9.88	0.007	1,222
	average					1,414				9.88	0.003	1,001

Table G5.1 Complete Results for PITT1 Using Extrapolated Response Curves (continued)

Well	Tracer	Residence Time (ft³)	Average Flow (ft³/h)	Mass Recovered (g)	Rf	Swept Volume (ft³)	K_W (LG/LL)	S_W	K_N (LG/LL)	V_N (ft³)	S_N	mg/kg (approx.)
TT5A	CH_4	808	30	36.9	1.00				0			
	CF_2H_2	1,294	30	28.4	1.60		1.7	0.255	0		0.025	
	C_7F_{14}	1,013	30	6.9	1.25	591		0.255	8.83	12.30	0.021	3,526
	C_8F_{16}	1,327	30	6.7	1.64	581		0.255	15.56	17.18	0.030	5,011
	C_9F_{18}	1,732	30	4.0	2.14	592		0.255	24.1	19.99	0.034	5,710
	$C_{10}F_{18}$	2,584	30	2.6	3.20	530		0.255	70.9	11.89	0.022	3,796
	average					573				15.34	0.027	4,508
TT5B	CH_4	766	20	22.1	1.00				0			
	CF_2H_2	893	20	16.3	1.17		1.7	0.088	0		0.009	
	C_7F_{14}	787	20	3.9	1.03	382		0.088	8.83	1.09	0.003	499
	C_8F_{16}	837	20	3.7	1.09	362		0.088	15.56	1.95	0.005	945
	C_9F_{18}	1,165	20	2.2	1.52	376		0.088	24.1	7.26	0.019	3,380
	$C_{10}F_{18}$	1,620	20	1.3	2.11	311		0.088	70.9	4.39	0.014	2,474
	average					343				5.82	0.010	2,927
TT5C	CH_4	54	1	1.0	1.00				0			
	CF_2H_2	73	1	0.8	1.34		1.7	0.163	0		0.021	
	C_7F_{14}	65	1	0.2	1.21	29		0.163	8.83	0.56	0.019	3,297
	C_8F_{16}	78	1	0.2	1.44	28		0.163	15.56	0.63	0.023	3,963
	C_9F_{18}	99	1	0.1	1.83	29		0.163	24.1	0.81	0.028	4,822
	$C_{10}F_{18}$	151	1	0.1	2.79	20		0.163	70.9	0.40	0.021	3,550
	average					26				0.60	0.023	3,908

Well	Tracer	Residence Time (ft³)	Average Flow (ft³/h)	Mass Recovered (g)	Rf	Swept Volume (ft³)	K_w (L$_G$/L$_L$)	S_w	K_N (L$_G$/L$_L$)	V_N (ft³)	S_N	mg/kg (approx.)
TT6A	CH_4	654	30	21.5	1.00				0			
	CF_2H_2	859	30	17.3	1.31		1.7	0.155	0		0.005	
	C_7F_{14}	673	30	4.3	1.03	255		0.155	8.83	0.73	0.003	495
	C_8F_{16}	654	30	4.2	1.00	253		0.155	15.56	0.01	0.000	10
	C_9F_{18}	782	30	2.5	1.20	251		0.155	24.1	1.71	0.007	1,181
	$C_{10}F_{18}$	1,159	30	1.7	1.77	245		0.155	70.9	2.24	0.009	1,578
	average					248				1.97	0.005	1,379
TT6B	CH_4	975	30	22.7	1.00				0			
	CF_2H_2	1,360	30	18.2	1.39		1.7	0.183	0		0.027	
	C_7F_{14}	1,153	30	4.2	1.18	393		0.183	8.83	6.52	0.017	2,848
	C_8F_{16}	1,310	30	4.1	1.34	384		0.183	15.56	6.78	0.018	3,034
	C_9F_{18}	1,614	30	2.5	1.66	396		0.183	24.1	8.58	0.022	3,717
	$C_{10}F_{18}$	2,941	30	1.6	3.02	370		0.183	70.9	8.35	0.023	3,879
	average					386				7.56	0.022	3,370
TT6C	CH_4	1,228	30	55.1	1.00				0			
	CF_2H_2	1,611	30	45.9	1.31		1.7	0.154	0		0.005	
	C_7F_{14}	1,272	30	12.3	1.04	1,379		0.154	8.83	4.71	0.003	592
	C_8F_{16}	1,307	30	12.3	1.06	1,376		0.154	15.56	4.76	0.003	600
	C_9F_{18}	1,603	30	7.0	1.31	1,351		0.154	24.1	14.28	0.011	1,829
	$C_{10}F_{18}$	2,228	30	4.8	1.81	1,285		0.154	70.9	12.34	0.010	1,662
	average					1,318				13.31	0.006	1,746
					SPV Sum	5,875				V_N Sum 80.44	0.014 Ave. S_N	

Table G5.2 Complete Results for PITT2 Using Extrapolated Response Curves

Well	Tracer	Residence Time (ft³)	Average Flow (ft³/h)	Mass Recovered (g)	Rf	Swept Volume (ft³)	K_W (Lc/LL)	S_W	K_N (Lc/LL)	V_N (ft³)	S_N	mg/kg (approx.)
TT4A	CH_4	787	31	33.4	1.00				0			
	CF_2H_2	943	31	17.4	1.20		1.7	0.104	0		0.011	
	C_7F_{14}	787	31	8.3	1.00	429		0.104	7.43	0.03	0.000	13
	C_8F_{16}	827	31	10.3	1.05	449		0.104	12.70	1.62	0.004	632
	C_9F_{18}	959	31	4.4	1.22	451		0.104	19.35	4.52	0.010	1,751
	$C_{10}F_{18}$	1,390	31	4.6	1.77	434		0.104	55.3	5.33	0.012	2,145
	average					445				3.82	0.008	1,509
TT4B	CH_4	1,203	30	45.3	1.00				0			
	CF_2H_2	1,388	30	27.4	1.15		1.7	0.082	0		0.013	
	C_7F_{14}	1,314	30	11.4	1.09	923		0.082	7.43	10.35	0.011	1,969
	C_8F_{16}	1,446	30	13.6	1.30	933		0.082	12.70	13.40	0.014	2,521
	C_9F_{18}	1,635	30	5.7	1.36	898		0.082	19.35	15.02	0.017	2,933
	$C_{10}F_{18}$	2,089	30	7.2	1.74	1,046		0.082	55.3	12.62	0.012	2,117
	average					~50				12.85	0.014	2,385
TT4C	CH_4	1,330	30	47.9	1.00				0			
	CF_2H_2	1,724	30	28.5	1.30		1.7	0.148	0		0.004	
	C_7F_{14}	1,482	30	12.4	1.11	1,201		0.148	7.43	15.49	0.013	2,234
	C_8F_{16}	1,618	30	14.7	1.22	1,196		0.148	12.70	17.08	0.014	2,474
	C_9F_{18}	1,817	30	6.4	1.37	1,202		0.148	19.35	19.03	0.016	2,740
	$C_{10}F_{18}$	2,223	30	7.9	1.67	1,378		0.148	55.3	14.10	0.100	1,773
	average					1,244				16.43	0.012	2,305

Well	Tracer	Residence Time (ft^3)	Average Flow (ft^3/h)	Mass Recovered (g)	Rf	Swept Volume (ft^3)	K_W (L_G/L_L)	S_W	K_N (L_G/L_L)	V_N (ft^3)	S_N	mg/kg (approx.)
TT5A	CH_4	455	30	27.6	1.00				0			
	CF_2H_2	795	30	15.9	1.75		1.7	0.297	0		0.025	
	C_7F_{14}	562	30	7.1	1.23	288		0.297	7.43	6.19	0.021	3,606
	C_8F_{16}	647	30	8.3	1.42	284		0.297	12.70	6.38	0.023	3,780
	C_9F_{18}	775	30	3.4	1.70	273		0.297	19.35	6.72	0.025	4,134
	$C_{10}F_{18}$	1,360	30	3.9	2.99	286		0.297	55.3	6.97	0.024	4,092
	average					**283**				**6.56**	**0.024**	**3,903**
TT5B	CH_4	983	20	35.2	1.00				0			
	CF_2H_2	1,171	20	19.7	1.19		1.7	0.099	0		0.020	
	C_7F_{14}	993	20	7.9	1.01	791		0.099	7.43	0.98	0.001	216
	C_8F_{16}	1,260	20	9.4	1.28	807		0.099	12.70	15.77	0.020	3,411
	C_9F_{18}	1,394	20	3.9	1.42	777		0.099	19.35	14.81	0.019	3,330
	$C_{10}F_{18}$	2,191	20	4.5	2.23	834		0.099	55.3	16.34	0.020	3,421
	average					**802**				**15.64**	**0.016**	**3,387**
TT5C	CH_4	60	1	0.8	1.00				0			
	CF_2H_2	95	1	0.4	1.59		1.7	0.252	0		0.025	
	C_7F_{14}	75	1	0.2	1.25	33		0.252	7.43	0.80	0.024	4,121
	C_8F_{16}	88	1	0.2	1.47	29		0.252	12.70	0.79	0.027	4,545
	C_9F_{18}	107	1	0.1	1.79	29		0.252	19.35	0.84	0.029	4,941
	$C_{10}F_{18}$	166	1	0.1	2.78	25		0.252	55.3	0.59	0.023	3,955
	average					**29**				**0.75**	**0.026**	**4,391**

Table G5.2 Complete Results for PITT2 Using Extrapolated Response Curves (continued)

Well	Tracer	Residence Time (ft³)	Average Flow (ft³/h)	Mass Recovered (g)	Rf	Swept Volume (ft³)	K_W (L_G/L_L)	S_W	K_N (L_G/L_L)	V_N (ft²)	S_N	mg/kg (approx.)
TT6A	CH_4	342	30	15.4	1.00				0			
	CF_2H_2	469	30	8.6	1.37		1.7	0.177	0		0.010	
	C_7F_{14}	367	30	3.8	1.07	97		0.177	7.43	0.78	0.008	1,386
	C_8F_{16}	391	30	4.4	1.14	96		0.177	12.70	0.87	0.009	1,563
	C_9F_{18}	437	30	1.8	1.28	90		0.177	19.35	1.04	0.012	1,994
	$C_{10}F_{18}$	605	30	1.9	1.77	87		0.177	55.3	0.98	0.011	1,941
	average					92				0.92	0.010	1,721
TT6B	CH_4	830	30	44.3	1.00				0			
	CF_2H_2	986	30	26.9	1.19		1.7	0.098	0		0.010	
	C_7F_{14}	827	30	10.9	1.00	612		0.098	7.43	-0.30	0.000	-86
	C_8F_{16}	963	30	13.6	1.16	651		0.098	12.70	7.29	0.011	1,961
	C_9F_{18}	986	30	5.5	1.19	604		0.098	19.35	5.22	0.009	1,514
	$C_{10}F_{18}$	1,383	30	6.1	1.67	628		0.098	55.3	6.75	0.011	1,879
	average					624				6.42	0.008	1,785
TT6C	CH_4	1,177	30	46.9	1.00				0			
	CF_2H_2	1,465	30	28.5	1.25		1.7	0.125	0		0.005	
	C_7F_{14}	1,241	30	11.6	1.05	963		0.125	7.43	6.15	0.006	1,113
	C_8F_{16}	1,369	30	14.2	1.16	991		0.125	12.70	11.03	0.011	1,937
	C_9F_{18}	1,283	30	5.9	1.09	942		0.125	19.35	3.81	0.004	706
	$C_{10}F_{18}$	1,525	30	7.1	1.30	1,059		0.125	55.3	4.93	0.005	811
	average					989				6.48	0.006	1,142
						SPV Sum 5,458				V_N Sum 69.87	0.014	Ave. S_N

Appendix H

Engineering Drawings of Hypothetical Designs

Appendix H Listing

Drawing S1: Site 1 Layout, RF Heating Soil Vapor Extraction
Drawing S2: Site 2 Layout, RF Heating Soil Vapor Extraction
Drawing P1: Process Flow Diagram, Vapor Extraction System with RF Heating
Drawing P2: Piping and Instrumentation Diagram, Soil Vapor Extraction System with RF Heating

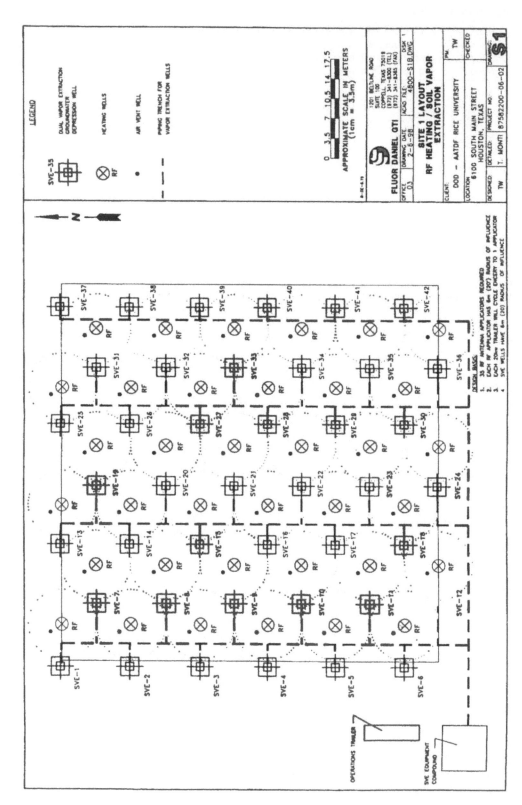

Drawing S1: Site 1 layout, RF heating soil vapor extraction.

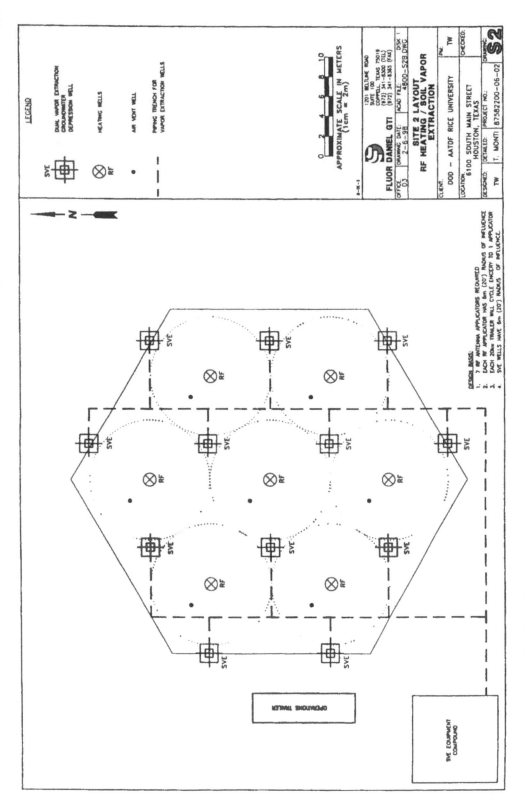

Drawing S2: Site 2 layout RF heating soil extraction.

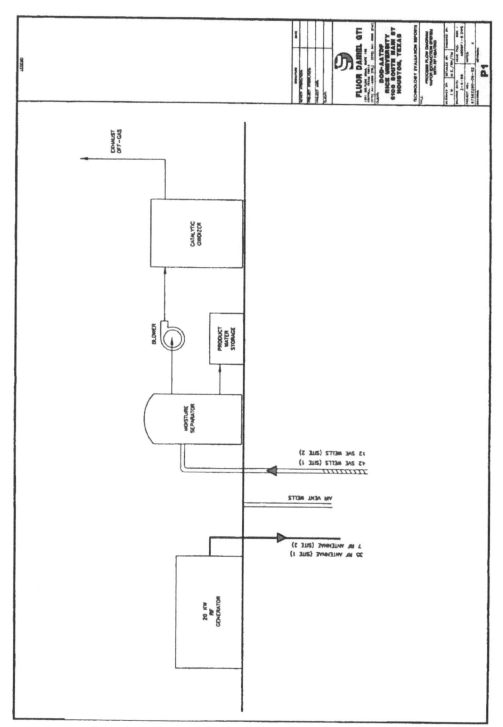

Drawing P1: Process flow diagram, vapor extraction system with RF heating.

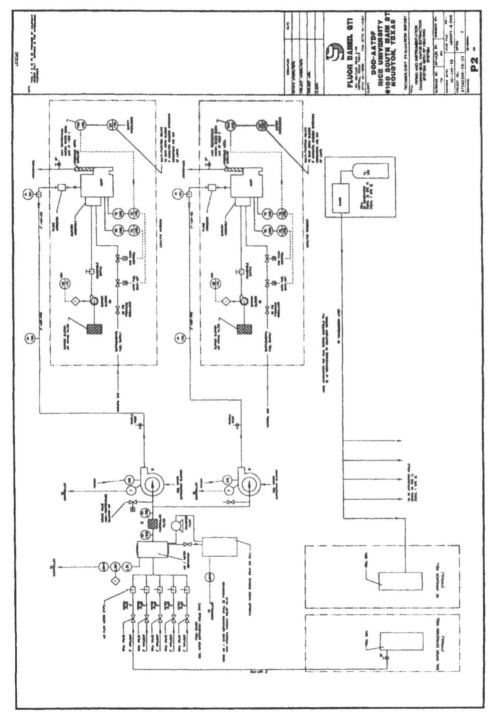

Drawing P2: Piping and instrumentation diagram, soil vapor extraction system with RF heating.

Appendix I

Equipment Catalog Cut-Sheet References

Appendix I Listing

Table I1 Equipment Manufacturers and Specifications

Item	Specifications/Information	Manufacturer
RF generator	KAI Mobile 20-kW Trailer (see layout) Control and diagnostic computer RF antenna and transmission lines Fiber optic thermometer	KAI Technologies, Inc. 170 West Road, Suite 4 Portsmouth, NH 03801 Phone: 603-431-2266 Fax: 603-431-4920 Mr. Raymond Kasevich Mr. Michael C. Marley

SVE blowers
(with control
equipment)
	ROOTS (Dresser)
	ROOTS Universal RAI
	  Positive-displacement blower
	  (Model URAI 56)
	  14.9-kW (20-hp) TEFC motor
	  12,750 slpm (450 scfm)
	  47.4 kPa (14 in. Hg) vacuum
		Diversified Remediation Controls, Inc.
		Charlotte NC Branch
		Charlotte, NC
		Phone: 704-843-4664
		Fax: 704-843-4661

Catalytic oxidation
unit
	Baker gas-fired oxidation system
	Case 1: Model 750 scfm (21,240 slpm)
	  ROOTS Blower Model URAI 76
	  14,500 (500 scfm)
	  47.4 kPa (14 in. Hg) vac
	  Maximum fuel supplement:
	  1,875,000 Btu/h
	  Moisture knock-out tank
	Case 2: Model 500 scfm (14,200 slpm)
	  ROOTS Blower Model URAI 68
	  14,500 (500 scfm)
	  47.4 kPa (14 in. Hg) vac
	  Maximum fuel supplement:
	  1,250,000 Btu/h
	  Moisture knock-out tank
		Baker Furnace, Inc.
		1015 East Discovery Lane
		Anaheim, CA 92801
		Phone: 714-491-9293
		Fax: 714-491-8221

Water transfer
pump
	Myers Centri-Thrift Pump
	  Model 100MM
	  0.74 kW (1 hp)
	  ~190 Lpm (~50 gpm)
	  Discharge pressure gauge
	  ISTA Model 1720 Water Meter
		F.E. Myers, A Pentair Co.
		1101 Myers Parkway
		Ashland, OH 44805
		Phone: 419-289-1144
		Fax: 419-289-6658

Air-water separator	Main condensate tank
	  Auto-relief valve
	  Auto-dilution valve
	  Ambient air filter
	  Vacuum gauge
	  High level switch
	  Solenoid valve
	Secondary collection tank
	  High level switch
	  Low level switch
	  Vacuum gauge
	Water transfer pump	

Transformers	12,470 – 480/277 V, 3-phase, 4-wire
	  Main RF unit transformer
	  Case 1: 1000 KVA
	  Case 2: 300 KVA
	  Weatherproofed enclosure
	  Meter-to-measure power consumption
	480 – 240/120 V, single phase, 3-wire
	  Service transformer for trailer unit
	  25 KVA
	  NEMA 3R enclosure	

Figure I1   KAI trailer-mounted, 10-kW generator unit.

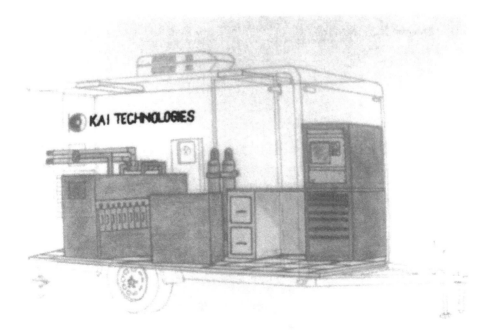

Figure I2   Basic layout of 10-kW RF generator unit

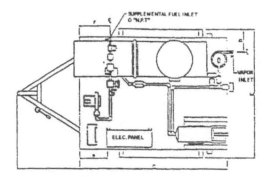

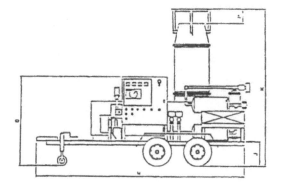

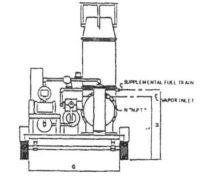

Figure I3   Thermal catalytic oxidizer.

Appendix J

Mass & Heat Balance Calculations for the Three Hypothetical Design Cases

Appendix J Listing and Description of Tables

Item	Title	Description
Calculations	Example calculations for Table J2, Case 1	Example calculations Case 1 in Table J2
Calculations	Example calculations for Table J3, Case 1	Example calculations for Case 1 in Table J3
Table J1	RF-SVE hypothetical design case studies and site conditions	1. Description of three hypothetical cases for RF-SVE remediation at two sites 2. Conceptual area and volume of each site 3. General site conditions
Table J2	RF-SVE design basis for hypothetical cases 1, 2, and 3	For each case: 1. Operating assumptions for RF-SVE system 2. Operating assumptions for SVOC contaminant removal
Table J3	Heat balance for hypothetical cases 1, 2, and 3	For each case: 1. Heat balance input data 2. Heat requirements for wet and dry soil 3. Heat requirements for water vapor 4. Heat for SVE air 5. Heat loss to the surroundings 6. Total heat requirements

EXAMPLE CALCULATIONS FOR TABLE J2, CASE 1

Case 1 — SVE System

Air extraction wells = 42
Maximum dry-airflow per extraction well = 750 slpm
AIR_{total} = maximum volumetric dry airflow
AIR_{total} = (42 wells)(750 slpm per well)(60 min/h)(24 h/day)/(1,000 L/m³) = 45,360 m³/day
AIR_{mass} = maximum mass airflow = (45,360 m³/day)(1.204 kg/m³) = 54,610 kg/day

Case 1 — RF-SVE System

Data from Table J1:
  Soil density = 1,788 kg/m³
  Initial soil moisture = 10.0% dwb
  Final soil moisture = 1.0% dwb
Data from Table J2:
  Total RF antennae (RFA) = 36
  kW per RF antenna (kW/RFA) = 20.0
  Applicator efficiency (RF energy to heat) = 92.5%
  Efficiency of AC power to RF energy = 72.0%

1. Heat wet soil to 100°C
  Total heat content and mass of soil moisture from Table J3, heat balances:
  $H_{wet soil}$, total heat content of wet soil at 100°C = 5,215 million kJ = 1,448,594 kWh
  Volume of soil to be treated from Table J1 = 24,843 m³
  Mass of soil (dry basis) = (24,843 m³)(1,778 kg/m³) = 44,171,547 kg
  Total mass of moisture in soil = (44,171,547 kg)(10.0/100%) = 4,417,155 kg
  Total mass of soil moisture to be removed = 4,417,155 kg − (44,171,547 kg)(1.0%/100%)
  Total mass of soil moisture to be removed = 3,975,439 kg

  Operating up to 100°C:
  Air extraction wells = 42
  Density of air at 20°C = 1.204 kg/m³
  Set airflow per well = 300 slpm
  Set humidity of air at 100°C = 0.2000 kg water/kg dry air

Change the value (**128 days**) until agreement with "days" value in 4.
T_{total}, Total time to treat site at 100°C = **128 days (iterative value with "days" in 4).**

AIR_{mass} = (300 Lpm)(60 min/h)(24 h/day)(1.204 kg/m³)(42 wells)/(1,000 L/m³) = 21,845 kg/day
Mass of dry air to remove water = (21,845 kg/day)(128 days) = *2,796,036 kg*
Mass of soil moisture removed at 100°C = 2,796,036 kg)(0.2000 kg/kg) = 559,207 kg water
Check: mass of dry air = (559,207 kg)/(0.2000 kg/kg) = *2,796,036 kg*

2. Remove water from wet soil
Operating at 100°C:
Airflow per well = 750 slpm
Air extraction wells = 42
Density of air at 20°C = 1.204 kg/m³

Change the value (**173 days**) until agreement with "days" value in 4.
T_{total}, total time to treat site at 100°C = **173 days (iterative value with "days" in 4).**

Total mass of soil moisture to be removed = 3,975,439 kg
Mass of soil moisture to be removed at 100°C = (3,975,439 − 559,207 kg) = 3,416,232 kg
AIR_{mass} = (750 Lpm)(60 min/h)(24 h/day)(1.204 kg/m³)(42 wells)/(1,000 L/m³) = 54,613 kg/day
Mass of dry air to remove water = (173 days)/(54,613 kg/day) = *9,447,542 kg*
Adjust humidity until the *check* value for mass of dry air equals calculated value
Humidity of air at 100°C = *0.3616* kg water/kg dry air
Check: mass of dry air = (3,416,232 kg)/(0.3616 kg/kg) = *9,447,545 kg*

3. Heat dry soil from 100–131.5°C
Operating from 100–131.5°C:
Airflow per well = 600 L/min
Air extraction wells = 42
Density of air at 20°C = 1.204 kg/m³

Change the value (**35 days**) until agreement with "days" value in 4.
T_{total}, total time to treat site at 131.5°C = **35 days (iterative value with "days" in 4).**

AIR_{mass} = (600 Lpm)(60 min/h)(24 h/day)(1.204 kg/m³)(42 wells)/(1,000 L/m³) = 43,688 kg/day
Mass of soil moisture to be removed = 3,975,439 kg
Mass of soil moisture removed to 131.5°C = (3,975,439 − 559,207 − 3,416,232 kg) = 0 kg
Mass of dry air to remove water = (35 day)/(43,688 kg/day) = 1,529,082 kg dry air
Humidity of air from 100–131.5°C = *0.0000* kg water/kg dry air

4. Heat wet soil, heat dry air, and remove water-SVOC
Operating up to 100°C
Total heat content from Table J3, heat balances
H_{total}, total heat requirement from RF system = 7,359 million kJ = 2,044,230 kWh
T_{total}, total time to treat site to 100°C
T_{total} = (2,044,230 kWh)/[(36 RFA)(20 kW/RFA)(92.5%/100%)]/(24 h/day)
T_{total} = **128 days (match # days in 1.)**

Operating at 100°C
Total heat content from Table J3, heat balances
H_{total}, total heat requirement from RF system = 9,953 million kJ = 2,764,612 kWh
T_{total}, total time to treat site at 100°C
T_{total} = (2,764,612 kWh)/[(36 RFA)(20 kW/RFA)(92.5/100)]/(24)
T_{total} = **173 days (match # days in 2).**

Heat dry soil from 100°C to 131.5°C
Total heat content from Table J3, heat balances
$H_{dry\ soil}$, total heat content of dry soil at 131.5°C = 1,433,502 kWh
$H_{dry\ soil}$, total heat content of dry soil at 100°C = 1,009,855 kWh
$H_{dry\ soil}$, heat dry soil from 100–131.5°C = (1,433,502 − 1,009,855 kWh)
$H_{dry\ soil}$ = 423,647 kWh
H_{total}, total heat requirement from RF system = 2,011 million kJ = 558,498 kWh
$T_{dry\ soil}$, time to heat dry soil from 100–131.5°C
$T_{dry\ soil}$ = (558,498 kWh)/[(36 RFA)(20 kW/RFA)(92.5/100)]/(24)
$T_{dry\ soil}$ = **35 days (match # days in 3).**

Case 1 — SVOC Contaminant Removal

Mass of SVOC removed in air at 131.5°C
Mass of soil (dry basis) = 44,171,547 kg
Initial SVOC concentration = 8,000 mg/kg
Final SVOC concentration = 500 mg/kg
Mass SVOC removed in air = (44,171,547 kg)(8,000 − 500 mg/kg)/10^6 = 331,287 kg

SVOC removal rates per extraction well
$T_{100°C}$, total time to treat site up to 100°C = 128 days
Mass of soil moisture removed up to 100°C = 559,207 kg
Mass of soil moisture to be removed = 3,975,439 kg
Mass of dry air to remove water = 2,796,036 kg
Operating at 15–100°C:
   Rate = (331,287 kg)(559,207/3,975,439)/[(42 wells)(128 days)] = 8.67 kg/day per well
   Content = (8.67 kg/day per well)/{2,796,036 kg/[(128 days)(42 wells)]}
   Content = 0.01668 kg/kg dry air
$T_{100°C}$, total time to treat site up to 131.5°C = (173 + 35 days) = 208 days
Mass of dry air to remove water and SVOC = (9,447,542 + 1,529,082 kg)
Mass of dry air to remove water and SVOC = 10,976,624 kg
Operating at 100–131.5°C:
   Rate = [331,287 kg − (8.67 kg/day/well)(128 days)][42 wells]/[(208 days)(42 wells)]
   Rate = 32.6 kg/day per well
   Content = (32.6 kg/day per well)/[10,976,624 kg/[(208 days)(42 wells)]]
   Content = 0.02593 kg/kg air

EXAMPLE CALCULATIONS FOR TABLE J3, CASE 1

Case 1 — Heat Requirement for Wet Soil

C_p, heat capacity of dry soil at 15°C
  Use soil heat capacity (Cp) in Table C1
  C_p = 0.777 kJ/kg-°C + (15 − 0°C)/(25 − 0°C)(830 − 0.777 kJ/kg-°C)
  C_p = 0.8090 kJ/kg-°C

C_p, heat capacity of dry soil at 100°C
  Use soil heat capacity (Cp) in Table C1
  C_p = 0.9444 kJ/kg-°C

C_p, heat capacity of dry soil at 131.5°C
  Use soil heat capacity (Cp) in Appendix C, Table C1
  C_p = 0.9740 kJ/kg-°C + (131.5 − 125°C)/(150 − 125°C)(1.000 − 0.974 kJ/kg-°C)
  C_p = 0.9807 kJ/kg-°C

H_l, heat content of moisture at 15 and 100°C
  Use enthalpy table for liquid water (H_l) in Table C5
  H_l = 48.8 kJ/kg + (15.0 − 11.86°C)/(16.86 − 11.86°C)(69.7 − 48.8 kJ/kg)
  H_l = 61.9 kJ/kg at 15°C
  H_l = 419.5 kJ/kg at 100°C

$H_{dry\ soil}$, total heat content of dry soil at 100°C
  $H_{dry\ soil\ 100°C}$ = [(44,171,547 kg)(0.9444 kJ/kg-°C)(100°C) − (44,171,547 kg)(0.8090 kJ/kg-°C)(15°C)]/(1 million
  kJ/1.0 million kJ) = 3,635.5 million kJ
  $H_{dry\ soil\ 100°C}$ = (3,635,500 kJ)(1,000 J/kJ)/(3,600,000 J/kWh) = 1,1,009,955 kWh

$H_{dry\ soil}$, total heat content of dry soil at 131.5°C
  $H_{dry\ soil\ 131.5°C}$ = [(44,171,547 kg)(0.9807 kJ/kg-°C)(131.5°C) − (44,171,547 kg)(0.8090 kJ/kg-°C)(15°C)]/(1 million
  kJ/1.0 million kJ) = 5,160.6 million kJ
  $H_{dry\ soil\ 131\ 5°C}$ = (5,160,600,000 kJ)(1,000 J/kJ)/(3,600,000 J/kWh) = 1,433,502 kWh

H_{water}, total heat content of moisture at 100°C
  H_{water} = (4,417,155 kg)(419.5 − 61.9 kJ/kg)/(1 million kJ/1.0 million kJ)
  H_{water} = 1,579.5 million kJ
  H_{water} = (1,579,500 kJ)(1,000 J/kJ)/(3,600,000 J/kWh) = 438,740 kWh

$H_{wet\ soil}$, total heat content of wet soil at 100°C
$H_{wet\ soil\ 100°C}$ = 3,635.5 million kJ + 1,579.5 million kJ = 5,214.9 million kJ
$H_{wet\ soil\ 100°C}$ = (5,197,600,000 kJ)(1,000 J/kJ)/(3,600,000 J/kWh) = 1,448,594 kWh

Case 1 — Heat Requirements for Water Vapor

Mass of soil moisture to be removed = (44,171,547 kg)(10/100% − 1/100%) = 3,975,439 kg

Heat wet soil to 100°C
Mass of soil moisture removed heating to 100°C: Table J2
Mass of soil moisture removed at 100°C = 559,207 kg
H_{fg} = heat of vaporization of water, Table C5 = 2,256 kJ/kg
$H_{water\ vapor}$, heat content of water vapor at 100°C
$H_{water\ vapor\ (100°C)}$ = (559,207 kg)(2,256 kJ/kg)/(1 million kJ/1.0 million kJ) = 1,262 million kJ
$H_{water\ vapor\ (100°C)}$ = (1,262,000,000 kJ)(1,000 J/kJ)/(3,600,000 J/kWh) = 350,452 kWh

Remove Water and SVOC from Wet Soil
Mass of soil moisture removed heating at 100°C: Table J2
Mass of soil moisture removed at 100°C = (3,975,439 − 559,207 kg) = 3,416,232 kg
$H_{water\ vapor\ 100°C}$, heat content of water vapor at 100°C
$H_{water\ vapor\ 100°C}$ = (3,416,232 kg)(2256 kJ/kg)/(1,000,000) = 7,707 million kJ
$H_{water\ vapor\ 100°C}$ = (7,707,000,000 kJ)(1,000 J/kJ)/(3,600,000 J/kWh) = 2,140,934 kWh
$H_{total\ vapor}$ = total heat content of water vapor
$H_{total\ vapor}$ = (1,262 million kJ + 7,707 million kJ) = 8,969 million kJ
$H_{total\ vapor}$ = (8,969,000,000 kJ)(1,000 J/kJ)/(3,600,000 J/kWh) = 2,491,386 kWh

Case 1 — Heat SVE Air

Heat wet soil to 100°C
Ambient air at 20°C (Range 15–30°C)
Mass of SVE air at 100°C = 2,796,036 kg
Cp_{air}, heat capacity of air at 100 and 20°C from Table C6
Cp_{air} (20°C) = 1.0210 kJ/kg-°C; Cp_{air} (100°C) = 1.0147 kJ/kg-°C
H_{air}, heat content of dry air at 100°C
H_{air} = (2,796,036 kg)[(100°C)(1.0147 kJ/kg-°C) − (20°C)(1.0210 kJ/kg-°C)] = 227.7 million kJ
H_{air} = (227,700,000 kJ)(1,000)/(3,600,000 J/kWh) = 63,239 kWh

Remove water and SVOC from wet soil
Mass of SVE air at 100°C = 9,447,542 kg
Cp_{air}, heat capacity of air at 100 and 20°C from Table C6
Cp_{air} (20°C) = 1.0210 kJ/kg-°C; Cp_{air} (100°C) = 1.0147 kJ/kg-°C
H_{air}, heat content of dry air at 100°C
H_{air} = (9,447,542 kg)[(100°C)(1.0147 kJ/kg-°C) − (20°C)(1.0210 kJ/kg-°C)] = 769.2 million kJ
H_{air} = (765,800,000 kJ)(1,000)/(3,600,000 J/kWh) = 213,679 kWh

Heat dry soil from 100–131.5°C
Mass of SVE air at 131.5°C = 1,529,082 kg
Cp_{air}, heat capacity of air at 131.5 and 20°C from Table C6, properties of air
Cp_{air} (20°C) = 1.0210 kJ/kg-°C; Cp_{air} (131.5°C) = 1.0025 kJ/kg-°C
H_{air}, heat content of dry air at 131.5°C
H_{air} = (1,529,082 kg)[(131.5°C)(1.0025 kJ/kg-°C) − (20°C)(1.0210 kJ/kg-°C)]
H_{air} = 174.5 million kJ
H_{air} = (174,500,000 kJ)(1,000)/(3,600,000 J/kWh) = 48,461 kWh

Case 1 — Heat Loss to the Surroundings

RF-SVE heat loss to the surroundings = 0.1220 kJ per kJ_{total} (total heat input basis)

Heat wet soil to 100°C
Heat content of wet soil from Table J3
$H_{wet\ soil}$ (100°C) = 5,215 million kJ
H_{total}, total heat requirement from RF system = 7,359 million kJ = 2,044,230 kWh
$T_{100°C}$, total time to treat site to 100°C = 128 days
T_{total}, total time to treat site = 336 days
H_{sur}, heat loss to surroundings = (0.1220 kJ/kJ_{total})(7,359 million kJ) − (0.1220 kJ/kJ_{total})(5,215 million)
    (128 days)/(336 days)
H_{sur}, heat loss to surroundings = *655 million kJ* = 182,030 kWh

Remove water and SVOC from wet soil

Heat content of wet and dry soil from Table J3

$H_{wet\ soil}$ at 100°C = 5,215 million kJ

$H_{dry\ soil}$ at 131.5°C = 5,161 million kJ

H_{total}, total heat requirement from RF system = 9,953 million kJ = 2,764,612 kWh

$T_{100°C}$, total time to treat site at 100°C = 173 days

$T_{131.5°C}$, total time to treat site up to 131.5°C = 35 days

T_{total}, total time to treat site = 336 days

H_{sur}, heat loss to surroundings = $(0.1220\ kJ/kJ_{total})(9,953$ million kJ) + $(0.1220\ kJ/kJ_{total})(5,215$ million) (173 days)/(336 days) − $(0.1220\ kJ/kJ_{total})(5,161$ million)(35 days)/(336 days)

H_{sur}, heat loss to surroundings = *1,476 million kJ* = 410,000 kWh

Heat dry soil from 100–131.5°C

Heat content of wet and dry soil from Table J3

$H_{wet\ soil}$ at 100°C = 5,215 million kJ

$H_{dry\ soil}$ at 131.5°C = 5,161 million kJ

H_{total}, total heat requirement from RF system = 2,011 million kJ = 558,498 kWh

$T_{131.5°C}$, total time to treat site up to 131.5°C = 35 days

T_{total}, total time to treat site = 336 days

H_{sur}, heat loss to surroundings = $(0.1220\ kJ/kJ_{total})(2,011$ million kJ) + $(0.1220\ kJ/kJ_{total})(5,161$ million) (35 days)/(336 days)

H_{sur}, heat loss to surroundings = *311 million kJ* = 86,389 kWh

Case 1 — Total Heat Requirements

Heat wet soil to 100°C

$H_{wet\ soil}$, total heat content of wet soil = 5,215 million kJ

H_{water} vapor, total heat content of water vapor = 1,262 million kJ

H_{air}, heat content of dry air at 100°C = 228 million kJ

H_{sur}, heat loss to surroundings = *655 million kJ* (iterative value with H_{sur} above)

Change the value (655 million kJ) until agreement with H_{sur} above

H_{total}, total heat requirement from RF system = 7,359 million kJ

Remove water and SVOC from wet soil at 100°C

H_{water} vapor, total heat content of water vapor = 7,707 million kJ

H_{air}, heat content of dry air at 100°C = 769 million kJ

H_{sur}, heat loss to surroundings = *1,476 million kJ* (iterative value with H_{sur} above)

Change the value (1,476 million kJ) until agreement with H_{sur} above

H_{total}, total heat requirement from RF system = 9,953 million kJ

Heat dry soil from 100–131.5°C

H_{dry}, heat dry from 100–131.5°C = 1,525 million kJ

$H_{water\ vapor}$, total heat content of water vapor = 0 million kJ

H_{air}, heat content of dry air at 100°C = 174 million kJ

H_{sur}, heat loss to surroundings = *311 million kJ* (iterative value with H_{sur} above)

Change the value (310 million kJ) until agreement with H_{sur} above

H_{total}, total heat requirement from RF system = 2,011 million kJ

Table J1 RF-SVE Hypothetical Design Case Studies and Site Conditions

---

Cost Study Cases

Case 1. 4,290 m² Site 1
1. Heat wet soil to 100°C
   Remove small portion of water & light-end SVOC
2. Operate at 100°C
   Remove remaining water & light-end SVOC
3. Heat dry soil areas to 131.5°C
   Remove heavy-end SVOC
4. Final SVOC content <500 mg/kg

Case 2. 817 m² Site 2
1. Heat wet soil to 100°C
   Remove small portion of water & light-end SVOC
2. Operate at 100°C
   Remove remaining water & light-end SVOC
3. Heat dry soil areas to 131.5°C
   Remove heavy-end SVOC
4. Final SVOC content <500 mg/kg

Case 3. 817 m² Site 2
1. Heat wet soil to 100°C
   Remove a small portion of the water and some SVOC while heating to 100°C
2. Operating at 100°C
   Oxidize both light-end and heavy-end SVOC by hydrous pyrolysis oxidation (HPO)
   Remove SVOC by SVE
3. Final SVOC <500 mg/kg

Area and Volume of Sites

Site 1

Radius of each RF zone	6.10 m
Length of treatment zone	66.00 m
Width of treatment zone	65.00 m
Number of RF zones	36
Direct area of RF zones	4,203 m²
Conceptual area	4,290 m²
Thickness of contamination	5.79 m
Direct volume of RF Zones	24,339 m³
Conceptual volume	24,843 m³

Site 2

Radius of each RF zone	6.10 m
Number of RF zones	7
Direct area of RF zones	817 m²
Conceptual area	847 m²
Thickness of contamination	5.79 m
Direct volume of RF zones	4,733 m³
Conceptual volume	4,906 m³

Site Conditions

Soil has similar physical properties to the Kirtland AFB site.	
Initial SVOC	*8,000* mg/kg (av value) (10,000 mg/kg max value)
Initial moisture	10 %
Final moisture	1 %
Soil density	1,778 kg/m³ (dry basis) (111 lb/ft³)
Soil porosity	34 %
Ambient temp	15–30 °C
Initial soil temp	15 °C
Water table depth	80 m
Depth to contaminated soil	0 m
SSTL for SVOC	500 mg/kg (less than value)
No free NAPL	
No smear zone	
Include impermeable barrier at site	

Note: Conversion factors, 3.28084 ft = 1.0 m; 4,046.86 m² = 1.0 acre; 16.018 lb/ft³ = 1.0 kg/m³; 0.02832 m³ = 28.32 L = 1.0 ft³; 3.6 × 10⁶ J = 1.0 kWh.

Table J2   RF-SVE Design Basis for Hypothetical Cases 1, 2, and 3

Case 1 — SVE System	
Time to heat soil and remove water-SVOC	336 days
Air injection wells	36
Air extraction wells	42
Standard (std) air temperature	20 °C
Standard (std) air pressure	101.3 kPa
Maximum airflow per injection well	875 std L/min (Lpm)
Maximum airflow per extraction well	750 std L/min (Lpm)
Density of dry air at 20°C and 101.3 kPa	1.204 kg/m^3

Case 1 — RF-SVE System	
Number of RF antenna wells	36
Total RF antennae (RFA)	36
Kilowatts per RF antenna (kW/RFA)	20
Antenna efficiency (RF energy to heat)	92.5 %
Efficiency of AC power to RF energy	72 %

1. **Heat wet soil to 100°C**

Mass of soil moisture to be removed	3,975,439 kg	
$H_{wet\ soil}$, total heat content of wet soil at 100°C	5,215 million kJ	1,448,594 kWh

Operating up to 100°C:

$T_{100°C}$, total time to treat site up to 100°C	*128 days (change #)*
Humidity of air (RF-SVE demonstration data)	*0.2000* kg water/kg dry air
Mass of soil moisture removed at 100°C	559,207 kg of water
Mass of dry air to remove water at 100°C	2,796,036 kg dry air — airflow = *300* Lpm
Check	*2,796,036*

2. **Remove water and SVOC from wet soil**
Operating at 100°C:

$T_{100°C}$, total time to treat site at 100°C	*173 days (change #)*
Humidity of air (RF-SVE demonstration)	*0.3616* kg water/kg dry air
Mass of soil moisture removed	3,416,232 kg of water
Mass of dry air to remove water	9,447,542 kg dry air — airflow = 750 Lpm
Check	*9,447,545* 90.00%

3. **Heat dry soil from 100–131.5°C**
Operating from 100–131.5°C:

$T_{131°C}$, total time to treat site up to 131.5°C	*35 days (change #)*
Humidity of air (RF-SVE demonstration)	*0.0000* kg water/kg dry air
Mass of soil moisture removed	0 kg of water
Mass of dry air to remove water	1,529,082 kg dry air — airflow = *600* Lpm

4. **Heat soil, heat dry air, and remove water-SVOC**
Operating to heat wet soil up to 100°C

H_{total}, total heat requirement from RF system	7,359 million kJ	2,044,230 kWh
$T_{>100\ °C}$, total time to treat site to 100°C	*128 days (match above #)*	

Operating at 100°C to remove water-SVOC

H_{total}, total heat requirement from RF system	9,953 million kJ	2,764,612 kWh
$T_{at\ 100°C}$, total time to treat site at 100°C	*173 days (match above #)*	

5. **Heat dry soil from 100–131.5°C**

$H_{dry\ soil\ at\ 131°C}$, heat content of dry soil at 131.5°C	5,160.6 million kJ	1,433,502 kWh
$H_{dry\ soil\ at\ 100°C}$, heat content of dry soil at 100°C	3,635.5 million kJ	1,009,855 kWh
$H_{dry\ soil}$, heat dry soil from 100–131.5°C	1,525.1 million kJ	423,647 kWh
H_{total}, total heat requirement from RF system	2,010.6 million kJ	558,498 kWh
$T_{dry\ soil}$, time to heat soil from 100–131.5°C	*35 days (match above #)*	

Table J2   RF-SVE Design Basis for Hypothetical Cases 1, 2, and 3 (continued)

Case 1 — SVOC Contaminant Removal

Mass of SVOC removed in air

Mass of soil (dry basis)	44,171,547 kg
Initial SVOC concentration	8,000 mg/kg
Final SVOC concentration	500 mg/kg
Mass of SVOC removed in air	331,287 kg

SVOC Removal Rates per Extraction Well

Operating from 15–100°C	8.7 kg/day	0.01668 kg/kg of air
Operating from 100–131.5°C	32.6 kg/day	0.02593 kg/kg of air

SVOC Removal Rate Assumptions
1. Demonstration rate of SVOC removal for 1 extraction well
   Airflow at ~280 std Lpm
   SVOC removal rate = ~2.0 kg/day
2. Demonstration SVOC averaged ~900 mg/kg and Site 1 contains 8000 mg/kg.
3. Demonstration SVOC was mixed with high concentrations of MRO that averaged ~6650 mg/kg.
4. Final average SVOC (mg/kg) is similar to Site-1's SSTL value of 500 mg/kg.

Case 2 — SVE System

Time to heat soil and remove water-SVOC	**332 days**
Air injection wells	7
Air extraction wells	12
Standard (std) air temperature	20 °C
Standard (std) air pressure	101.3 kPa
Maximum dry airflow per injection well	1,029 std L/min (Lpm)
Maximum dry airflow per extraction well	600 std L/min (Lpm)
Density of dry air at 20°C and 101.3 kPa	1.204 kg/m^3

Case 2 — RF-SVE System

RF antenna wells	7
Total RF antennae (RFA)	7
Kilowatts per RF antenna (kW/RFA)	20
Antenna efficiency (RF Energy to Heat)	92.5 %
Efficiency of AC power to RF energy	72 %

1. Heat wet soil to 100 °C

Mass of soil moisture to be removed	785,079 kg	
$H_{wet\ soil}$, total heat content of wet soil at 100°C	1,030 million kJ	286,072 kWh

Operating up to 100°C:

$T_{100°C}$, total time to treat site up to 100°C	*144 days (change #)*
Humidity of air (RF-SVE demonstration data)	*0.2000* kg water/kg dry air
Mass of soil moisture removed at 100°C	179,745 kg of water
Mass of dry air to remove water at 100°C	898,726 kg dry air — airflow = *300* Lpm
Check	*898,726*

2. Remove water and SVOC from wet soil
Operating at 100°C:

$T_{100°C}$, total time to treat site at 100°C	*152 days (change #)*
Humidity of air (RF-SVE demonstration)	0.3170 kg water/kg dry air
Mass of soil moisture removed	605,334 kg of water
Mass of dry air to remove water	1,897,310 kg dry air — airflow = *600* Lpm
Check	*1,909,787* 90.00%

3. Heat dry soil from 100–131.5°C
Operating from 100–131.5°C:

$T_{131°C}$, total time to treat site up to 131.5°C	*36 days (change #)*
Humidity of air (RF-SVE demonstration)	*0.0000* kg water/kg dry air
Mass of soil moisture removed	0 kg of water
Mass of dry air to remove water	449,363 kg dry air — airflow = *600* Lpm

4. Heat soil, heat dry air, and remove water-SVOC
   Operating to heat wet soil up to 100°C

H_{total}, total heat requirement from RF system	1,611 million kJ	447,377 kWh
$T_{>100°C}$, total time to treat site at 100°C	144 days (match # above)	

   Operating at 100–remove water-SVOC

H_{total}, total heat requirement from RF system	1,703 million kJ	473,165 kWh
$T_{at\ 100°C}$, total time to treat site at 100°C	152 days (match # above)	

5. Heat dry soil from 100–131.5°C

$H_{dry\ soil\ at\ 131°C}$, heat content of dry soil at 131.5°C	1,019.1 million kJ	283,091 kWh
$H_{dry\ soil\ at\ 100°C}$, heat content of dry soil at 100°C	717.9 million kJ	199,428 kWh
$H_{dry\ soil}$, heat dry soil from 100–131.5°C	301.2 million kJ	83,663 kWh
H_{total}, total heat requirement from RF system	397.3 million kJ	110,349 kWh
$T_{dry\ soil}$, time to heat soil from 100–131.5 °C	36 days (match # above)	

Case 2 — SVOC Contaminant Removal

Mass of SVOC removed in air

Mass of soil (dry basis)	8,723,100 kg	
Initial SVOC concentration	8,000 mg/kg	
Final SVOC concentration	500 mg/kg	
Mass of SVOC removed in air	65,423 kg	

SVOC removal rates per extraction well

Operating from 15–100 °C	8.67 kg/day	0.01667 kg/kg of air
Operating from 100–131.5 °C	22.39 kg/day	0.02153 kg/kg of air

SVOC removal rate assumptions
1. Demonstration rate of SVOC removal for 1 extraction well
   Airflow at ~280 std Lpm
   SVOC removal rate = ~2.0 kg/day
2. Demonstration SVOC averaged ~900 mg/kg and Site 1 contains 8000 mg/kg
3. Demonstration SVOC was mixed with high concentrations of MRO that averaged ~6650 mg/kg
4. Final average SVOC (mg/kg) is similar to Site-1's SSTL value of 500 mg/kg

Case 3 — SVE System

Time to Heat Soil and Remove Water-SVOC	*365 days (change #)*
Air injection wells	7
Air extraction wells	12
Standard (std) air temperature	20 °C
Standard (std) air pressure	101.3 kPa
Maximum airflow per injection well	1,029 std L/min (Lpm)
Maximum airflow per extraction well	600 std L/min (Lpm)
Density of air at 20°C and 101.3 kPa	1.204 kg/m³

Case 3 — RF-SVE System

	Period 1	Period 2
RF antenna wells	7	7
Total RF antennae (RFA)	4	*2 (change #)*
kW per RF antenna (kW/RFA)	20.0	*11.4 (change #)*
Antenna efficiency (RF Energy to Heat)	92.5%	92.5%
Efficiency of AC power to RF energy	72.0%	72.0%

1. Heat wet soil to 100°C

Mass of soil moisture	785,079 kg	
Water produced by oxidation reaction	44,233 kg	
Mass of available water	829,312 kg	
$H_{wet\ soil}$, total heat requirement for wet soil	1,030 million kJ	286,072 kWh

   Operating up to 100°C:

$T_{wet\ soil}$, total time to heat wet soil up to 100°C	*179 days (change #)*
Humidity of air (RF-SVE demonstration data)	*0.200 kg water/kg dry air (change #)*

Table J2   RF-SVE Design Basis for Hypothetical Cases 1, 2, and 3 (continued)

Mass of soil moisture removed up to 100°C	22,343 kg of water
Mass of dry air to remove water up to 100°C	111,717 kg dry air — airflow = 30 Lpm
Check	*111,717*

2. Remove water and SVOC from wet soil
   Operating at 100°C:

$T_{100°C}$, total time to treat site at 100°C	186 days (fixed #)
Humidity of air (RF-SVE demonstration data)	0.224 kg water/kg dry air (change #)
Mass of soil moisture removed at 100°C	521,036 kg of water
Mass of dry air to oxidize SVOCs at 100°C	488,759 kg of Air
Utilization of O_2 from air for oxidation	21 % (change #)
Mass of dry air to oxidize SVOCs at 100°C	2,321,709 kg dry air — airflow = *600* Lpm
Check	*2,321,708*

3. Heat wet soil, heat dry air, and remove water-SVOC
   Operating to heat wet soil up to 100°C

H_{total}, total heat requirement from RF system	1,145 million kJ	318,115 kWh
$T_{100°C}$, total time to heat wet soil to 100°C	179 days (match above #)	

Operating at 100°C to remove water-SVOC

H_{requir}water vapor, air, loss to surroundings	1,543 million kJ	428,701 kWh
H_{oxid}, net heat added by SVOC oxidation	−1,543 million kJ	-428,701 kWh
H_{total}, total heat requirement from RF system	0 million kJ	0 kWh
$T_{at\ 100°C}$, total time to treat site by RF heat	0 days	

Case 3 — SVOC Contaminant Removal

Mass of SVOC removed in air

Mass of soil (dry basis)	8,723,100 kg
Initial SVOC concentration	8,000 mg/kg
Final SVOC concentration	500 mg/kg
Amount of SVOC removed	94 %
Mass of SVOC removed in air	65,423 kg

SVOC removal rates per extraction well

Operating from 15–100°C (by SVE)	7.61 kg/day
Operating at 100°C (by oxidation and SVE)	21.98 kg/day

SVOC rate of removal

Operating from 15–100°C (by SVE)	10.47 mg/day-kg soil
Operating at 100°C (by oxidation and SVE)	30.24 mg/day-kg soil

SVOC removal rate assumptions
1. Demonstration rate of removal for 1 extraction well
   Airflow at ~280 std Lpm
   SVOC removal rate = ~2.0 kg/day
2. Demonstration SVOC averaged ~900 mg/kg and Site 1 contains 8000 mg/kg
3. Demonstration SVOC was mixed with high concentrations of MRO
   that averaged ~6650 mg/kg
4. We estimated that only ~50% of the SVOC compounds can be removed by RF-SVE
   operating up to 100°C (Period 1) and at 100°C (Period 2)

Case 3 — Heat, Reactants, and Products of Reaction — SVOC Contaminants

1. SVOC oxidation reaction for pentadecane
   $C_{15}H_{32} + 23O_2 = 15CO_2 + 16H_2O$

	Mol Wt	ΔH_f^o
$(C_{15}H_{32})$ = 212.42 g/mol		−860.0 kJ/g-mol
$23(O_2)$ = 735.97 g/mol		0.0 kJ/g-mol
$15(CO_2)$ = 660.15 g/mol		−393.5 kJ/g-mol
$16(H_2O)$ = 288.24 g/mol		−294.6 kJ/g-mol

ΔH_f^o(reaction) = −45,931 kJ/kg $C_{15}H_{34}$
ΔH_c^o(combustion) = −47,658 kJ/kg $C_{15}H_{34}$

2. SVOCs: heat of oxidation for n-alkanes

	Mol Wt	ΔH_f^0
$(C_{12}H_{26})$ =	170.34 g/mol	−47,830 kJ/kg alkane
$(C_{15}H_{32})$ =	212.42 g/mol	−47,658 kJ/kg alkane
$(C_{17}H_{36})$ =	240.47 g/mol	−47,577 kJ/kg alkane
$(C_{19}H_{40})$ =	268.53 g/mol	−47,513 kJ/kg alkane
$(C_{20}H_{42})$ =	282.55 g/mol	−47,485 kJ/kg alkane
Average =		**−47,613 kJ/kg alkane**

3. SVOCs: oxygen (air) requirements and water produced @ 50% oxidized

	kg O_2 per kg alkane	kg H_2O per kg alkane	kg CO_2 per kg alkane
$(C_{12}H_{26})$ =	3.48	1.375	3.100
$(C_{15}H_{32})$ =	3.46	1.357	3.108
$(C_{17}H_{36})$ =	3.46	1.348	3.111
$(C_{19}H_{40})$ =	3.46	1.342	3.114
$(C_{20}H_{42})$ =	3.45	1.339	3.115
Average =	3.46	1.352	3.110
Average =	**14.94 kg air/kg**		

4. SVOC rate of removal — operating at 100°C

Mass of SVOC oxidized @ 50% of total	32,712 kg
$T_{100°C}$, total time to treat site at 100°C	186 day
Oxidation rate operating at 100°C	20.16 mg/day-kg soil
SVE removal rate operating at 100°C	10.08 mg/day-kg soil

5. SVOCs: reaction summary @ 50% oxidized

Total heat evolved =	−1,557 million kJ	
Total O_2 required =	113,245 kg	79,275 m³
Total air required =	488,759 kg	377,499 m³
Total water produced =	44,233 kg	
Total CO_2 produced =	101,724 kg	51,775 m³

6. Heat capacities of oxidation reaction gases

T, K = °C + 273.14
1.0 calorie (cal) = 4.184 J
Cp O_2: 8.27 + 0.000258*T - 187,700/T², cal/degree-mol
Cp CO_2: 10.34 + 0.002617*T - 195,500/T², cal/degree-mol
Cp H_2O: 1.0 cal/degree-g

ΔCp (20–100°C)
0.918 kJ/degree-kg
0.942 kJ/degree-kg
4.184 kJ/degree-kg

7. Heat content of oxidation reactants and products

$\Delta H_R° O_2$:	−8.31 million kJ
$\Delta H_R° CO_2$:	7.67 million kJ
$\Delta H_R° H_2O$:	14.81 million kJ
difference	14.16 million kJ

Table J3   Heat Balances for Hypothetical Cases 1, 2, and 3

Case 1 — Heat Balance Input Data

Av soil temperature	15.0 °C
Boiling point of water	100.0 °C
Av final temperature	131.5 °C
Vol of soil	24,843 m³
Mass of soil (dry basis)	44,171,547 kg
Mass of soil (dry basis)	44,172 t
Total mass of moisture in the soil	4,417,155 kg
Mass of soil moisture to be removed	3,975,439 kg
C_p, heat capacity of dry soil at 131.5°C	0.9807 kJ/kg-°C
C_p, heat capacity of dry soil at 100°C	0.9444 kJ/kg-°C
C_p, heat capacity of dry soil at 15°C	0.8090 kJ/kg-°C
H_l, heat content of soil moisture at 100°C	419.5 kJ/kg
H_l, heat content of soil moisture at 15°C	61.9 kJ/kg

Case 1 — Heat Requirements for Wet and Dry Soil

H_{soil}, total heat content of dry soil at 100°C	3,635.5 million kJ	1,009,855 kWh
H_{water}, total heat content of moisture at 100°C	1,579.5 million kJ	438,740 kWh
$H_{wet\ soil}$, total heat content of wet soil at 100°C	**5,214.9 million kJ**	**1,448,594 kWh**
H_{soil}, total heat content of dry soil at 131.5°C	5,160.6 million kJ	
$H_{dry\ soil}$, total heat content of dry soil at 131.5°C	**5,160.6 million kJ**	**1,433,502 kWh**

Case 1 — Heat Requirements for Water Vapor

Mass of soil moisture to be removed	3,975,439 kg of water	
Heat wet soil to 100°C		
Mass of soil moisture removed at 100°C	559,207 kg of water	
H_{lg}, heat of vaporization for water at 100°C	2,256 kJ/kg	
$H_{water\ vapor}$, heat content of water vapor at 100°C	1,262 million kJ	350,452 kWh
Remove water and SVOC from wet soil		
Mass of soil moisture removed at 100°C	3,416,232 kg of water	
H_{lg}, heat of vaporization for water at 100°C	2,256 kJ/kg	
$H_{water\ vapor}$, heat content of water vapor at 100°C	7,707 million kJ	2,140,934 kWh
$H_{total\ vapor}$, total heat content of water vapor	**8,969 million kJ**	**2,491,386 kWh**

Case 1 — Heat SVE Air

Heat wet soil to 100°C		
Heat air from 20–100°C		
$T_{>100°C}$, total time to treat site up to 100°C	128 days	
Humidity of SVE air at 100°C	0.2000 kg water/kg dry air	
Mass of dry SVE air	2,796,036 kg	
Cp_{air}, heat capacity of air at 100°C	1.0147 kJ/kg-°C	
Cp_{air}, heat capacity of air at 20°C	1.0025 kJ/kg-°C	
H_{air}, heat content of dry air up to 100°C	227.7 million kJ	63,239 kWh
Remove water and SVOC from wet soil		
Heat air from 20–100°C		
$T_{at\ 100°C}$, total time to treat site at 100°C	173 days	
Humidity of SVE air at 100°C	0.3616 kg water/kg dry air	
Mass of Dry SVE air	9,447,542 kg	
Cp_{air}, heat capacity of air at 100°C	1.0147 kJ/kg-°C	
Cp_{air}, heat capacity of air at 20°C	1.0025 kJ/kg-°C	
H_{air}, heat content of dry air at 100°C	769.2 million kJ	213,679 kWh
Heat dry soil from 100–131.5°C		
Heat air from 20–131.5°C		
$T_{>131°C}$, total time to treat site up to 131.5°C	35 days	
Humidity of SVE air at temperature up to 131.5°C	0.0000 kg water/kg dry air	
Mass of dry SVE air	1,529,082 kg	
Cp_{air}, Heat Capacity of Air at 131.5°C	1.0201 kJ/kg-°C	
Cp_{air}, Heat Capacity of Air at 20°C	1.0025 kJ/kg-°C	
H_{air}, heat content of dry air up to 131.5°C	174.5 million kJ	48,461 kWh

Case 1 — Heat Loss to the Surroundings

RF-SVE heat loss to the surroundings	0.1220 kJ per kJ$_{total}$ (Total Heat Input Basis)	

Heat wet soil to 100°C
Operating up to 100°C
H_{sur} excludes heat loss by wet soil

H_{sur} heat loss to surroundings	655 million kJ	182,030 kWh

Remove water and SVOC from wet soil
Operating at 100°C
Remove light-end SVOC
H_{sur} includes heat loss by wet soil

H_{sur} heat loss to surroundings	1,476 million kJ	410,104 kWh

Heat dry soil from 100–131.5°C
Operating up to 131.5°C
Remove heavy-end SVOC
H_{sur} includes heat loss by dry soil

H_{sur} heat loss to surroundings	311 million kJ	86,365 kWh

Case 1 — Total Heat Requirements

Heat wet soil to 100°C

$H_{wet\ soil}$ total heat content of wet soil	5,215 million kJ	1,448,594 kWh
$H_{water\ vapor}$ total heat content of water vapor	1,262 million kJ	350,452 kWh
H_{air} heat content of dry air at 100°C	228 million kJ	63,239 kWh
H_{sur} heat loss to surroundings	655 million kJ	181,944 kWh
H_{total} total heat requirement from RF system	**7,359 million kJ**	**2,044,230 kWh**
Check	*7,359 million kJ*	

Remove water and SVOC from wet soil

$H_{water\ vapor}$ total heat content of water vapor	7,707 million kJ	2,140,934 kWh
H_{air} heat content of dry air at 100°C	769 million kJ	213,679 kWh
H_{sur} heat loss to surroundings	1,476 million kJ	410,000 kWh
H_{total} total heat requirement from RF system	**9,953 million kJ**	**2,764,612 kWh**
Check	*9,953 million kJ*	

Heat dry soil from 100–131.5°CI

H_{dry} heat dry soil from 100–131.5 °C	1,525 million kJ	423,647 kWh
$H_{water\ vapor}$ total heat content of water vapor	0 million kJ	0 kWh
H_{air} heat content of dry air from 100–131.5 °C	174 million kJ	48,461 kWh
H_{sur} heat loss to surroundings	311 million kJ	86,389 kWh
H_{total} total heat requirement from RF system	**2,011 million kJ**	**558,498 kWh**
Check	*2,011 million kJ*	

H_{total} project heat requirement from RF system	**19,322 million kJ**	**5,367,340 kWh**
Check	*0.1264*	

Case 2 — Heat Balance Input Data

Av soil temp	15.0 °C
Boiling point of water	100.0 °C
Av final temp	131.5 °C
Vol of soil	4,906 m³
Mass of soil (dry basis)	8,723,100 kg
Mass of soil (dry basis)	8,723 t
Total mass of moisture in the soil	872,310 kg
Mass of soil moisture to be removed	785,079 kg
Cp, heat capacity of dry soil at 131.5°C	0.9807 kJ/kg-°C
Cp, heat capacity of dry soil at 100°C	0.9444 kJ/kg-°C
Cp, heat capacity of dry soil at 15°C	0.8090 kJ/kg-°C
H_l, heat content of soil moisture at 100°C	419.5 kJ/kg
H_l, heat content of soil moisture at 15°C	61.9 kJ/kg

Case 2 — Heat Requirements for Wet and Dry Soil

H_{soil} total heat content of dry soil at 100°C	717.9 million kJ	199,428 kWh
H_{water} total heat content of moisture at 100°C	311.9 million kJ	86,643 kWh
$H_{wet\ soil}$ total heat content of wet soil at 100°C	**1,029.9 million kJ**	**286,072 kWh**

Table J3   Heat Balances for Hypothetical Cases 1, 2, and 3 (continued)

$H_{dry\ soil}$, total heat content of dry soil at 131.5°C	1,019.1 million kJ	283,091 kWh
$H_{dry\ soil}$, total heat content of dry soil at 131.5°C	1,019.1 million kJ	283,091 kWh

Case 2 — Heat Requirements for Water Vapor

Mass of soil moisture to be removed	785,079 kg of water	
Heat wet soil to 100°C		
Mass of soil moisture removed at 100°C	179,745 kg of water	
H_{lg}, heat of vaporization for water at 100°C	2,256 kJ/kg	
$H_{water\ vapor}$ heat content of water vapor at 100°C	406 million kJ	112,645 kWh
Remove water and SVOC from wet soil		
Mass of soil moisture removed at 100°C	605,334 kg of water	
H_{lg}, heat of vaporization for water at 100°C	2,256 kJ/kg	
$H_{water\ vapor}$ heat content of water vapor at 131.5°C	1,366 million kJ	379,359 kWh
$H_{total\ vapor}$, total heat content of water vapor	1,771 million kJ	492,005 kWh

Case 2 — Heat SVE Air

Heat wet soil to 100°C		
Heat air from 20–100°C		
$T_{>100°C}$, total time to treat site up to 100°C	144 days	
Humidity of SVE air at 100°C	0.2000 kg water/kg dry air	
Mass of dry SVE air	898,726 kg	
Cp_{air}, heat capacity of air at 100°C	1.0147 kJ/kg-°C	
Cp_{air}, heat capacity of air at 20°C	1.0025 kJ/kg-°C	
H_{air}, heat content of dry air up to 100°C	73.2 million kJ	20,327 kWh
Remove water and SVOC from wet soil		
Heat air from 20–100°C		
$T_{at\ 100°C}$, total time to treat site at 100°C	152 days	
Humidity of SVE air at 100°C	0.3170 kg water/kg dry air	
Mass of dry SVE air	1,897,310 kg	
Cp_{air}, heat capacity of air at 100°C	1.0147 kJ/kg-°C	
Cp_{air}, heat capacity of air at 20°C	1.0025 kJ/kg-°C	
H_{air}, heat content of dry air at 100°C	151.7 million kJ	42,139 kWh
Heat dry soil to 131.5°C		
Heat air from 20–131.5°C		
$T_{>131°C}$, total time to treat site up to 131.5°C	36 days	
Humidity of SVE air at up to 131.5°C	0.0000 kg water/kg dry air	
Mass of dry SVE air	449,363 kg	
Cp_{air}, heat capacity of air at 131.5°C	1.0201 kJ/kg-°C	
Cp_{air}, heat capacity of air at 20°C	1.0025 kJ/kg-°C	
H_{air}, heat content of dry air up to 131.5°C	50.1 million kJ	13,908 kWh

Case 2 — Heat Loss to the Surroundings

RF-SVE heat loss to the surroundings	*0.0900* kJ per kJ$_{total}$ (Total Heat Input Basis)	
Heat wet soil to 100°C		
Operating up to 100°C		
H_{sur} excludes heat loss from wet soil		
H_{sur} heat loss to surroundings	102 million kJ	28,465 kWh
Remove water and SVOC from wet soil		
Operating at 100°C		
Remove light-end SVOC		
H_{sur} includes heat loss from wet soil		
H_{sur} heat loss to surroundings	186 million kJ	51,618 kWh
Heat dry soil from 100–131.5°C		
Operating up to 131.5°C		
Remove heavy-end SVOC		
H_{sur} includes heat loss from dry soil		
H_{sur} heat loss to surroundings	46 million kJ	12,697 kWh

Case 2 — Total Heat Requirements

Heat wet soil to 100°C

$H_{wet\,soil}$, total heat content of wet soil	1,030 million kJ	286,072 kWh
$H_{water\,vapor}$, total heat content of water vapor	406 million kJ	112,645 kWh
H_{air}, heat content of dry air to 100°C	73 million kJ	20,327 kWh
H_{sur}, heat loss to surroundings	102 million kJ	28,333 kWh
H_{total}, total heat requirement from RF system	**1,611 million kJ**	**447,377 kWh**
Check	1,611 million kJ	

Remove water and SVOC from wet soil

$H_{water\,vapor}$, total heat content of water vapor	1,366 million kJ	379,359 kWh
H_{air}, heat content of dry air at 100°C	152 million kJ	42,139 kWh
H_{sur}, heat loss to surroundings	186 million kJ	51,667 kWh
H_{total}, total heat requirement from RF system	**1,703 million kJ**	**473,165 kWh**
Check	1,703 million kJ	

Heat dry soil from 100–132.5°C

$H_{dry\,soil}$, heat dry soil from 100–131.5°C	301 million kJ	83,663 kWh
$H_{water\,vapor}$, total heat content of water vapor	0 million kJ	0 kWh
H_{air}, heat content of dry air from 100–131.5°C	50 million kJ	13,908 kWh
H_{sur}, heat loss to surroundings	46 million kJ	12,778 kWh
H_{total}, total heat requirement from RF system	**397 million kJ**	**110,349 kWh**
Check	397 million kJ	

H_{total}, **project heat requirement from RF system**	**3,711 million kJ**	**1,030,891 kWh**
Check	0.0900	

Case 3 — Heat Balance Input Data

Av soil temp	15.0 °C
Boiling point of water	100.0 °C
Av final temp	100.0 °C
Vol of soil	4,906 m³
Mass of soil (dry basis)	8,723,100 kg
Mass of soil (dry basis)	8,723 t
Total mass of moisture in the soil	872,310 kg
Mass of soil moisture to be removed	785,079 kg
Cp, heat capacity of dry soil at 100°C	0.9444 kJ/kg-°C
Cp, heat capacity of dry soil at 15°C	0.8090 kJ/kg-°C
H_l, heat content of soil moisture at 100°C	419.5 kJ/kg
H_l, heat content of soil moisture at 15°C	61.9 kJ/kg

Case 3 — Heat Requirements for Wet and Dry Soil

H_{soil}, total heat content of dry soil at 100°C	717.9 million kJ	199,428 kWh
H_{water}, total heat content of moisture at 100°C	311.9 million kJ	86,643 kWh
$H_{wet\,soil}$, total heat content of wet soil at 100°C	1,029.9 million kJ	286,072 kWh

Case 3 — Heat Requirements for Water Vapor

Mass of soil moisture to be removed	785,079 kg of water	
Mass of water produced by SVOC oxidation	44,233 kg of water	
Total mass of water to be removed	829,312 kg of water	

Heat wet soil to 100°C

Mass of soil moisture removed at 100°C	22,343 kg of water	
H_{lg}, heat of vaporization for water at 100°C	2,256 kJ/kg	
$H_{water\,vapor}$ heat content of water vapor at 100°C	50 million kJ	14,002 kWh

Remove water and SVOC from wet soil

Mass of soil moisture removed at 100°C	521,036 kg of water	
H_{lg}, heat of vaporization for water at 100°C	2,256 kJ/kg	
$H_{water\,vapor}$ heat content of water vapor at 100°C	1,176 million kJ	326,530 kWh
$H_{total\,vapor}$ total heat content of water vapor	1,226 million kJ	340,533 kWh

Table J3   Heat Balances for Hypothetical Cases 1, 2, and 3 (continued)

Case 3 — Heat SVE Air to 100°C

Heat wet soil to 100°C
    Heat air from 20–100°C
    $T_{>100°C}$, total time to treat site up to 100°C   179 days
    Humidity of SVE air at 100°C   0.2000 kg water/kg dry air
    Mass of dry SVE air   111,717 kg
    Cp_{air}, heat capacity of air at 100°C   1.0147 kJ/kg-°C
    Cp_{air}, heat capacity of air at 20°C   1.0025 kJ/kg-°C
    H_{air}, heat content of dry air at 100°C   **9.1 million kJ**   **2,527 kWh**
Remove water and SVOC from wet soil
    Heat air from 20–100°C
    $T_{at\ 100°C}$, total time to treat site at 100°C   186 days
    Humidity of SVE air at 100°C   0.2244 kg water/kg dry air
    Mass of dry SVE air   2,321,709 kg
    Cp_{air}, heat capacity of air at 100°C   0.9444 kJ/kg-°C
    Cp_{air}, heat capacity of air at 20°C   0.8090 kJ/kg-°C
    H_{air}, heat content of dry air at 100°C   **181.7 million kJ**   **50,471 kWh**

Case 3 — Heat Loss to the Surroundings

The heat loss to the surroundings is based on the heat balance for the RF-SVE demonstration.

RF-SVE heat loss to the surroundings   0.0900 kJ per kJ_{total} (total heat input basis)
$H_{wet\ soil}$, total heat content of wet soil   1,030 million kJ   286,072 kWh

Heat wet soil to 100°C
    H_{sur} excludes heat loss by wet soil
    H_{sur}, heat loss to surroundings while heating   **56 million kJ**   **15,514 kWh**
Remove water and SVOC from wet soil
    H_{sur} includes heat loss by wet soil
    H_{sur}, heat loss to surroundings at 100°C   **186 million kJ**   **51,700 kWh**

Case 3 — Total Heat Requirements

Heat wet soil to 100°C

$H_{wet\ soil}$, total heat content of wet soil	1,030 million kJ	286,072 kWh
$H_{water\ vapor}$, total heat content of water vapor	50 million kJ	14,002 kWh
H_{air}, heat content of dry air at 100°C	9 million kJ	2,527 kWh
H_{sur}, heat loss to surroundings	56 million kJ	15,514 kWh
H_{total}, total heat requirement from RF system	**1,145 million kJ**	**318,115 kWh**
Check	1,145 million kJ	

Remove water and SVOC from wet soil

$H_{water\ vapor}$, total heat content of water vapor	1,176 million kJ	326,530 kWh
H_{air}, heat content of dry air at 100°C	182 million kJ	50,471 kWh
H_{sur}, heat loss to surroundings	186 million kJ	51,700 kWh
H_{oxid}, net heat added by SVOC oxidation	−1,543 million kJ	−428,701 kWh
H_{total}, total heat requirement from RF system	**0 million kJ**	**0 kWh**
Check	0 million kJ	

H_{total}, project heat requirement from RF system	**1,145 million kJ**	**746,816 kWh**
Check	0.0900	

Appendix K

Cost Estimates for the Three
Hypothetical Design Cases

Appendix K Listing

Table K1 Case 1 Cost Summary Sheet (Large Site)

Preliminary Cost Estimate
RF-SVE Full-Scale Remediation

Before Treatment Costs

WBS #[a]	Description	Unit	Units	Unit Cost ($)	Total Cost ($)
33 01	Mobilization/preparatory work	Lump	1	20,000	20,000
33 02	Premonitoring, sampling, analysis, assessments	N.A.	Assumed Completed		
33 03	Site work (clearing/utilities install/pads), surveying	Lump	1	40,960	40,960
33 03	Engineering [SVE design only]	Lump	1	17,438	17,438
33 05	Markup (see itemized cost breakdown)	%	15		9,144
	Subtotal				87,542

Treatment Costs (includes equipment cost/installation, RF design, and RF-SVE system operation)

Thermal Treatment — WBS# 33 14
Physical Treatment — WBS# 33 13

WBS #[a]	Treatment Cost Descriptions	Unit	Units	Unit Cost ($)	Total Cost ($)
33 13-01	SVE system equipment and installation	Lump	1	417,550	417,550
33 13-02	SVE operating costs	Lump	1	220,011	220,011
33 14-01	RF equipment, design, and installation	Lump	1	150,325	150,325
33 14-02	RF operating costs	Lump	1	3,380,936	3,380,936
	Markup — selected subcontract, capital, O&M costs	%	15		184,095
	Subtotal				4,352,917

After Treatment Costs

WBS #[a]	Description	Unit	Units	Unit Cost ($)	Total Cost ($)
33 17	Decontamination and decommissioning	Lump	1	7,000	7,000
33 18	Disposal	Lump	1	7,500	7,500
33 20	Site restoration	Lump	1	5,000	5,000
33 21	Demobilization	Lump	1	4,000	4,000
33 91	Confirmatory sampling	Lump	1	57,880	57,880
33 92	As-built drawings/final closure report	Lump	1	3,600	3,600
	Markup (see itemized cost breakdown)	%	15		3,525
	Subtotal				88,505
	Total Remediation Cost				4,528,964
	Check Total Costs				4,528,964

Itemized Breakdown of Costs

Labor (RF Heating System Design & Installation Oversight)

WBS #*	Title	Duty	Hours	Rate $/h	Total Cost ($)
33 14-01	Project manager (1)	Management	100	135	13,500
33 14-01	RF supplier's engineer (1)	RF system design/technical review	1	Lump	30,000
33 14-01	Technician (2)	RF system installation oversight	560	57	31,920
33 14-01	Lead engineer (1)	RF system installation	125	85	10,625
33 14-01	Per diem	2 people onsite at all times @ ($)80/day per person	1	Lump	4,480
		Subtotal			90,525

Sub-Contract: Remediation System (RF & SVE) Installation

WBS #*	Description	Notes	Unit	Units	Unit Cost ($)	Total Cost ($)
33 13-01	42 2" SVE wells @ 20 ft each (complete)		LF	840	40	33,600
33 13-01	36 2" Air vent wells @ 20 ft each (passive)		LF	720	30	21,600
33 13-01	Trenching (4 @ 240ft + 24 ft to SVE+ 24 × 15 ft + 18 × 25 ft)	1	LF	2,010	25	50,250
33 13-01	Subsurface utilities (includes steel piping)		LF	2,010	15	30,150
33 13-01	Backfill, compact, & resurface (with same soils)		LF	2,010	20	40,200
33 13-01	Waste disposal (well cuttings)	2	Lump	1	5,000	5,000
33 03	Surveying for installation of all systems		Lump	1	5,000	5,000
33 03	Compound for SVE system		Lump	1	5,000	5,000
33 14-01	Install 36 RF heating wells	1	EA	36	800	28,800
33 14-01	Set structures/complete connections		Lump	1	16,000	16,000
33 07	Install vapor treatment system (incl. elect., gas hook-up)	3	Lump	1	5,000	5,000
33 13-01	Provide & install well vaults		EA	42	500	21,000
33 13-01	Install wellhead manifolding		EA	42	200	8,400
33 01	Mobilization (incl. electrical service install)		Lump	1	20,000	20,000
33 03	Site fence & 2 gates (complete)		LF	1,500	8	12,000
33 03	Per diem (4 persons) @ ($)80/day per person		day	28	320	8,960
33 03	Permitting		Lump	1	10,000	10,000
33 17	Decontamination (incl. steam cleaner rental ($)150/day)		Lump	1	7,000	7,000
33 18	Waste disposal (site debris)		Lump	1	7,500	7,500
33 20	Site restoration		Lump	1	5,000	5,000
33 21	Demobilization		Lump	1	4,000	4,000
33 90	Health & safety plan/RF-trained personnel monitoring		Lump	1	5,000	5,000
	Markup		%	15		
	Subtotal					401,879

Note: (1) RF lines will be installed in SVE trenches, assume no saw cuts into concrete; (2) assume nonhazard disposal @ $30/yd; (3) install only, no equipment costs. Assume 28 days (1 month) for RF system installation.

Table K1 Case 1 Cost Summary Sheet (Large Site) (continued)

Labor: SVE Well Design & Installation Oversight (Install 42 SVE wells)

WBS #ᵃ	Title	Duty	Hours	Rate $/h	Total Cost ($)
33 13-01	Project manager (1)	Management	20	135	2,700
33 05	Senior engineer/hydro (1)	System Design/Technical Review	150	116	17,438
33 13-01	Staff engineer (1)	Oversight	32	75	2,400
33 13-01	Staff geologist (2)	Install Wells (10 wells per day)	70	75	5,250
33 92	Staff engineer (1)	As-built Drawings/Final Closure Report	48	75	3,600
	Subtotal				31,388

Capital Cost: RF Heating/SVE System Equipment

WBS #ᵃ	Description	Notes	Unit	Units	Unit Cost ($)	Total Cost ($)
33 14-01	RF generator system (rental — see operating costs)	1	EA	1	0	0
33 14-01	Heating monitoring points/equipment	2	Lump	1	15,000	15,000
33 13-01	14.9-kW blowers for SVE system (12,750 Lpm [450 scfm])	3	EA	3	15,000	45,000
33 13-01	Piping and valves (within compound)		Lump	1	8,000	8,000
33 13-01	Valves, fittings, and gauges with SVE wells		Lump	1	10,000	10,000
33 13-01	Central controller (Elect. for blowers, Cat-Ox)		EA	1	5,000	5,000
33 13-01	Electrical transformers (1,000 KVA)	4	Lump	1	22,000	22,000
33 07	Catalytic oxidation units (21,240 Lpm [750 scfm]) each		EA	2	51,000	102,000
33 13-01	Markup		%	15		31,050
	Subtotal					238,050
	Total system install					761,842

Note: (1) RF equipt. rental assumed operating, not capital cost; (2) cost of RF transmission lines, antennae, fiber optic thermometers; (3) SVE system includes moisture separator, transfer pump, particle filters, appurtenances; (4) 1,000-KVA transformer required for 36 RF units (includes material and labor to install).

Operation & Maintenance (O&M) — Labor: (O&M Oversight)

WBS #ᵃ	Title	Duty	Hours	Rate $/h	Total ($)
BOTH	Project manager	Management	290	135	39,150
BOTH	Technicians (2)	Site work (includes sampling labor)	2000	57	114,000
BOTH	Staff engineer	Data Interpretation	500	75	37,500
BOTH	Health & safety officer	Health & safety on site	220	116	25,520
BOTH	Responsible engineer	Engineer of record, PE	280	116	32,480
	Subtotal				248,650

Note: Assume 336 Days (11 months) as time for remediation.

RF/SVE System O & M Costs

WBS #*	Description	Notes	Unit	Quantity	Unit Cost ($)	Total Cost ($)
33 14-02	36 20-kW RF generator/trailer units @ ($)6,000/month		Month	12	216,000	2,592,000
33 14-02	Heating system routine maintenance/mat'ls	1	Lump	1	10,000	10,000
33 14-02	Quarterly reporting		ea	4	5,000	20,000
33 14-02	Electricity for RF heating system @ ($)80/1,000 kWh	2	kWh	7,454,639	80	596,371
33 13-02	Regulatory sampling	3	Month	12	200	2,400
33 13-02	Electricity for SVE system @ ($)80/1,000 kWh		kWh	18,080	80	1,446
33 13-02	Propane gas (for Cat-Ox.)	4	Month	12	1,500	18,000
33 13-02	FID (monitor once/week)		Week	52	100	5,200
33 13-02	Office trailer/telephone/toilet		Month	12	450	5,400
33 13-02	SVE/vapor abatement system maintenance		Month	12	2,500	30,000
33 13-02	Travel (2 technicians) @ ($)550/trip		Month	12	1,100	13,200
33 13-02	Per diem (2 persons on site) @ ($)80/day per person		Day	333	160	53,280
33 13-02	Markup (excluding RF rental)		%	15		113,295
	Subtotal					**3,460,592**
	Total O & M costs/year					**3,460,592**

Note: (1) Includes maintaining transmission lines, adjusting temperatures, replacing mat'ls, SVE flow; (2) 7,454,600 kWh RF energy required for 336 days of operation, assumed AC to RF energy 72%; (3) based on TX regulations, data interpretation only (not sampling labor); (4) propane & tank available at $1/gal.

Confirmatory Sampling

WBS #*	Title	Duty	Hours	Rate $/h	Total ($)
33 91	Project manager (1)	Management/final report	120	135	16,200
33 91	Technicians (2)	Site work (includes sampling labor)	200	57	11,400
33 91	Staff engineer (1)	Data interpretation/final report	200	75	15,000
33 91	Staff geologist (2)	Install monitor wells (10 wells per day)	80	75	6,000
33 91	Responsible engineer (1)	Engineer of record, PE	80	116	9,280
	Subtotal				**57,880**

Note: Completed 1 year after remediation. Assume 40 confirmatory borings were required.

* Categorized according to interagency WBS# cost elements.

Table K1 Case 1 Cost Summary Sheet (Large Site) (continued)

Project Cost Scenarios

Case	Volume or Mass of Soil		
	m³	yd³	tons
1	24,843	32,493	48,700
2	4,903	6,374	9,610
3	4,903	6,374	9,610

Description of Cost Scenario

			Unit Costs	
	Cost ($)	$/m³	$/yd³	$/ton
Case 1 — Total project cost Table J1	4,528,964	182.30	139.38	93.00
Case 2 — Total project cost Table J2	1,411,620	287.91	221.47	146.89
Case 3 — Total project cost Table J3	1,024,652	208.98	160.75	106.62
Case 2 to Case 3 comparison: decrease in RF energy requirement	91,072	18.57	14.29	9.48
Case 2 to Case 3 comparison: decrease in the number of RF units	300,000	61.19	47.07	31.22
Case 2 to Case 3: add above unit costs		79.76	61.35	40.69
Case 1: compare leased RF units to capitalized RF units	1,815,000	109.24	83.52	55.73
Case 2: compare leased RF units to capitalized RF units	353,000	215.91	166.08	110.16
Case 3: compare leased RF units to capitalized RF units	118,000	184.92	142.24	94.34

Capital Costs

$$A/P_{i,n} = (i)(1 + i)^n/[(1 + i)^n - 1]$$

i = Interest rate = 7%
n = Number of years for capital write-off = 5 years
$P_{i,n}$ = Capital expenditure for 20 kW RF Unit = ($)70,000 per unit
A = Annual capital-recovery payment for capital expenditure $P_{i,n}$

Case	Total Lease ($)	$P_{i,n}$ ($)	A* per Unit	Total Capital Cost ($)	Payoff Years	Capital Cost per Year[b] ($)	Cost RF Units	Difference ($)
1	2,592,000	70,000	18,780	2,520,000	3.24	777,000	36	1,815,000
2	504,000	70,000	18,780	490,000	3.25	151,000	7	353,000
3	204,000	70,000	18,780	280,000	3.26	86,000	4	118,000

a Includes a 10% contingency; b includes a 15% markup.

Table K2 Case 2 Cost Summary Sheet (Small Site)

Preliminary Cost Estimate
RF-SVE Full-Scale Remediation

Before Treatment Costs

WBS #*	Description	Unit	Units	Unit Cost ($)	Total Cost ($)
33 01	Mobilization/preparatory work	Lump	1	12,000	12,000
33 02	Pre-monitoring, sampling, analysis, assessments	N.A.	Assumed completed		
33 03	Site work (clearing/utilities install/pads), surveying	Lump	1	26,860	26,860
33 05	Engineering [SVE design only]	Lump	1	13,950	13,950
	Markup (see itemized cost breakdown)	%	15		5,829
	Subtotal				58,639

Treatment Costs (includes equipment cost/installation, RF design, and RF-SVE system operation)

Thermal Treatment — WBS# 33 14
Physical Treatment — WBS# 33 13

WBS #*	Treatment Cost Descriptions	Unit	Units	Unit Cost ($)	Total Cost ($)
33 13-01	SVE system equipment and installation	Lump	1	146,670	146,670
33 13-02	SVE operating costs	Lump	1	201,765	201,765
33 14-01	RF equipment, design and installation	Lump	1	79,180	79,180
33 14-02	RF operating costs	Lump	1	806,668	806,668
	Markup — selected subcontract, capital, O&M costs	%	15		62,953
	Subtotal				1,297,236

After Treatment Costs

WBS #¹	Description	Unit	Units	Unit Cost ($)	Total Cost ($)
33 17	Decontamination and decommissioning	Lump	1	2,500	2,500
33 18	Disposal	Lump	1	5,000	5,000
33 20	Site restoration	Lump	1	5,000	5,000
33 21	Demobilization	Lump	1	3,000	3,000
33 91	Confirmatory sampling	Lump	1	35,520	35,520
33 92	As-built drawings/final closure report	Lump	1	2,400	2,400
	Markup (see itemized cost breakdown)	%	15		2,325
	Subtotal				55,745
	Total remediation cost				$1,411,620
	Check total cost				$1,411,620

Table K2   Case 2 Cost Summary Sheet (Small Site) (continued)

Itemized Breakdown of Costs
Labor (RF Heating System Design & Installation Oversight)

WBS #[1]	Title	Duty	Hours	Rate ($)/h	Total Cost ($)
33 14-01	Project manager (1)	Management	60	135	8,100
33 14-01	RF supplier's engineer (1)	RF system design/technical review	1	Lump	20,000
33 14-01	Technician (2)	RF system installation oversight	300	57	17,100
33 14-01	Lead engineer (1)	RF system installation	100	85	8,500
33 14-01	Per diem	2 people on site at all times @ ($)80/day per person	1	Lump	2,880
	Subtotal				56,580

Note: Assume 18 (1 month) days for RF system installation.

Sub-Contract: Remediation System (RF & SVE) Installation

WBS #[1]	Description	Notes	Unit	Units	Unit Cost ($)	Total Cost ($)
33 13-01	12 2" SVE wells @ 20 ft each (complete)		LF	240	40	9,600
33 13-01	7 2" Air vent wells @ 20 ft each (passive)		LF	140	30	4,200
33 13-01	Trenching (2@135 ft + 1@100 ft to SVE + 6@22 ft + 6@13 ft)	1	LF	580	25	14,500
33 13-01	Subsurface utilities (includes steel piping)		LF	580	15	8,700
33 13-01	Backfill, compact, & resurface (with same soils)		LF	580	20	11,600
33 13-01	Waste disposal (well cuttings)	2	Lump	1	2,000	2,000
33 03	Surveying for installation of all systems		Lump	1	4,000	4,000
33 03	Compound for SVE system		Lump	1	3,500	3,500
33 14-01	Install 7 RF heating wells	1	Ea	7	800	5,600
33 14-01	Set structures/complete connections		Lump	1	9,000	9,000
33 07	Install vapor treatment system (incl. elect/gas hook-up)	3	Lump	1	3,000	3,000
33 13-01	Provide & install well vaults		Ea	12	500	6,000
33 13-01	Install wellhead manifolding		Ea	12	200	2,400
33 01	Mobilization (incl. electrical service install)		Lump	1	12,000	12,000
33 03	Site fence & 2 gates (complete)		LF	700	8	5,600
33 03	Per diem (4 persons) @ ($)80/day per person		Day	18	320	5,760
33 03	Permitting		Lump	1	8,000	8,000
33 17	Decontamination (incl. steam cleaner rental ($150/day))		Lump	1	2,500	2,500
33 18	Waste disposal (site debris)		Lump	1	5,000	5,000
33 20	Site restoration		Lump	1	5,000	5,000
33 21	Demobilization		Lump	1	3,000	3,000
33 90	Health & safety plan/RF-trained personnel monitoring		Lump	1	5,000	5,000
	Markup		%	15		20,394
	Subtotal					156,354

Note: (1) RF lines will be installed in SVE trenches, assume no saw cuts into concrete; (2) assume nonhazard disposal @ ($)30/yd; (3) install only, no equipment costs

Labor: SVE Well Design & Installation Oversight (Install 12 SVE wells).

WBS #a	Title	Duty	Hours	Rate $/h	Total Cost ($)
33 13-01	Project manager (1)	Management	12	135	1,620
33 05	Senior engineer/hydro (1)	System design/technical review	120	116	13,950
33 13-01	Staff engineer (1)	Oversight	20	75	1,500
33 13-01	Staff geologist (2)	Install wells (10 wells per day)	30	75	2,250
33 92	Staff engineer (l)	As-built drawings/final closure report	32	75	2,400
	Subtotal				21,720

Capital Cost: RF Heating/SVE System Equipment

WBS #a	Description	Notes	Unit	Units	Unit Cost ($)	Total Cost ($)
33 14-01	RF generator system (rental — see operating costs)	1	Ea	1	0	0
33 14-01	Heating monitoring points/equipment	2	Lump	1	8,000	8,000
33 13-01	14.9-kW blower for SVE system (12,750 Lpm [450 scfm])	3	Ea	1	15,000	15,000
33 13-01	Piping and valves (within compound)		Lump	1	5,000	5,000
33 13-01	Valves, fittings, and gauges with SVE wells		Lump	1	4,000	4,000
33 13-01	Central controller (elect. for blowers, Cat-Ox)		Ea	1	3,000	3,000
33 13-01	Electrical transformer (300 KVA)	4	Lump	1	12,300	12,300
33 07	Catalytic oxidation unit (14,150 Lpm [500 scfm])		Ea	1	40,000	40,000
33 13-01	Markup		%	15		13,095
	Subtotal					100,395
	Total system install					335,049

Note: (1) RF equipment rental assumed operating, not capital cost; (2) cost of RF transmission lines, antennae, fiber optic thermometers; (3) SVE system includes moisture separator, transfer pump, particle filters, appurtenances; (4) 300-KVA transformer required for 7 RF units, includes material and labor to install.

Operation & Maintenance (O&M)— Labor: (O&M Oversight)

WBS #a	Title	Duty	Hours	Rate $/h	Total ($)
BOTH	Project manager	Management	290	135	39,150
BOTH	Technicians (2)	Site work (includes sampling labor)	2000	57	114,000
BOTH	Staff engineer	Data interpretation	500	75	37,500
BOTH	Health & safety officer	Health & safety on site	220	116	25,520
BOTH	Responsible engineer	Engineer of record, PE	280	116	32,480
	Subtotal				248,650

Note: Assume 333 Days (11 months) as time for remediation.

Table K2 Case 2 Cost Summary Sheet (Small Site) (continued)

RF/SVE System O & M Costs

WBS #ᵃ	Description	Notes	Unit	Quantity	Unit Cost ($)	Total Cost ($)
33 14-02	7 20-kW RF generators/trailers @ ($)6,000/month)		Month	12	42,000	504,000
33 14-02	Heating system routine maintenance/mat'ls	1	Lump	1	5,000	5,000
33 14-02	Quarterly reporting		Ea	4	5,000	20,000
33 14-02	Electricity for RF heating system @ ($)80/1,000 kWh	2	kWh	1,431,793	80	114,543
33 13-02	Regulatory sampling	3	Month	12	200	2,400
33 13-02	Electricity for SVE system @ ($)80/1,000 kWh		kWh	8,000	80	640
33 13-02	Propane gas (for Cat-Ox.)	4	Month	12	1,500	18,000
33 13-02	FID (monitor once/week)		Day	52	100	5,200
33 13-02	Office trailer/telephone/toilet		Month	12	450	5,400
33 13-02	SVE/vapor abatement system maintenance		Month	12	1,000	12,000
33 13-02	Travel (2 technicians) @ ($)550/trip		Month	12	1,100	13,200
33 13-02	Per diem (2 persons on site) @ ($)80/day per person		Day	340	160	54,400
33 13-02	Markup (excluding RF rental)		%	15		37,618
	Subtotal					**792,401**
	Total O & M costs/year					**1,041,051**

Note: (1) Includes maintaining transmission lines, adjusting temperatures, replacing materials, SVE flow; (2) 1,431,800 kWh RF energy required for 333 days of operation; assumed AC to RF energy 72%; (3) based on TX regulations, data interpretation only (not sampling labor); (4) propane & tank available at ($)1/gal.

Confirmatory Sampling

WBS #ᵃ	Title	Duty	Hours	Rate $/h	Total ($)
33 91	Project manager	Management/final report	50	135	6,750
33 91	Technicians (2)	Site work (includes sampling labor)	80	57	4,560
33 91	Staff engineer	Data interpretation/final report	200	75	15,000
33 91	Staff geologist (2)	Install monitor wells (10 wells per day)	30	75	2,250
33 91	Responsible engineer	Engineer of record, PE	60	116	6,960
	Subtotal				**35,520**

Note: Completed 1 year after remediation. Assume 15 confirmatory borings were required.

ᵃ Categorized according to interagency WBS # cost elements

Table K3   Case 3 Cost Summary Sheet (Hydrous Pyrolysis Oxidation)

Preliminary Cost Estimate
RF-SVE Full-Scale Remediation

Before Treatment Costs

WBS #ᵃ	Description	Unit	Units	Unit Cost ($)	Total Cost ($)
33 01	Mobilization/preparatory work	Lump	1	12,000	12,000
33 02	Premonitoring, sampling, analysis, assessments	N.A.	Assumed completed		
33 03	Site work (clearing/utilities install/pads), surveying	Lump	1	26,860	26,860
33 05	Engineering [SVE design only]	Lump	1	13,950	13,950
	Markup (see itemized cost breakdown)	%	15		5,829
	Subtotal				**58,639**

Treatment Costs (includes equipment cost/installation, RF design, and RF-SVE system operation)

Thermal Treatment — WBS# 33 14
Physical Treatment — WBS# 33 13

WBS #ᵃ	Treatment Cost Descriptions	Unit	Units	Unit Cost ($)	Total Cost ($)
33 13-01	SVE system equipment and installation	Lump	1	138,670	138,670
33 13-02	SVE operating costs	Lump	1	206,098	206,098
33 14-01	RF equipment, design, and installation	Lump	1	79,180	79,180
33 14-02	RF operating costs	Lump	1	435,305	435,305
	Markup — selected subcontract, capital, O&M costs	%	15		51,014
	Subtotal				**910,268**

After Treatment Costs

WBS #ᵃ	Description	Unit	Units	Unit Cost ($)	Total Cost ($)
33 17	Decontamination and decommissioning	Lump	1	2,500	2,500
33 18	Disposal	Lump	1	5,000	5,000
33 20	Site restoration	Lump	1	5,000	5,000
33 21	Demobilization	Lump	1	3,000	3,000
33 91	Confirmatory sampling	Lump	1	35,520	35,520
33 92	As-built drawings/final closure report	Lump	1	2,400	2,400
	Markup (see itemized cost breakdown)	%	15		2,325
	Subtotal				**55,745**
	Total remediation cost				**1,024,652**
	Check total cost				**1,024,652**

Table K3   Case 3 Cost Summary Sheet (Hydrous Pyrolysis Oxidation) (continued)

Itemized breakdown of costs
Labor (RF Heating System Design & Installation Oversight)

WBS #*	Title	Duty	Hours	Rate $/h	Total Cost ($)
33 14-01	Project manager (1)	Management	60	135	8,100
33 14-01	RF supplier's engineer (1)	RF system design/technical review	1	Lump	20,000
33 14-01	Technician (2)	RF system installation oversight	300	57	17,100
33 14-01	Lead engineer (1)	RF system installation	100	85	8,500
33 14-01	Per diem	2 people on site at all times @ ($)80/day per person	1	Lump	2,880
	Subtotal				**56,580**

Sub-Contract: Remediation System (RF & SVE) Installation

WBS #*	Description	Notes	Unit	Units	Unit Cost ($)	Total Cost ($)
33 13-01	12 2" SVE wells @ 20 ft each (complete)		LF	240	40	9,600
33 13-01	7 2" Air vent wells @ 20 ft each (passive)		LF	140	30	4,200
33 13-01	Trenching (2@135 ft + 1@100 ft to SVE+ 6@22 ft + 6@13 ft)	1	LF	580	25	14,500
33 13-01	Subsurface utilities (includes steel piping)		LF	580	15	8,700
33 13-01	Backfill, compact, & resurface (with same soils)		LF	580	20	11,600
33 13-01	Waste disposal (well cuttings)	2	Lump	1	2,000	2,000
33 03	Surveying for installation of all systems		Lump	1	4,000	4,000
33 03	Compound for SVE system		Lump	1	3,500	3,500
33 14-01	Install 7 RF heating wells	1	Ea	7	800	5,600
33 14-01	Set structures/complete connections		Lump	1	9,000	9,000
33 07	Install vapor treatment system (incl. elect/gas hook-up)	3	Lump	1	3,000	3,000
33 13-01	Provide & install well vaults		Ea	12	500	6,000
33 13-01	Install wellhead manifolding		Ea	12	200	2,400
33 01	Mobilization (incl. electrical service install)		Lump	1	12,000	12,000
33 03	Site fence & 2 gates (complete)		LF	700	8	5,600
33 03	Per diem (4 persons) @ ($)80/day per person		day	18	320	5,760
33 03	Permitting		Lump	1	8,000	8,000
33 17	Decontamination (incl. steam cleaner rental ($)150/day)		Lump	1	2,500	2,500
33 18	Waste disposal (site debris)		Lump	1	5,000	5,000
33 20	Site restoration		Lump	1	5,000	5,000
33 21	Demobilization		Lump	1	3,000	3,000
33 90	Health & safety plan / RF-trained personnel monitoring		Lump	1	5,000	5,000
	Markup		%	15		20,394
	Subtotal					**156,354**

Note: (1) RF lines will be installed in SVE trenches, assume no saw cuts into concrete; (2) assume nonhazard disposal @ $30/yd; (3) install only, no equipment costs. Assume 18 days (1 month) for RF system installation.

Labor: SVE Well Design & Installation Oversight (Install 12 SVE wells)

WBS #*	Title	Duty	Hours	Rate ($/h)	Total Cost ($)
33 13-01	Project manager (1)	Management	12	135	1,620
33 05	Senior engineer/hydro (1)	System design/technical review	120	116	13,950
33 13-01	Staff engineer (1)	Oversight	20	75	1,500
33 13-01	Staff geologist (2)	Install wells (10 wells/day)	30	75	2,250
33 92	Staff engineer (I)	As-built drawings/final closure report	32	75	2,400
	Subtotal				21,720

Capital Cost: RF Heating/SVE System Equipment

WBS #*	Description	Notes	Unit	Units	Unit Cost ($)	Total Cost ($)
33 14-01	RF generator system (rental — see operating costs)	1	Ea	1	0	0
33 14-01	Heating monitoring points/equipment	2	Lump	1	8,000	8,000
33 13-01	7.5-kW blower for SVE system (7,075 Lpm [250 scfm])	3	Ea	1	12,000	12,000
33 13-01	Piping and valves (within compound)		Lump	1	5,000	5,000
33 13-01	Valves, fittings, and gauges with SVE wells		Lump	1	4,000	4,000
33 13-01	Central controller (elect. for blowers, Cat-Ox)		Ea	1	3,000	3,000
33 13-01	Electrical transformer (300 KVA)	4	Lump	1	12,300	12,300
33 07	Catalytic oxidation unit (7,075 Lpm [250 scfm])		Ea	1	35,000	35,000
33 13-01	Markup		%	15		11,895
	Subtotal					91,195
	Total system install					325,849

Note: (1) RF equipment rental assumed operating, not capital cost; (2) cost of RF transmission lines, antennae, fiber optic thermometers; (3) SVE system includes moisture separator, transfer pump, particle filters, appurtenances; (4) 300-KVA transformer required for 7 RF units, includes materials and labor to install.

Operation & Maintenance (O&M) — Labor: (O&M Oversight)

WBS #*	Title	Duty	Hours	Rate $/h	Total ($)
BOTH		Management	290	135	39,150
BOTH	Technicians (2)	Site work (includes sampling labor)	2080	57	118,560
BOTH	Staff engineer	Data interpretation	500	75	37,500
BOTH	Health & safety officer	Health & safety on-site	220	116	25,520
BOTH	Responsible engineer	Engineer of record, PE	280	116	32,480
	Subtotal				253,210

Note: Assume 365 days (12 months) as time for remediation.

Table K3 Case 3 Cost Summary Sheet (Hydrous Pyrolysis Oxidation) (continued)

RF/SVE System O & M Costs

WBS #ᵃ	Description	Notes	Unit	Quantity	Unit Cost ($)	Total Cost ($)
33 14-02	4 & 1, 20-kW RF generators/trailers (($)6,000/month/unit)	1	Month	7 & 6	6,000	204,000
33 14-02	Heating system routine maintenance/mat'ls		Lump	1	5,000	5,000
33 14-02	Quarterly reporting		Ea	5	5,000	25,000
33 14-02	Electricity for RF heating system @ ($)80/1,000 kWh	2	kWh	441,878	80	35,350
33 13-02	Regulatory sampling	3	Month	12	200	2,400
33 13-02	Electricity for SVE system @ ($)80/1,000 kWh		kWh	8,667	80	693
33 13-02	Propane gas (for Cat-Ox.)	4	Month	12	1,500	18,000
33 13-02	FID (monitor once/week)		Day	52	100	5,200
33 13-02	Office trailer/telephone/toilet		Month	13	450	5,850
33 13-02	SVE/vapor abatement system maintenance		Month	13	1,000	13,000
33 13-02	Travel (2 technicians) @ ($)550/trip		Month	13	1,100	14,300
33 13-02	Per diem (2 persons on site) @ ($)80/day per person		Day	340	160	54,400
33 13-02	Markup (excluding RF rental)		%	15		26,879
	Subtotal					410,073
	Total O & M costs/year					663,283

Note: (1) Includes maintaining transmission lines, adjusting temperatures, replacing materials, SVE flow; (2) 441,900 kWh RF energy required for 365 days of operation, assumed AC to RF energy 72%; (3) based on TX regulations, data interpretation only (not sampling labor); (4) propane & tank available at $1/gal.

Confirmatory Sampling

WBS #ᵃ	Title	Duty	Hours	Rate ($)/h	Total ($)
33 91	Project manager	Management/final report	50	135	6,750
33 91	Technicians (2)	Site work (includes sampling labor)	80	57	4,560
33 91	Staff engineer	Data interpretation/final report	200	75	15,000
33 91	Staff geologist (2)	Install monitor wells (10 wells/day)	30	75	2,250
33 91	Responsible engineer	Engineer of record, PE	60	116	6,960
	Subtotal				35,520

Note: Completed 1 year after remediation. Assume 15 confirmatory borings required.

ᵃ Categorized according to interagency WBS # cost elements.

References

Monograph References

Allison, S.B., G.A. Pope, and K. Sepehrnoori. 1991. Analysis of field tracers for reservoir description, *J. Pet. Sci. Eng.*, 5, 173.

API Recommended Practice 40: Recommended Practices for Core Analysis. 1996. American Petroleum Institute (API).

ASTM: Soil and Rock (I): D420 – D4914, vol. 04.08. 1996. American Society for Testing and Materials (ASTM), Philadelphia.

Balshaw-Biddle, K., C.L. Oubre, and C.H. Ward, Eds., Steam and electroheating remediation of tight soils, AATDF Monograph, CRC Press LLC, Boca Raton, FL (in press).

Brown & Root Environmental (BRE), Final site specific health and safety plan for enhanced soil vapor extraction with radio frequency heating demonstration. 1996a. Project Work Plan, Project No. 6736, Oak Ridge, TN.

Brown & Root Environmental (BRE), Permeability test data: enhanced soil vapor extraction with radio frequency heating demonstration. 1996b. Project Letter Report, Project No. 6736, Oak Ridge, TN.

Brown & Root Environmental (BRE) 1997. Private communication.

Chang, Y.B., Development and application of an equation of state compositional simulator. 1990. Ph.D. dissertation, University of Texas, Austin.

Daniel, D.E., J.A. Pearce, and J.J. Bowders. 1995. Enhanced soil vapor extraction with radio frequency heating, Phase 2 Work Plan, prepared for DOD/AATDF.

Datta-Gupta, A. and M.J. King. 1995. A semianalytic approach to tracer flow modeling in heterogeneous permeable media, *Adv. Water Resour.*, 18, 9.

Delshad, M., G.A. Pope, and K. Sepehrnoori. 1996. A compositional simulator for modeling enhanced aquifer remediation formulation, *J. Contaminant Hydrol.*, 23, 303.

Dev, H. 1986. Radio frequency enhanced in-situ decontamination of soils contaminated with halogenated hydrocarbons, Proc. U.S. EPA 12th Ann. Res. Symp., Publ. No. EPA/600/9-86/022, 402-412.

Dev, H. and D. Downey. 1988. Zapping hazwastes, *Civ. Eng.*, 43.

Dev, H., J. Enk, G. Sresty, and J. Bridges. 1989. In-situ decontamination by radio-frequency heating — field test, Final Report, USAF/SD Contract No. F04701-86-C-0002, 175p.

Dong, W.L. 1991. Computer controlled measurement system for complex permittivity over radio frequencies and microwave frequencies, M.S. thesis, University of Texas, Austin.

Dwarakanath, V. 1997. Characterization and remediation of aquifers contaminated by nonaqueous phase liquids using partitioning tracers and surfactants, Ph.D. dissertation, University of Texas, Austin.

ENVIROGEN. 1996. Results of modeling support for design of vapor extraction well system, Project Letter Report, Project No. 67461, Canton, MA, October 4.

Federal Remediation Technologies Roundtable (FRTR). 1995. Guide to documenting cost and performance for remedial technologies, prepared by member agencies of the FRTR, Washington, D.C., March.

Fluor Daniel GTI, 1998. Enhanced soil vapor extraction using radio frequency heating, AATDF Technology Evaluation Report, No. TR-98-10.

Hansen, T. L. 1997. Soil contaminated with hydrocarbons, Masters thesis, Technical University of Denmark.

Harneshaug, T. 1996. Permeability and saturation distributions from tracer test data, Masters thesis, University of Texas, Austin.

Harrington, R.F., *Time Harmonic Electromagnetic Fields*, McGraw-Hill, New York, 1961.

Incropera, F.P. and D.P. DeWitt. 1990. *Fundamentals of Heat and Mass Transfer*, John Wiley & Sons, New York, New York.

INTERA, Inc. 1997. Application of a NAPL partitioning interwell tracer test to estimate the spatial distribution and volume of NAPL beneath the chemical waste landfill, Sandia National Laboratories, New Mexico, Final Draft Report.

Jackson, R. E., F.J. Holzmer, M. Jin, J.T. Londergan, P.E. Mariner, K. Rotert, J., Studer, C.M. Young, and G.A. Pope. 1997. Partitioning tracer testing of alluvium to determine the spatial distribution and volume of residual TCE DNAPL, *Ground Water Monitoring Remeditation*, vol./pg.

Jin, M., M. Delshad, V. Dwarakanath, D.C. McKinney, G.A. Pope, K. Sepehrnoori, and C.E. Tilburg. 1995. Partitioning tracer test for detection, estimation and remediation performance assessment of subsurface nonaqueous phase liquids, *Water Resour. Res.*, 31(5), 1201.

Kasevich, R., R. Holmes, D. Faust, and R. Beleski. 1993. Radio frequency heat enhances contaminant removal, *Soils*, 18.

Kasevich, R.S., D. Wiberg, M. Johnson, and S. Price. 1996. Enhanced removal of gasoline constituents from the capillary fringe utilizing radio frequency heating, Proc. 11th Annual Conf. Contaminated Soils, University of Massachusetts at Amherst, October 21 to 24.

Khan, S.A. 1992. An expert system to aid in compositional simulation of miscible gas flooding, Ph.D. dissertation, University of Texas, Austin.

Knauss, K.G., R.D. Aines, M.J. Dibley, R.N. Leif, and D.A. Mew. 1997. Hydrous pyrolysis/oxidation: in-ground thermal destruction of organic contaminants, Lawrence Livermore National Laboratory, pre-print (UCRL-JC-126636) for AIChE 1997 Spring Meeting, March 10 to 12.

Knauss, K.G., M.J. Dibley, R.N. Leif, D.A. Mew, and R.D. Aines. 1998. Aqueous oxidation of trichloroethane (TCE): a kinetic and thermodynamic analysis, Lawrence Livermore National Laboratory, Preprint (UCRL-JC-129932) for the First Int. Conf. Remediation of Chlorinated and Recalcitrant Compounds, May 18 to 21.

Kubaschewski, O., C.B. Alcock, and P.J. Spencer. 1993. *Materials Thermochemistry*, Pergamon Press, New York, New York.

Kurihara, M., 1995. Development of Three Dimensional Streamline Model (UTSTREAM) and its application, Ph.D. dissertation, University of Texas, Austin.

Leif, R.N., R.D. Aines, and K.G. Knauss. 1997. Hydrous pyrolysis of pole treating chemicals: (A) Initial measurement of hydrous pyrolysis rates for naphthalene and pentachlorophenol; (B) solubility of flourene at temperatures up to 150°C, Lawrence Livermore National Laboratory, preprint (UCRL-CR-129838) for the Southern California Edison Company, November 15.

Leif, R.N., K.G. Knauss, D.A. Mew, and R.D. Aines. 1998. Destruction of 2,2,3-trichlorobiphenyl in aqueous solution by hydrous pyrolysis/oxidation (HPO), Lawrence Livermore National Laboratory (UCRL-ID-129837), November 25.

Lowe, D.F., C.L. Oubre, and C.H. Ward, Eds. *Surfactants and Cosolvents for NAPL Remediation (A Technology Practices Manual)*, AATDF Monograph, Lewis Publishers, Boca Raton, Florida.

Newmark, R.L. and R.D. Aines. 1997. Dumping pump and treat: rapid cleanups using thermal technology, Lawrence Livermore National Laboratory, preprint (UCRL-JC-126637) for AIChE 1997 Spring Meeting, March 10 to 12.

Parks, R.D. 1949. *Examination and Valuation of Mineral Property*, Addison-Wesley Press, Inc., Cambridge, Massachusetts.

Parkhomenko, E.I. 1967. *Electrical Properties of Rocks*, translated and edited by George V. Keller, Plenum Press, New York.

Perry, R.H. and D.W. Green. 1984. *Perry's Chemical Engineering Handbook*, 6th ed., McGraw-Hill, New York.

Saxena, S.K., N. Chatterjee, Y. Fei, and G. Shen. 1993. *Thermodynamic Data on Oxides and Silicates*, Springer-Verlag, New York.

Tang, J.S., 1992. Interwell tracer test to determine residual oil saturation to waterflood at Judy Creek BHL "A" pool, *J. Can. Pet. Tech.*, 31 (8), 61.

U.S. Department of Energy (DOE) Office of Environmental Management, Office of Technology Development. 1995. Six phase soil heating, Report No. USDOE/EM-0272, April.

U.S. Department of Energy (DOE) Office of Environmental Management, Office of Technology Development. 1994. VOCs in non-arid soils, integrated demonstration, Report No. USDOE/EM-0135P.

U.S. Environmental Protection Agency (EPA). 1992. *Test Methods for Evaluating Solid Waste: Physical/Chemical Methods,* 3rd ed., USEPA/SW-846, U.S. EPA, Office of Solid Waste and Emergency Response, Washington, D.C.

U.S. Environmental Protection Agency (EPA). 1994a. Demonstration bulletin on IITRI radio frequency heating technology, USEPA/540/R-94/527a, Office of Research and Development, Cincinnati, OH.

U.S. Environmental Protection Agency (EPA). 1994b. Demonstration bulletin on KAI radio frequency heating technology, USEPA/540/R-94/528a, Office of Research and Development, Cincinnati, OH.

U.S. Environmental Protection Agency (EPA). 1995. *Test Methods for Evaluating Solid Waste,* 3rd ed., USEPA/SW846, USEPA Office of Solid Waste and Emergency Response, Washington, D.C.

U.S. Environmental Protection Agency (EPA). 1995a. SITE technology capsule: IITRI radio frequency heating technology, USEPA/540/R-94/527a, Office of Research and Development, Cincinnati, OH.

U.S. Environmental Protection Agency (EPA). 1995b. SITE technology capsule: KAI radio frequency heating technology, USEPA/540/R-94/528a, Office of Research and Development, Cincinnati, OH.

Weston, Roy R., Inc. 1992. Final Rocky Mountain Arsenal In-Situ Radio Frequency Heating/Vapor Extraction Pilot Test Report, Vol. 1, prepared for U.S. Army Material Command, Rocky Mountain Arsenal, Document No. 5300-01-12-AAFP, Lakewood, CO.

Whitley, G.A., G.A. Pope, D.C. McKinney, B.A. Rouse, and P.E. Mariner. 1995. Vadose zone nonaqueous phase liquid characterization using partitioning gas tracer, Proc. 3rd Int. Symp. In-Situ and On-Site Bioreclamation, San Diego, CA, 211.

Whitley, G.A. 1997. An investigation of partitioning tracer for characterization of nonaqueous phase liquids in the vadose zone, Dissertation, University of Texas, Austin.

Young, C.M., R.E. Jackson, M. Jin, J.T. Londergan, P.E. Mariner, G.A. Pope, and F.J. Anderson. 1999. Characterization of a TCE DNAPL zone in partitioning tracers, *Ground Water Monitoring & Remediation,* 19(1), 84-94.

Related References

American Public Health Association. 1995. *Standard Methods for Examination of Water and Wastewater,* 19th ed., American Public Health Association.

Brock, T. D., M. Madigan, J. Martinko, and J. Parker. 1994. *Biology of Microorganisms,* Prentice-Hall, Englewood Cliffs, NJ.

Cohen, R.M., J.W. Mercer, and J. Mathews. 1993. *DNAPL Site Evaluation,* C.K. Smoley, Ed., CRC Press, Boca Raton, FL.

Engineering Science Inc. 1981. Kirtland AFB Installation Restoration Program (IRP) Phase 1 — Records Search Hazardous Materials Disposal Sites.

Hickey, W.J., 1995. In-situ respirometry: field methods and implications for hydrocarbon biodegradation in subsurface soils, *J. Environ. Qual.,* 24, 583.

Hinchee, R.E., S.K. Ong, R.N. Miller, D.C. Downey, and R. Frandt. 1992. Test Plan and Technical Protocol for a Field Treatability Test for Bioventing, Environmental Services Office, Air Force Center for Environmental Excellence (AFCEE), Brooks Air Force Base, TX.

Hinchee, R.E. and S.K. Ong. 1992. A rapid in-situ respiration test for measuring aerobic biodegradation rates of hydrocarbons in soil, *J. Air Waste Manage. Assoc.,* 42 (10), 1305.

Hutzler, N.J., B.E. Murphy, and J.S. Gierke. 1989. State of technology review: soil vapor extraction systems, U.S. EPA Risk Reduction Engr. Laboratory, Cincinnati, OH, Publ. No. USEPA/600/2-89/024 (87 pages).

Jarosch, T.R., R.J. Beleski, and D. Faust. 1994. Final report: in-situ radio frequency heating demonstration, prepared for the U.S. Department of Energy, Contract No. DE-AC09-89SR18035.

Joss, C.J. and A.L. Baehr. 1997. Documentation of a computer program to simulate two-dimensional axisymmetric air flow in the unsaturated zone, U.S. Geological Survey Open-File Report 97-588, p. 106.

Kirtland Air Force Base. 1996. Appendix II, Phase 2 SAP, Phase II RCRA Facility Investigation for Appendix II Solid Waste Management Units (SWMUs) Final Draft Sampling and Analysis Plan, Kirtland Air Force Base, NM.

Mayer, A.S. and C.T. Miller. 1990. The influence of porous medium characteristics and measurement scale on pore-scale distributions of residual non-aqueous phase liquids, *J. Contaminant Hydrol.*, 6, 107.

McCarthy, D.F. 1993. *Essentials of Soil Mechanics and Foundation Engineering*, 4th ed., Prentice-Hall, Upper Saddle River, NJ.

Phelan, J.M. and S.W. Webb. 1994. Thermal enhanced vapor extraction systems — designs, application, performance prediction, including contaminant behavior, for presentation at the 33rd Annu. Hanford Symp. Health and the Environment, Pasco, WA.

Phelan, J.M. and H. Dev. 1995. Application of thermal enhanced vapor extraction system technology on a complex chemical waste mixture, Sandia National Laboratories, Albuquerque, NM.

Studer, J.E., P. Mariner, M. Jin, G.A. Pope, D. McKinney, and R. Fate. 1996. Application of a PITT to support DNAPL remediation at Sandia National Labs chemical waste landfill, Proc. Superfund/Hazwaste West Conf., Las Vegas, NV.

Sugiura, K., M. Ishihara, T. Shimachi, and S. Harayama. 1997. Physical properties and biodegradability of crude oil, *Environ. Sci. Technol.*, 31, 45.

U.S. Air Force, 1995. Stage 2B final RCRA facility investigation, Kirtland Air Force Base, NM.

U.S. Environmental Protection Agency (EPA). 1988. Compendium of Methods for Determination of Toxic Organic Compounds in Ambient Air, EPA-600/4-84-041, Quality Assurance Division of the Environmental Monitoring Systems Laboratory, Research Triangle Park, NC.

U.S. Geological Survey (USGS) Water Resources Division. 1993b. RCRA facility investigation (RFI) stage 2A, Vol. 1 Tech. Inf. Rep. Kirtland AFB, NM.

U.S. Geological Survey (USGS) Water Resources Division. 1994a. Analytical results from an investigation of six sites on Kirtland Air Force Base, NM, Open File Rep. 94-547.

U.S. Geological Survey (USGS) Water Resources Division. 1994b. Analytical data informal technical information report (ITIR) stage 2B soils, Kirtland Air Force Base, NM.

Venosa, A.D, M.T. Suidan, B.A. Wrenn, K.L. Strohmeier, J.R. Haines, B.L. Eberhart, D. King, and E. Holder. 1996. Bioremediation of an experimental oil spill on the shoreline of Delaware Bay, *Environ. Sci. Technol.*, 30, 1764.

Wight, G.D. 1994. *Fundamentals of Air Sampling*, Lewis Publishers, Boca Raton, FL, Chapter 12.

Index

Milton Keynes UK
Ingram Content Group UK Ltd.
UKHW052020071024
449327UK00027B/2360